Nanomaterials for Energy Applications

Nanomaterials for Energy Applications provides readers with an in-depth understanding of advanced nanomaterials and their applications in energy generation and utilization concepts. It focuses on emerging nanomaterials and applications in various energy-related fields.

- Describes nanomaterials for use in photovoltaic cells, solid state lighting, fuel cells, electrochemical batteries, electrochemical capacitors, super-conductors, hydrogen storage, and photocatalysts.
- Focuses on commercial and economic aspects.
- Includes case studies drawn from practical research.

This book is aimed at researchers, advanced students, and practicing engineers in the disciplines of materials, mechanical, electrical, and related fields of engineering.

Emerging Materials and Technologies

Series Editor: Boris I. Kharissov

The *Emerging Materialsand Technologies* series is devoted to highlighting publications centered on emerging advanced materials and novel technologies. Attention is paid to those newly discovered or applied materials with potential to solve pressing societal problems and improve quality of life, corresponding to environmental protection, medicine, communications, energy, transportation, advanced manufacturing, and related areas.

The series takes into account that, under present strong demands for energy, material, and cost savings, as well as heavy contamination problems and worldwide pandemic conditions, the area of emerging materials and related scalable technologies is a highly interdisciplinary field, with the need for researchers, professionals, and academics across the spectrum of engineering and technological disciplines. The main objective of this book series is to attract more attention to these materials and technologies and invite conversation among the international R&D community.

For more information about this series, please visit: www.routledge.com/Emerging-Materials-and-Technologies/book-series/CRCEMT

Nanomaterials for Energy Applications

Edited by
L. Syam Sundar
Shaik Feroz
Faramarz Djavanroodi

CRC Press
Taylor & Francis Group
Boca Raton London New York

CRC Press is an imprint of the
Taylor & Francis Group, an **informa** business

First edition published 2024
by CRC Press
2385 Executive Center Drive, Suite 320, Boca Raton, FL 33431

and by CRC Press
4 Park Square, Milton Park, Abingdon, Oxon, OX14 4RN

CRC Press is an imprint of Taylor & Francis Group, LLC

ISBN: 978-1-032-42901-4 (hbk)
ISBN: 978-1-032-42904-5 (pbk)
ISBN: 978-1-003-36482-5 (ebk)

DOI: 10.1201/9781003364825

Typeset in Times New Roman
by MPS Limited, Dehradun

Contents

Preface

Cinemas, television shows and entertainment hubs frequently give the idea that the only truly significant factors in the creation of modern civilization were religions, emperors and monarchs, wars and treaties, invasions and uprisings, and sickness and famine. But it is also possible to make the case that the accessibility of usable energy has been just as significant, if not more so. The use of fire, animals and basic mechanical tools like the wheel, lever and pulley all had a significant impact on the lives of early humans. After simpler mechanisms came more sophisticated ones like internal combustion and steam engines as well as cutting-edge electrical and electronic technologies. All these inventions have made it possible for humanity to continuously increase its usage of energy, which has led to better living conditions and economic prosperity.

Industrialized nations became aware of their wasteful energy usage during the 1970s oil shortages and price shocks, and people learned about the advantages of conservation. Nations started programs to promote conservation in transportation, industry, residential, and commercial sectors. Energy efficiency increased significantly as a result of these measures. Oil was the primary source of energy at the time we started the forerunner to this book, more than ten years ago, and its price was falling in terms adjusted for inflation. With fresh oil and gas discoveries and enhanced extraction techniques, major oil producers guaranteed a steady supply of oil for the next 40 years.

Energy issues were no longer on most people's radar due to the reduction in oil prices in the 1990s and the apparent oil surplus. On the spot market, crude oil was being sold for less than \$20 a barrel, or roughly the same price as it was in the 1920s after accounting for inflation. Nonetheless, the issues of where and how our society will get the energy to meet its needs in the future remained crucial.

For those of us who work in the field of energy research, this truth is very clear. To move a new technology from its initial implementation to an usable scale, changes in the pattern of energy use must be made over a long period of time. Even if we continue to use oil at our current rate for 40 years, and even though we are likely to uncover new sources, we will eventually reach a point where the anticipated increases in demand will exceed our capacity to produce oil at the required rate. At that point, we'll require more resources. The development of new energy sources has historically taken a very long time, in part because doing so frequently necessitates creating new infrastructure for its extraction, distribution, and usage. We can no longer ignore the looming energy crisis, which is why it is critical that we address the energy issue right away. Due to the abolition of fossil fuels, one can think about energy storage systems. Nowadays, due to the advancement in nanotechnology, advanced nanomaterials are synthesized and those can be used as high energy storage materials.

The foundation of nanoscience and nanotechnology is nanomaterials. Over the past few years, research and development activity in the vast and interdisciplinary field of nanostructure science and technology has exploded globally. It has the potential to fundamentally alter the processes used to make materials and products, as well as the variety and type of functionalities that may be accessible. It already has a large commercial influence and this impact will undoubtedly grow in the future.

This book discusses how nanoparticles can be used for energy storage. The book aids in gradually building up understanding of the use of a wide variety of nanoparticles for energy storage applications beginning with the synthesis and characterization of high energy nanomaterials. The book discusses sophisticated characterization techniques as well as numerous physical, chemical, and hybrid methods of nanomaterial production and nanofabrication. Chapters on the different uses of nanoscience and nanotechnology are included. Students, researchers and academicians studying the physical and material sciences can get benefit from it because of how easily it is written.

Editor Biographies

L. Syam Sundar is currently working at Prince Mohammed Bin Fahd University, Kingdom of Saudi Arabia. Dr. L. Syam Sundar obtained his doctorate in the field of Energy Systems from Jawaharlal Nehru Technological University, Hyderabad, India in 2010, Postdoctoral Research Fellow from the University of Aveiro in 2016, M. Tech. in Thermal Engineering from Jawaharlal Nehru Technological University, Hyderabad, India in 2003 and B. Tech. in Mechanical Engineering from Andhra University, India in 1998.

Dr. L. Syam Sundar has expertise in thermal process engineering, power plant design and nanofluid heat transfer, synthesis of hybrid nanoparticles, thermophysical property analysis, CFD analysis and finite element method analysis. He has more than 110 publications to his credit in journals and conferences of international repute and supervised two Ph.D. research works with another two on-going. Dr. L. Syam Sundar is the top 120,000 or in the top 2% of the most cited researchers in the world throughout their career and in their scientific field, according to a Stanford University study signed by the team led by John P.A. Ioannidis.

Shaik Feroz is currently working at Prince Mohammed Bin Fahd University, Kingdom of Saudi Arabia. Dr. Feroz obtained his doctorate in the field of Chemical Engineering from Andhra University, India in 2004, Postdoctoral Research Fellow from Leibniz University, Germany in 2015, M. Tech. in Chemical Engineering from Osmania University, India in 1998, B. Tech. in Chemical Engineering from S.V. University, India in 1992 and Post Graduate Diploma in Environmental Studies from Andhra University in 2003. Dr. Feroz has expertise in process engineering, plant design and troubleshooting, quality control using advance analytical equipment, wastewater treatment, solar energy systems (PV & CSP) for desalination, hot water systems and water treatment, synthesis of nano photocatalysts, simultaneous treatment of wastewater and production of hydrogen, environmental impact assessment, design, delivery and management of chemical and process engineering programs and tailor-made industry-based training programs related to chemical and process engineering. Dr. Feroz has more than 230 publications to his credit in journals and conferences of international repute and supervised four Ph.D. research works with another four on-going. Dr. Feroz is associated as the Principal Investigator/ Co-Investigator to a number of research projects and also involved as "Technology transfer agent" by Innovation Research Center, Sultanate of Oman. He has made significant contribution as editorial member and peer reviewer for various reputed international journals and conferences. He was awarded the "Best Researcher" by the Caledonian College of Engineering for the year 2014. Dr. Feroz has a total experience of 30 years in teaching, research and industry.

Faramarz Djavanroodi received his Ph.D. from Imperial College, London in 1989. He is presently the Chair of Mechanical Engineering Department and Director of Patent Office at Prince Mohammad Bin Fahd University (PMU), Kingdom of

Saudi Arabia. During his career, he stayed at Imperial College, at Qassim University, and at IUST. Prof. Djavanroodi has been involved in many international collaborations, including the KACST and SASO-funded projects. Dr. Djavanroodi has been featured among the world's top 2% Scientists List 2020 and 2021 according to recent study published by Elsevier. His current fields of research include nanostructure materials, life assessment of high temperature components, creep/fatigue fracture, finite element analysis. Dr. Djavanroodi is currently serving on the editorial board of the *Journal of Mechanics of Advanced Materials and Structures*.

Contributors

Saroj Kumar Acharya
Department of Mechanical Engineering
Institute of Technical Education and
 Research
S 'O' A Deemed to be University
Bhubaneswar, Orissa, India

Nisar Ahmed
National University of Sciences and
 Technology
Islamabad, Pakistan

Sirajuddin Ahmed
Department of Environmental Science
 and Engineering
Jamia Millia Islamia
Central University
New Delhi, India

Syed Murtuza Ali
Department of Mechanical and
 Industrial Engineering
National University of Science
 and Technology
Muscat, Sultanate of Oman

Summan Aman
University of Gujrat
Gujrat, Pakistan

Kim Andrews
Prince Mohammad Bin Fahd University
Kingdom of Saudi Arabia

M.M. Armendáriz-Ontiveros
Instituto Tecnológico de Sonora
Sonora, Mexico

M. Waqar Ashraf
Department of Mathematics and Natural
 Sciences
Prince Mohammad Bin Fahd University
Al Khobar, Eastern Province
Kingdom of Saudi Arabia

Anita Bisht
Department of Chemistry
Govt. (PG) College
Kotdwar Pauri, Garhwal,
 Uttarakhand, India

Chinmay Deheri
Department of Mechanical Engineering
Institute of Technical Education and
 Research
S 'O' A Deemed to be University
Bhubaneswar, Orissa, India

Faramarz Djavanroodi
Department of Mechanical Engineering
Prince Mohammad Bin Fahd University
Al Khobar, Eastern Province, Kingdom
 of Saudi Arabia

Lioz Etgar
The Institute of Chemistry
The Hebrew University of Jerusalem
Jerusalem, Israel

Shaik Feroz
Department of Mechanical Engineering
Prince Mohammad Bin Fahd University
Al Khobar, Eastern Province, Kingdom
 of Saudi Arabia

G.A. Fimbres-Weihs
School of Chemical and Biomolecular
 Engineering
The University of Sydney
Sydney, Australia

V. Harshitha
Department of Chemistry
University College of Science
  and Department of Studies
  and Research in Organic Chemistry
Tumkur University
Tumakuru, Karnataka, India

Ashfaq Khan
Robert Gordon University
Aberdeen, United Kingdom

Mushtaq Khan
Department of Mechanical Engineering
Prince Mohammad Bin Fahd University
Al Khobar, Eastern Province
Kingdom of Saudi Arabia

Zuhaib Ali Khan
University of Engineering
  and Technology
Peshawar, Pakistan

Umm E Kulsoom
East China University of Science
  and Technology
Shanghai, People's Republic of China

Anuj Kumar
Department of Chemistry
Govt. (PG) College
Kotdwar Pauri, Garhwal,
  Uttarakhand, India

Pendyala Naresh Kumar
The Institute of Chemistry
The Hebrew University of Jerusalem
Jerusalem, Israel

Sourav Laha
Department of Chemistry
National Institute of
  Technology Durgapur
Durgapur, West Bengal, India

Abinash Mahapatro
Department of Mechanical Engineering
Institute of Technical Education and
  Research
S 'O' A Deemed to be University
Bhubaneswar, Orissa, India

M. Amin Mir
Department of Mathematics and Natural
  Sciences
Prince Mohammad Bin Fahd University
Al Khobar, Eastern Province
Kingdom of Saudi Arabia

Bhagiratha Mishra
Department of Material Science
Altafuel Private Limited
Bokhara, Nagpur, Maharashtra, India

Subhradeep Mistry
Department of Chemistry
Hemvati Nandan Bahuguna
  Garhwal University (A Central
  University)
SRT Campus, Badshithaul,
  Uttarakhand, India

Ehsan Ullah Mughal
East China University of Science
  and Technology
Shanghai, People's Republic of China

Asim Mushtaq
Zhejiang Sci-Tech University
Hangzhou, People's Republic of China

P.C. Nethravathi
Department of Chemistry
University College of Science
  and Department of Studies
  and Research in Organic Chemistry
Tumkur University
Tumakuru, Karnataka, India

Laila Noureen
Peking University Shenzhen Graduate
    School
Shenzhen, People's Republic
    of China

Binayak Pattanayak
Department of Mechanical Engineering
Institute of Technical Education and
    Research
S 'O' A Deemed to be University
Bhubaneswar, Orissa, India
Chamba, Uttarakhand, India

Shailendra Prakash
Department of Chemistry
Govt. (PG) College
Kotdwar Pauri, Garhwal,
    Uttarakhand, India

Hummera Rafique
East China University of Science
    and Technology
Shanghai, People's Republic of China

S.G. Salinas-Rodriguez
Water Supply, Sanitation and
    Environmental Engineering
    Department
IHE Delft Institute for Water Education
Delft, Netherlands

S. Sethi
School of Chemical and Biomolecular
    Engineering
The University of Sydney
Sydney, Australia

Mehreen Shah
Department of Environmental Science
    and Engineering
Jamia Millia Islamia
Central University
New Delhi, India

Ananthakumar Soosaimanickam
R&D Division
Intercomet S.L.
Madrid, Spain

L. Syam Sundar
Department of Mechanical Engineering
Prince Mohammad Bin Fahd University
Al Khobar, Eastern Province, Kingdom
    of Saudi Arabia

D. Suresh
Department of Chemistry
University College of Science
    and Department of Studies
    and Research in
    Organic Chemistry
Tumkur University
Tumakuru, Karnataka, India

Muhammad Tayyab
East China University of Science and
    Technology
Shanghai, People's Republic of China
    and
University of Gujrat
Gujrat, Pakistan

Muhammad Zeeshan Zahir
University of Engineering
    and Technology
Peshawar, Pakistan

Acronyms and Abbreviations

PSC	Perovskite solar cells
SSL	Solid-state lighting
UV	Ultraviolet
UV-Vis	Ultraviolet-visible
PV	Photovoltaic
VB	Valence band
CB	Conduction band
CdTe	Cadmium telluride
a-Si	Amorphous silicon
CIGS	Copper indium gallium diselenide
SnO_2	Tin(IV) oxide
TCO	Transparent conducting oxide
CdS	Cadmium sulfide
i-ZnO	Intrinsic zinc oxide
AZO	Aluminum-doped ZnO
MgF_2	Magnesium fluoride
$CuInSe_2$	Copper indium selenide
CIS	Copper indium selenium
KOH	Potassium hydroxide
CVD	Chemical vapor deposition
TiO_2	Titanium dioxide
DSSC	Dye-sensitized solar cells
CNT	Carbon nanotubes
Nb_2O_5	Niobium pentoxide
Ru	Ruthenium
QDSSCs	quantum dot sensitized solar cells
LED	Light emitting diode
NWs	Nanowires
PSC	Perovskite solar cells
FTO	Fluorine-doped tin oxide
ITO	Indium tin-doped oxide
BHJ	Bulk heterojunction
HTL	Hole transport layers
SiNWs	Silicon nanowires
PVD	Physical vapor deposition
RIE	Reactive ion etching (RIE)
MCEE	Metal catalyzed electroless etching
MJSC	Multi-junction solar cells
PCE	Photovoltaic conversion efficiency
Voc	Open circuit potential (V)
J	Short circuit current (mA/cm^2) and

FF	Fill factor
AR	Anti reflection coating
EL	Emitter layer
BL	Base layer
CL	Confinement layer
BZO	Boron-doped zinc oxide
BL	Buffer layer
TJ	Tunnel junction
PSCs	Perovskite solar cells
PCE	Power conversion efficiency
2D	Two-dimensional
3D	Three-dimensional
PbI_2	Lead iodide
QW	Quantum well
RP	Ruddlesden-Popper
DJ	Dion-Jacobson
ACI	Alternating cations in the interlayer space
MQWs	Multiple well widths
BAAC	Molten-state n-butylacetate
FTIR	Fourier-transform infrared spectra
GABr	Guanidinium bromide
BzDAI	Benzyl diammonium iodide
DipI	Dipropylammonium iodide
HRTEM	High-resolution transmission electron microscope
GIWAXS	Grazing incidence wide-angle X-ray scattering
ThMA	2-thiophenemethylammonium
BA	Butyl ammonium
HR-SEM	High-resolution scanning electron microscopy
B-GUA	*B*-guanidinopropionic acid
GAI	Guanidinium iodide
5-AVAI	5-ammoniumvaleric acid iodide
XRD	X-ray diffraction
FBA	4,4,4-trifluorobutylammonium
FPEA	4-fluorophenylethylammonium
PL	Photoluminescence
PNCs	Perovskite nanocrystals
LEDs	Light-emitting diodes
PLQY	Photoluminescent quantum yield
PL	Photoluminescence
SQ	Shockley-Queisser
PQDs	Perovskite quantum dots
LARP	Ligand-assisted reprecipitation
QDSSCs	Quantum-dot sensitized solar cells
EtoAc	Ethyl acetate
MeOAc	Methyl acetate
LBL	Layer-by-layer

PTAA	Poly-triarylamine
CsOAc	Cesium acetate
GPC	Gel permeation chromatography technique
EDTA	Ethylene diamine tetracetic acid
TMSI	Trimethylsulfonium iodide
TOP	Trioctylphosphine
TBI	Tertiary-butyl iodide
EXFAS	X-ray absorption fine structure
MPA	3-mercaptopropionic acid
HTL	Hole-transporting polymer
KPFM	Kelvin probe force microscopy
TBAB	Tetrabutyl ammonium bromide
TMSI	Trimethyl silyl iodide
QRs	Quantum rods
R2R	Roll-to-roll
TiO_2	Titanium dioxide
Pt	Platinum
PECs	Photoelectrochemical cells
PbS	Lead sulphide
PANI	Polyaniline
ATO	Antimony-doped tin oxide
HER	Hydrogen evolution reaction
OER	Oxygen evolution reaction
CF	Copper foam
CVO	Copper-vanadium oxides
HER	Hydrogen evolution reaction
OER	Oxygen evolution reaction
MOFs	Metal-organic frameworks
EWS	Electrochemical water splitting
BET	Brunauer–Emmett–Teller
ZIF	Zeoliticimidazolate framework
CNTs	Carbon nanotubes
PXRD	Powder X-ray diffraction
TEM	Transmission electron microscopy
SEM	Scanning electron microscopy
AFM	Atomic force microscopy
HAADF	High-angle annular dark-field imaging
ICP-MS	Inductively coupled plasma mass spectrometry
EDS	Energy-dispersive X-ray spectroscopy
EDXA	Energy dispersive X-ray analysis
EDXMA	Energy dispersive X-ray microanalysis
XPS	X-ray photoelectron spectroscopy
BHT	Benzenehexathiol
TF	Thin film
LB	Langmuir-Blodgett
RDE	Rotating disk electrode

LDHs	Layered double hydroxides
HER	Hydrogen evolution reaction
HEP	Hydrogen evolution photocatalysts
CB	Conduction band
VB	Valence band
NFs	Nanofibers
TDs	Titanium dioxide dots
RHE	Reversible hydrogen electrode
SZO	Silver-doped zinc oxide
CZO	Copper-doped zinc oxide
PHE	Photocatalytic hydrogen evolution
SBUs	Secondary building units
MOL	Metal-organic layer
MOFs	Metal organic frameworks
ESR	Electron spin resonance
HER	Hydrogen evolution reactions
DBPs	Disinfection by-products
PPCPs	Pharmaceuticals and personal care products
CPs	Conducting polymers (CPs)
CPNs	Conducting polymers nanocomposites
TiO_2	Titanium dioxide
SnO_2	Tin dioxide
XRD	X-ray diffraction analysis
OPW	Oil produced water
DO	Dissolved oxygen
COD	Chemical oxygen demand
TDS	Total dissolved solids
TGA	Thermal gravimetric analysis
TOC	Total organic carbon
TP	Treatment period
HSs	Hollow spheres
LiBs	Lithium-ion batteries
$LiMn_2O_4$	Lithium manganese oxide
SEI	Solid-electrolyte Interphase
LCO	Lithium cobalt oxide
NCA	Nickel cobalt aluminum oxide
LCP	Lithium cobalt phosphate
LMO	Lithium manganese oxide
LFP	Lithium iron phosphate
LTS	Lithium titanium sulphide
MO	Metal oxide
PE	Polyethylene
PP	Polypropylene
$LiFP_6$	Lithium hexafluorophosphate
$LiAsF_6$	Lithium hexafluoroarsenate monohydrate
$LiClO_4$	Lithium perchlorate

$LiBF_4$	Lithium tetrafluoroborate
SWNTs	Single-walled nanotubes
MWNTs	Multi-walled carbon nanotubes
CNTs	Carbon nanotubes
NiO	Nickel oxide
AAO	Anodic aluminum oxide
SEI	Solid electrolyte interface
EVs	Electric vehicles
PEO	Polyethylene oxide
LOBs	Lithium-oxygen batteries
CNTs	Carbon nanotubes
HOPG	Highly oriented pyrolytic graphite
nGr	Nano-graphite
GNP	Graphene nanoplatelets
GO	Graphene oxide
CVD	Chemical vapor deposition
LPE	Liquid-phase exfoliation
PAHs	Polycyclic aromatic hydrocarbons
GNR	Graphene nano ribbon
THF	Tetrahydrofuran
PMMA	Polymethylmethacrylate
LIBs	Lithium-ion batteries
EVs	Electric vehicles
HEVs	Hybrid electric vehicles
MCMB	Meso-carbon microbead
VGCF	Vapor grown carbon fiber
MCF	Mesophase-pitch-based carbon fiber
MAG	Massive artificial graphite
ECs	Electrochemical capacitors
EDLCs	Electrochemical double layer capacitors
SC	Super capacitor
SRAM	Static random-access memory
CNTs	Carbon nanotubes
CVD	Chemical vapour deposition
GN	Graphene nanosheets
XRD	X-ray diffraction (XRD)
SEM/EDX	Scanning electron microscopy/energy-dispersive analysis
HRTEM	High-resolution transmission electron microscopy
IR	Infrared spectra
BET	Brunauer-Emmett-Teller
TGA	Thermogravimetric analysis
CV	Cyclic voltammetry
PTFE	Polyvinylidenedifluoride
SCE	Saturated calomel electrode
PLD	Pulsed laser deposition
NPs	Nanoparticles

MOFs	Metal organic frameworks
PdHx	Palladium hydride
Tc	Critical temperature
Hc	Crucial area
LHC	Large hadron collider
MRI	Magnetic resonance imaging
BCS	Bardeen, Cooper, and Schrieffer
RO	Reverse osmosis
PRO	Pressure retarded osmosis
SWRO	Seawater reverse osmosis
TOC	Total organic carbon
MOS	Microorganisms
TF	Thin film
SEC	Specific energy consumption
ERD	Energy recovery device
CNTs	Carbon nanotubes
MB	Methylene blue
CR	Congo red

1 Introduction

L. Syam Sundar, Shaik Feroz, and Faramarz Djavanroodi
Prince Mohammad Bin Fahd University, Kingdom of Saudi Arabia

In this modern era, our society faces a serious energy crisis and demand due to an increase in human population and urbanization. Energy consumption ranges from small-scale electronic gadgets to high power consuming industrial equipment. To maintain the ever-increasing energy demand, various energy resources are being exploited with advanced and new technologies. The efficient conversion and storage performance of an advanced technology depend on their material properties. It was recently discovered that nanostructuring of the device components leads to enhanced efficiency in terms of robustness and reliability of energy conversion and storage systems. Moreover, nanostructured materials have attracted great interest due to their unique physicochemical and electrochemical properties. Hence, the utilization of such materials in nanodimensions will create an enormous impact on the efficiency of various energy conversion and storage devices [1].

Nanomaterials, both organic and inorganic, owing to their characteristic properties such as quantum size effect, high surface-to-volume ratio, tunable surfaces, and easy functionalization with desired chemical moieties, have already proven their value in various technological applications. It is noteworthy to mention that with the growing research in the field of nanotechnology, scientists have developed various methods for their synthesis and characterization. Nanomaterials and nanostructures provide unique mechanical, electrical, and optical properties and have played an important role in recent advances in energy-related applications. Nanomaterials have played an essential role in developing novel energy generation devices. The unusual quantum effect at nanoscale benefits electron transport and band engineering in nanomaterials, which brings about an excellent performance for devices [2].

Nanomaterials can be metallic-based nanoparticles (ferrites, chromates, aluminates, bismutates, etc.) or carbon oxides (carbon nanotubes, graphenes, graphene oxides, etc.). The tailoring of the shape, size, and size distribution of nanoparticles, and the properties of hybrid nanoparticles, is achieved through different synthesis routes by modifying parameters such as pH, concentration of reactants, dopants, or stirring speed. Some of these methods are complex, involving the use of reduction agents with little to no impact on the environment and needing a longer reaction time or a high processing temperature to complete crystallization. These tailored properties of nanoparticles make them suitable candidates for technological applications in energy conversion and storage technologies such as solar cells, fuel cells, secondary

DOI: 10.1201/9781003364825-1

batteries, supercapacitors and other self-powered systems, photocatalysis, photoluminescence, biosensors, catalysis, humidity sensors, permanent magnets, magnetic drug delivery, magnetic liquids, magnetic refrigeration, ceramic pigments, microwave absorbents, corrosion protection, water decontamination, photocatalysis, antimicrobial agents, or biomedicine (hyperthermia) [3–6].

Nanotechnology and its products (or nanomaterials) mostly involve in the applications of renewable energies (such as solar and hydrogen fuel cells and energy storage device), which result in nearly zero CO_2 emissions. Increasing the use and efficiency of renewable/ecofriendly energy resources will overcome the use of burning fossil fuel, and at the same time reducing the consumption of current fuels is one way to slow down and ultimately stop global warming. Advanced development in nanomaterials is still in progress, which can economically absorb carbon dioxide from air, capture toxic pollutants from water, and degrade solid waste into useful products. Nanomaterials are efficient catalysts and mostly recyclable [7].

Hydrogen energy has potential to solve energy problems of the planet earth giving it a sustainable and safe future, resulting in a clean planet. H_2 storage and release is a key challenge which is solved by metal hydrides that can absorb hydrogen in atomic form and release it easily by raising their temperature or pressure. Lots of important advances have been made during the last one decade for developing nanostructure materials with high volumetric and gravimetric hydrogen capabilities. There is an urgent need to develop low cost, safe, and inexpensive nanostructured hydride materials possessing high hydrogen content and fast desorbing properties at low temperature and pressure [7].

Solar cell materials are the current prospects for clean energy research to offer strong power outputs from low-cost raw materials that are relatively simple to process into working devices. Although the potential of the material (perovskite-based solar cells) is just starting to be understood, it has caught the attention of the world's leading solar researchers who are trying to commercialize it. For example, organic inorganic lead halide perovskite solar cells are contenders in the drive to provide a cheap and clean source of energy with electrical power conversion efficiencies of over 29%. The highly efficient photovoltaic materials are recognized for optically high absorption characteristic sand-balanced charge transport properties with long diffusion lengths. Nevertheless, there are lots of puzzles to unravel to understand the fundamental basis of such advanced materials. Perovskite solar cells (PSC) are economically viable from a viewpoint of efficiency; however, commercialization is still challenging because of the toxicity of lead, long-term stability, and cost-effectiveness. Future research directions will benefit from finding lead-free light-absorbing materials. However, the reported efficiencies are thus far too low to commercialize PSCs. The cost-effectiveness of the raw materials and the fabrication processes is a significant issue. High throughput fabrication strategies with reproducible materials and processes should be developed.

Moreover, mechanical functionalities such as flexibility, stretchable, and long-term stability properties will lead to making PSCs more economically viable [7]. Chapter 1 is the introduction about the nanomaterials for the energy applications. Chapters 2 and 3 of this book deal with advanced nanomaterials for photovoltaic cells.

Chapter 4 of this book contains the strategies and progress in the fabrication of metal halide perovskite nanocrystal solar cells.

Luminescence nanoparticles have tremendous potential in revolutionizing many interesting applications in today's emerging cutting-edge optical technology such as solid-state lighting. Solid-state lighting (SSL) relies on the conversion of electricity to visible white light using solid materials. SSL using any of the materials (inorganic, organic, or hybrid) has the potential for unprecedented efficiencies. The development of novel mercury-free inexpensive nanomaterials that convert longer wavelength ultraviolet (UV) to blue light eventually into white light are eco-friendly with improved luminous efficacy, energy savings, long lifetime, and low power consumption characteristics [8]. Chapter 5 of this book discusses the effect of nanomaterials on the production of hydrogen from the electrochemical process of water splitting.

Fuel cell progress always attracts a lot of interest in the development of high-performance materials with novel design and preparing technologies in which nanomaterials have played a critical role. For example, in low-temperature fuel cells, nearly half of the cost of the fuel cell is linked to the electrocatalyst cost. To reduce the cost, catalysts with novel nanostructure and high performance have been developed and applied in fuel cells. Nanomaterials that can operate at low temperature or high temperature with high efficiency were developed and employed successfully in fuel cells. Chapters 6 and 7 emphasize the application of advanced nanomaterials in fuel cells.

Electrochemistry and nanoscience (and/or nanotechnology) are interdisciplinary fields, both of which are gaining increasing importance in the development of high-performance and reliable alternative energy devices (conversion or storage). Electrochemical capacitors are a special class of electric energy storage devices that are based on non-faradaic and/or faradaic charging/discharging at the interface between an electrode and an electrolyte. Current trends are on using nanostructure advanced materials in the fabrication of electrochemical capacitors for enhancing their performance. Super conductivity is an important property of the material that will potentially change the electronics world. Researchers are now using principles of nanotechnology to increase the superconductivity temperature of many materials. Chapters 8–10 present the application of advanced nanomaterials for electrochemical cells, electrochemical capacitors, and super conductors. Chapter 11 of this book is focused on the use of advanced nanomaterials for lithium-ion rechargeable batteries.

Nanomaterials of metallic or semiconducting nature that can be excited by radiation in the UV–vis (ultraviolet-visible) range have become very popular. These materials differ in characteristics from micrometric-sized materials because nanometer-sized particles exhibit new and unique magnetic, electrical, optical, and catalytic properties. A good photocatalyst should be characterized by (i) the ability to absorb radiation from a wide spectral range of light, (ii) the appropriate position of the energy bands of the semiconductor about the redox reaction potentials, (iii) high mobility and long diffusion path of charge carriers, (iv) thermodynamic, electrochemical, and photoelectrochemical stability. Moreover, for the reactions involving the resulting photocarriers to occur efficiently, it is necessary to effectively prevent recombination by separating electron-hole pairs, and then their transport to the semiconductor surface. Meeting these requirements by semiconductors is very difficult. Therefore, efforts are being made to increase the efficiency of photo processes by changing the electron structure, surface morphology, and crystal structure of semiconductors. Nanostructure semiconductors play a vital role in enhancing the efficiency of the photocatalytic processes [9]. Chapters 12 and 13 discuss the application of advanced nanomaterials in photocatalytic processes. Advanced nanomaterial applications for SSL are presented in Chapter 14. Chapter 15 present the application of nanomaterials in the water treatment process.

REFERENCES

1. Balasingam, S.K., Nallathambi, K.S., Jabbar, M.H.A., Ramadoss, A, Kamaraj, S.K., and Kundu, M., Nanomaterials for electrochemical energy conversion and storage technologies, *Journal of Nanomaterials*, (2019) 1–2. https://doi.org/10.1155/2019/1089842

2. Wan, J., Song, T., Flox, C., Yang, J., Yang, Q.-H., and Han, X., Advanced nanomaterials for energy-related applications, *Journal of Nanomaterials*, (2015) 1–2. https://doi.org/10.1155/2015/564097

3. Dippong, T., Synthesis, physicochemical characterization and applications of advanced nanomaterials, *Materials*, 16 (2023) 1674.

4. Dippong, T., Levei, E.A., Cadar, O., Formation, structure and magnetic properties of $MFe_2O_4@SiO_2$ (M = Co, Mn, Zn, Ni, Cu) nanocomposites, *Materials*, 14 (2021) 1139.

5. Dippong, T., Cadar, O., Levei, E.A., Deac, I.G., Borodi, G., and Barbu-Tudoran, L., Influence of polyol structure and molecular weight on the shape and properties of $Ni0.5Co0.5Fe_2O_4$ nanoparticles obtained by sol-gel synthesis. *Ceramics International*, 45 (2019) 7458–7467.

6. Dippong, T., Levei, E.A., Cadar, O., Deac, I.G., Lazar, M., Borodi, G., and Petean, I., Effect of amorphous SiO_2 matrix on structural and magnetic properties of $Cu0.6Co0.4Fe_2O_4/SiO_2$ nanocomposites, *Journal of Alloys and Compounds*, 849 (2020) 156695.

7. Lal, C., Advanced nanomaterials for energy and environmental applications, *Nanotechnology and Advanced Material Science*, 3(1) (2020) 1–2.

8. Tiwari, S., and Yakhmi, J.V., Recent advances in luminescent nanomaterials for solid state lighting applications, *Defect and Diffusion Forum*, 361 (2015) 15–68.

9. Feliczak-Guzik, A., Nanomaterials as photocatalysts—synthesis and their potential applications, *Materials*, 16 (2023) 193.

2 Nanomaterials for Solar Cells

Ashfaq Khan
Robert Gordon University, UK

Mushtaq Khan
Prince Mohammad Bin Fahd University, Kingdom of Saudi Arabia

Nisar Ahmed
National University of Sciences and Technology (NUST), Pakistan

Zuhaib Ali Khan and Muhammad Zeeshan Zahir
University of Engineering and Technology (UET), Pakistan

2.1 OVERVIEW OF SOLAR ENERGY

2.1.1 WORLD ENERGY MIX

With rapid growth in population and industry, there is an ever-increasing demand for energy, in particular electrical energy. Hence, low-cost energy generation has become one of the biggest challenges in the industrial revolution [1]. Fossil fuels have been a traditional choice for energy due to their availability, transportability, energy density and particularly the incomplete realization of their environmental impacts. Fossil fuels including coal, natural gas and crude oil account for approximately 82% of the total global energy consumption [2]. The environmental impact of extensive and unchecked use of fossil fuels can eventually lead to droughts and floods, which, in turn, risk food security and thus can threaten the very existence of the human race [1]. Also, fossil energy has created a dependence between producers and consumers and any threat to the energy supply such as the 2022 Russian-Ukraine war creates a very volatile situation leading to a global economic recession [3]. Fossil fuels are also a finite and rapidly depleting energy resource and would eventually need to be replaced with a more sustainable energy resource.

2.1.2 SOLAR AS ALTERNATIVE TO FOSSIL FUEL

Several alternatives such as hydel power, nuclear energy and renewable sources (solar thermal, photovoltaic (PV), wind, geothermal and biomass energy) have been

DOI: 10.1201/9781003364825-2

explored in the past as a replacement to fossil fuels. Among all renewable sources, solar PV has received a lot of attention. Every hour, the sun irradiates enough energy to fulfill the global energy demand for an year [4].

Significant improvements in renewable energy technologies in conjunction with increased environmental awareness have encouraged widespread interest in renewable energy. Increase in efficiency of renewable systems and the Paris Agreement 2015 have also promoted the adoption of renewable energy sources. Although all the aforementioned renewable technologies have received interest, solar energy has been at the forefront due to its advantages [5].

2.1.3 ADVANTAGES OF SOLAR ENERGY

Earth receives solar energy flux of 1.36 KW/m^2 which makes solar energy one of the most widely available and economical sources [6]. Solar PV systems do not have any moving parts, leading to low maintenance costs and much higher reliability [7]. The effectiveness of the PV systems depends upon weather, but since a large portion of the earth's surface is viable for solar PVs, this is not a critical limitation for energy harnessing [8,9]. On top of this, the PV technology is free of grid and easily scalable [10,11].

2.1.4 WORKING OF PV SYSTEMS

PV systems consist of two materials with two energetically separate bands. The bandgap is defined as the distance between the valence band (VB) and the conduction band (CB). High energy photons from sunlight can excite the electrons from VB to CB and thus create a flow of electrons. In general, exposure to higher solar irradiation will produce a higher electron flow. The materials used for PV should have two distinct charge carriers: electrons and vacancies/holes. Since this characteristic is only available in semiconductors, they are the only suitable candidates for solar PVs. Metals have free-flowing electrons and insulators have very large bandgaps and thus they are both unsuitable for PV applications. The conductivity of the semiconductors is controlled by adding impurities, a process called doping. Doping can be carried out by materials that have higher concentrations of electrons or holes. Depending on the type of doping, the PV cells can be p-type or n-type doped. The performance of solar cells is analyzed by four major parameters. These include the photovoltaic conversion efficiency (PCE), open circuit potential (V_{oc}), short circuit current (J) and fill factor (FF). PCE corresponds to the percentage of sunlight being converted into electricity. J is the current density and FF determines the maximum power of the solar cells.

2.1.5 CLASSIFICATION OF SOLAR CELLS

Based on the synthesis process and materials, we can classify solar cells into three generations. First-generation solar cells are made from single-crystal and multi-crystal silicon wafers. Silicon is the most widely investigated and used material for solar cells. Silicon-based solar cells have efficiencies in the range of 15–24% and are the commercially available solar cells in the market. Producing monocrystalline solar cells is a high-precision process. Therefore, monocrystalline solar cells are expensive but also have a 4–8% higher power output as compared to multi-crystalline solar cells [12,13].

Thin-film (TF) technology is used for second-generation solar cells with cadmium telluride (CdTe) being the most commonly manufactured solar cell of this type. TF solar cells are relatively less expensive but they have lower efficiencies than the first-generation solar cells. The only exception is copper indium gallium diselenide (CIGS)-based solar cells, as their lab scale efficiency exceeded 23% [14]. Amorphous silicon (a-Si) was developed in the 1980s and a thin layer of silicon is deposited on substrates by plasma-enhanced chemical vapour deposition. These solar cells have a lower cost but also have low efficiency of around 6% [12]. CdTe is a cost-effective and chemically stable making it an ideal candidate for TF solar cells. It has a high absorption coefficient which gives it an efficiency in the range of 9–11%. However, cadmium is a heavy metal that can accumulate in plants and animals making it toxic and limiting its use [12,15].

CIGS are TF solar cells where copper indium gallium and selenium layers are deposited on either plastic or glass. These TFs can be deposited on flexible surfaces, which makes them more versatile. CIGS cells can be produced by physical vapor deposition, electrochemical coatings and solgel method [16]. The third-generation solar PVs are the latest development and currently being researched extensively. They were developed to address the limitations of high production and installation costs of first- and second-generation solar cells and to make higher-efficiency devices. These solar PVs are not silicon based. So far, they are less efficient than first- and second-generation solar PVs, but due to low cost, they provide lower cost per watt generated. Third-generation solar cells exploit the benefits of nanotechnology to produce lightweight, flexible, efficient and highly economical devices. Third-generation solar cells have greater compatibility with the large-scale implementation of PVs. This generation primarily includes organic solar cells, dye-sensitized solar cells and perovskite solar cells (PSCs) [17,18]. Solar cells are often classified in three generations as described in this section. However, in some literature, emerging technologies like hybrid solar cells are termed as the fourth generation of solar cells [19].

2.1.6 Nanomaterials in Solar Cells

Nanotechnology is defined as the design, characterization, production and application of materials, devices and systems by controlling shape and size in the nanoscale, where nanoscale includes the size in the range of 1–100 nm. As the size of systems or particles decreases from macro to micro, there is no significant change in the properties of the material. But once the sizes are decreased to the nanoscale, statistical mechanical effects as well as quantum mechanical effects become more prominent and there is a change in the electronic properties due to quantum size effect [15]. Depending on the size and structure of nanomaterials, bandgap, optical path, charge carrier recombination, etc. can be tailored. Nanostructures can improve the properties of current solar cells and can help create unique materials and designs that can better harness solar energy at a lower cost. In essence, materials behave differently on nanoscale and can be used for different well-suited applications [17].

Nanotechnology is a highly researched multi-disciplinary field and has proven its applications in a wide variety of areas including physics, biological sciences, materials, electronics and solar energy systems. The advantages offered by nanotechnology have been used to improve PV energy harvesting. Nanotechnology can improve the absorption and retention of solar radiation, reduce solar reflection, improve efficiency by using nanowires (NWs)/particles, use nano particles to develop non-silicon-based PV devices and boost the efficiency of PV systems by adding NWs/particles and nano coatings [4].

As discussed in the earlier sections, nanotechnology can add significantly to PV systems and there can be multiple classifications of nanostructured solar cells. The conventional silicon-based solar technology is commercially matured. Theoretically, no significant improvement in PCE of first-generation solar cells is possible due to Shockley–Queisser limit [20]. However, second- and third-generation-based solar cells are not well established yet and have huge research potential. In the coming sections, we will discuss second- and third-generation technologies and the prospect of nanomaterials to enhance the PCE and viability of these technologies.

2.2 NANOMATERIALS IN SECOND-GENERATION SOLAR TECHNOLOGY

2.2.1 CADMIUM TELLURIDE-BASED SOLAR CELL

As solar energy has become a viable option, not all metals have been used extensively in PV technology. Initially, lithium was considered a suitable material for the contacts in the electrodes that transport the electron–hole pairs. However, it was later discovered that lithium had a major drawback – it gradually infiltrated into the silicon, resulting in the p-n junction moving further away from the surface and decreasing the PV cell's ability to generate current at the same level of efficiency [21–23]. Due to this, lithium was no longer considered a feasible option and not suitable to be used with other metals. Therefore, researchers began using doped semiconductors as electrical contacts in place of lithium [24].

CdTe is a PV technology that works by absorbing sunlight and converting it into electricity through the use of a TF of CdTe [25,26]. After silicon, the second most often used solar cell material worldwide is CdTe. For a sizeable segment of the PV industry, specifically in large-scale systems, CdTe-based solar panels are the only TF PV technology that is more cost-effective than crystalline silicone PV [25]. The active layers in CdTe-based PV are only a few microns thick and therefore considered TF technology. CdTe is known for its potential to produce highly efficient and cost-effective TF solar cells [27].

Similar to tin(IV) oxide (SnO_2), CdTe solar cells also have a transparent conducting oxide (TCO) for light to pass through and effectively transfer current. These layers improve the electrical properties between the TCO and CdTe through the use of intermediate layers such as cadmium sulfide (CdS). The CdTe film, which is the primary layer for photo-conversion, absorbs the majority of visible light. The light absorbed in the CdTe layer is then converted into current and voltage through the help of the electric field formed by the CdTe, intermediate and TCO layers [26]. Due to the relative abundance of cadmium, it has not experienced the same dramatic price fluctuations as silicon during the past two years [28]. However, compared to the normal efficiencies of silicon solar cells, the present efficiency of CdTe solar panels (10.6%) is substantially lower.

2.2.2 COPPER INDIUM GALLIUM SELENIDE-BASED SOLAR CELLS

Copper indium gallium selenide (CIGS)-based solar cells are TF-based PVs and are classified as second-generation solar cells. CIGS solar cells employ a CIGS layer to absorb light and convert it into electric current. These devices traditionally used soda lime glass as a substrate material, but nowadays some manufacturers are using flexible substrates like polyamide as well. The assembly of the CIGS solar cell is shown in Figure 2.1. The first layer after the substrate is back contact layer. Molybdenum is usually employed as a back contact due to its refractory nature and good electrical conductivity. After that, p-type CIGS absorber layer is deposited followed by n-type buffer layer usually composed of CdS. The next layer is intrinsic zinc oxide i-ZnO layer which is capped by aluminum-doped ZnO (AZO) layer. To reduce the reflection losses, anti-reflection coating like magnesium fluoride (MgF_2) is often deposited as a top layer in CIGS solar cells [29].

The first CIGS TF-based solar cell was developed in 1976 with PCE of 4% using copper indium selenide ($CuInSe_2$) powder (deposited via evaporation). However, in 1981 there was a breakthrough when 9.4% efficiency was gained from copper indium selenium (CIS) solar cells by co-evaporating different elemental sources. The bandgap of CIS is 1.04 eV, and it was enhanced by replacing some of indium with gallium to form CIGS. If all indium is replaced by gallium, the bandgap will be 1.7 eV. Hence, the band can be effectively engineered by using the optimum ratio of

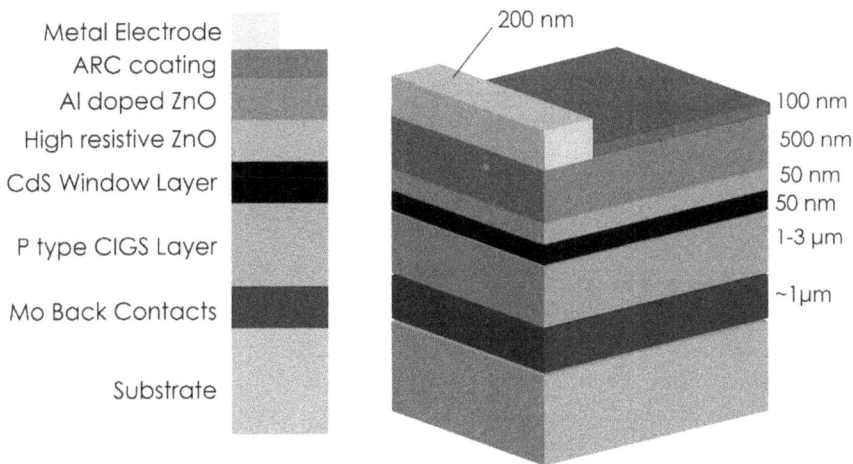

FIGURE 2.1 Schematic of CIGS-based solar cells.

indium to gallium. In recent years, by using rubidium and cesium instead of sodium, the best efficiency of 22.6% has been achieved [30]. Hence, these solar cells raise less health and environmental concerns as compared to CdTe-based solar cells.

These solar cells are in the very early stages of large-scale commercialization. The increase in the market share of CIGS-based solar cells is subjected to improvement in performance, uniformity and reliability. For CIGS production at an industrial scale, the three-stage co-evaporation process is employed. The efficiency of commercially prepared CIGS solar cells is about 16%, almost 6% lesser than that of lab-scale modules. This gap is due to the inhomogeneity of the CIGS layer in commercial solar cells. The efficiency gap can be downsized by optimizing the fabrication process to form a more homogenous film. Even with the 16% PCE, the commercial CIGS-based solar cells offer almost half the breakeven period as compared to commercially matured silicon-based solar cells.

2.2.3 Nano Texturing of Silicon for Solar Cells

Silicon is extensively used for solar cells due to its feasible bandgap. However, the smooth polished surface of silicon has substantial reflection and thus some of the solar energy is not fully harnessed. One of the ways to enhance the efficiency of solar cells is by texturing their surface using micro/nano textures that can improve light trapping and induce secondary reflections. Crystalline silicon with pyramid-shaped micro textures having nano roughness has been reported to have been used for PV applications with an efficiency of 17.5% [31]. These micro nano textures are generally produced by etching with potassium hydroxide (KOH)-based solutions but other methods such as plasma etching, CVD methods and laser texturing have also been used for this purpose [32,33]

2.3 NANOMATERIALS IN THIRD-GENERATION SOLAR TECHNOLOGY

2.3.1 TiO$_2$-Based Dye-Sensitized Solar Cells

In 1960, it was found that when organic dyes were exposed to light, they were able to generate electricity at oxide electrodes. Later in 1972, through understanding the process of photosynthesis, a mechanism was developed for the generation of electricity via dye-synthesized solar cells (DSSCs). The first DSSC called the "Gratzel cell" was invented in 1991 by sensitization of TiO$_2$ films utilizing a Ru biprydine complex and the PCE of 7% was achieved. This led to a wide interest in DSSCs as they provided a cheaper alternative to silicon. TiO$_2$ is one of the earliest and most widely used DSSCs developed due to its simplicity and ease of manufacturing under ambient conditions. It also has a large bandgap, suitable band edge levels and better cost. Specifically, the crystalline phase of TiO$_2$ "anatase" is most widely utilized due to its excellent charge transport characteristics [34–37]. As the name suggests, DSSCs contain dye which is used as sensitizer. In addition, it contains electrolyte and electrodes. The working electrode is made of nanostructured, mesoporous conductive oxide on the transparent surface generally fluorine-doped tin oxide conducting glass.

A non-aqueous electrolyte having a redox couple mediator is placed between the electrodes. Since the purpose of electrolyte is to exhibit efficient ionic transport without any hindrance, so it must have low viscosity and low resistivity [38].

Over the past 25 years, many researchers have come up with slight variations in the design. In a PV cell, the dye absorbs the light where excited electrons are transferred into the conduction band of TiO_2. The dye produces electrons and transfers them to TiO_2 and the holes to the electrolyte. The electrons from TiO_2 are then collected on the electrode and on to external load. At the counter electrode, the electrons reduce to restore the dye to the initial state and the entire cycle is repeated. Efficiency of the DSSCs is dependent upon many factors including the properties of TiO_2 powder and its generation, preparation of conductive film on the electrodes, electrolyte, transmission of the glass and type of dye. It has also been reported that efficiency can be increased by using hydrothermal or sol-gel to add carbon nanotubes (CNTs) in TiO_2 [39–41].

The PCE of TiO_2-based DSSCs can also be enhanced by modifying the morphology of TiO_2. Different nanostructures of titania such as NWs, nanotubes, nanorods and hollow spheres have been successfully synthesized. Table 2.1 shows the PCE of TiO_2 with different morphologies. Among all the materials, nanofibers are particularly effective at absorbing light in the visible range. Nanofibers of titanium dioxide (TiO_2) exhibit a high degree of light absorption in the visible region. This phenomenon is partially attributed to the extension of the optical path length, which arises from the scattering of light in a layer of these nanofibers. The extent of this scattering is influenced by the wavelength of the incident light, as well as the size, shape and orientation of TiO_2 nanofibers [42].

Although TiO_2 is proven to be a good photo anode, it still lacks the ideal conductivity. The electronic properties of TiO_2 can be enhanced by doping with different elements in very small amounts. Table 2.2 shows the performance of TiO_2 based on different elemental dopants.

Table 2.3 displays the PCEs of different metal oxide-based DSSCs. The table clearly shows that titanium oxide-based DSSCs are much more efficient than others. However, TiO_2-based DSSCs use ruthenium-based dyes which are relatively expensive. It has created a huge research gap and there is a potential to explore other metal oxides like ZnO, SnO_2 and niobium pentoxide (Nb_2O_5) in DSSCs as they are compatible with relatively economical dyes.

TABLE 2.1
Efficiency of DSSCs with Different Morphologies

Structure	PCE	Year	References
Nanofibers	10.3	2008	[43]
Nanowires	6.00	2017	[44]
Hollow spheres	10.34	2008	[45]
Nanotubes	4.8	2009	[46]

TABLE 2.2

PCEs of Doped TiO$_2$ Photoanodes DSSCs

Dopant	PCE	Year	References
2.5 mol% Nb-doped TiO$_2$	8.00	2011	[47]
0.1% Ce doped-TiO$_2$ 7.65	6.3	2012	[48]
0.2% Scandium (Sc) doped-TiO$_2$	9.9	2013	[49]
Ni-doped TiO$_2$	6.84	2014	[50]
N-doped TiO$_2$ multilayer photoelectrode	8.0	2017	[51]
S-doped TiO$_2$ nanoparticles	6.91	2018	[52]

TABLE 2.3

Efficiencies of Solar Cells Based on Metal Oxides Photo Anodes

PE Material	PCE	Year	References
ZnO	5.60	2008	[53]
ZnO NPs	5.31	2018	[54]
SnO$_2$	2.85	2016	[55]
Nb2O5 NPs	3.15	2016	[56]
TiO$_2$	10.2	2014	[50]
TiO$_2$	14.2	2015	[57]

2.3.2   ZnO-Based DSSCs

After DSSCs were discovered, significant research was directed towards creating high-performance solar cells with high wattage per unit area and reasonable costs. Although TiO$_2$ has been extensively studied due to its aforementioned properties, there has been huge interest in zinc oxide and tin oxide. Similar to TiO$_2$, these oxides must have a high surface area which is achieved by nano-structured and mesoporous surfaces [58] Most TiO$_2$-based DSSCs use ruthenium-based dyes which are expensive, so work was done to develop cheaper dyes including metal-complex porphyrin, oligothiophene, coumarin and indoline dyes [59,60].

Zinc oxide (ZnO) is of interest for DSSCs due to its physical properties and stability. Zinc oxide has excellent structural, electrical and optical properties along with good mechanical strength. While ZnO has similar optoelectronic properties to TiO$_2$, such as refractive index and conduction band edge, it possesses a greater mobility of electrons [61,62]. The electron diffusion coefficient of ZnO is significantly higher (about ten times) than that of TiO$_2$ in bulk form. This has resulted in the use of ZnO in different applications such as biosensors, transistors, lasers, piezoelectric devices and DSSCs. Also, ZnO is a direct band semiconductor and it can be converted into various crystalline forms to achieve different nanostructured anodes for DSSCs. Also, ZnO can be coated on all types of substrate with TCO and can form a variety of structures. This gives ZnO the ability to form flexible and tuneable cells at a low cost. Also, higher performance can be achieved with better identification of dye and redox couple [63–65].

Despite its advantages, ZnO suffers from chemical instability and lower efficiency. The efficiency of ZnO is less than TiO_2, but it could be attributed to the fact that ZnO has not been so extensively explored. Apart from low surface area which can limit its functionality, ZnO has inadequate chemical stability in both alkaline and acidic environments [66]. Nonetheless, the advantages of ZnO are significant and it has been considered as a viable alternative for TiO_2 for DSSCs but needs more research to fully explore its potential.

2.3.3 SnO_2 DSSCs

TiO_2 has been previously utilized as a photoanode material in DSSCs. However, its high surface area caused slow electron movement which led to high charge recombination [1,67]. ZnO is another commonly used photoanode material with a bandgap similar to TiO_2 but is chemically unstable with Ruthenium (Ru) dyes [68,69]. SnO_2 is one of the first materials that exhibited PV characteristics. SnO_2 has a lower conduction band (CB) than TiO_2 and its nanostructures are employed as photo anodes in DSSCs [4,70]. SnO_2 showed intrinsic lower CB results in a lower V_{OC} despite the higher J_{SC}. However, SnO_2 yields low photocurrent and has a lower overall efficiency compared to TiO_2. SnO_2 performance in DSSCs can be improved by doping multiple divalent metal oxides of Cu, Ni, Pb and Cd [5,71]. Doping improves the PCE compared to pure SnO_2-based DSSCs.

2.3.4 Nb_2O_5 DSSCs

In DSSCs, anatase has the best performance to date as a mesoporous layer. But niobium oxides have very interesting properties including relatively high conduction band edge, electron injection efficiency and chemical stability and can be a suitable replacement for anatase [72,73]. It can also be used as a top/under layer to prevent recombination losses [74]. Among all niobium oxides, niobium penta oxide (with oxidation state 5) is the most stable in terms of thermodynamics. Nb_2O_5 with 99.9% purity is used to manufacture fine ceramics that are utilized in the production of capacitors for optical lenses and electronics. Nb_2O_5 is a solid white power that is stable in air and insoluble in water but dissolved by the alkaline or acidic fluxes [75]. It has more than 15 structures, hence polymorphic in nature [76,77]. The Nb_2O_5 band-gap energy depends upon the crystalline phase. It can vary from 3.2 eV up to more resistive values of 5 eV [76]. The orthorhombic version also known as T-Nb_2O_5 has been most frequently utilized in solar energy applications due to its low annealing temperature [74,78]. In DSSCs, the annealing temperature is restricted by the use of conductive glass substrates (FTO). Above 600°C, both the electrical and optical properties of the substrates are affected. [74,79].

The forward and backward electron transfer rates must be carefully designed to achieve highly efficient devices [80]. To prevent recombination of injected electrons with the oxidized species, the dye in DSSC must be able to quickly regenerate after absorbing light and transferring an electron to Nb_2O_5. Electrons must efficiently reach external circuit through the semiconductor oxide. Nb_2O_5 film is very useful in making and improvement of PV devices by working as electron blocking layer which reduces device hysteresis [72,78,81,82].

2.3.5 QUANTUM DOT SOLAR CELLS

As discussed earlier, bandgap is the most important parameter for a material to be considered for PV applications. The ideal bandgap being 1.34 eV and for this reason silicon is extensively used as it has a band of 1.1 eV close to the ideal. The maximum theoretical efficiency of these solar cells, as limited by the Shockley–Queisser limit, is 33% [20]. However, there are ways to achieve higher efficiency by using tandem or multi-junction solar cells (MJSCs) with much higher efficiencies. Quantum dots are semiconducting particles with sizes smaller than Bohr excitation radius which gives them the unique characteristic of being quantum wells with tunable bandgaps. Usually, the bandgap is inversely proportional to the size of the particles and the bandgap can be adjusted by changing the particle size. Sun's high-energy photons are used to produce electricity as dictated by the Shockley–Queisser effect, but the low-energy photons are not utilized. Lower bandgap quantum dots can be used to utilize the low-energy photons to generate electricity. Tunability of the bandgap makes them ideal for PV applications [65].

Quantum dots are used on the anode of DSSCs and such systems are referred to as li. Their working principles remain the same as DSSCs [83]. The mesoporous semiconductor materials (such as TiO_2 and SnO_2) used for DSSC are coated with quantum dots. Charge is generated in the quantum dots under solar illumination that is transferred to the mesoporous semiconductor. All other aspects of energy harvesting remain the same as typical DSSCs. Quantum dots have been used with a variety of materials including CdSe, CdTe, CdS, ZnSe, Ag_2Se, Ag_2S, InAs, InP, In_2O_3, PbS, Si, graphene and many more. QDSSC has garnered great interest due to its low cost, ease of processability, tunability of bandgap and has a theoretical conversion efficiency of 44%, whereas solar cells have been fabricated to give an efficiency up to 16.6 % [84]. $CsPbI_3$-based QDSSCs have be reported to have an efficiency of 15.1% [85].

2.3.6 ORGANIC NANOCOMPOSITES FOR SOLAR CELLS

The PCE of commercial PVs are about 22–23%, which is 2–30% less than the finest laboratory crystalline silicon solar cells [86]. In 1992, invention of organic PVs led to decades of intense research in this field. P-type and N-type materials are combined and arranged to create nano 3D woven networks, according to the bulk heterojunction (BHJ) idea. The nanometer-sized domains are organized in an ideal way, which enables the efficient generation and transport of charge carriers towards the corresponding electrodes. Under illumination, photons are absorbed by materials that are either p-type or n-type, producing the corresponding extractions. The main benefit of BHJ PVs over previously known planar junction devices is the dramatic increase in interface of the p- and n-type materials [87].

A significant advancement in the understanding of organic solar cells was made through design and study of various organic semiconductor materials, mainly conjugated polymers, as light-collecting components. The blend of donor and acceptor components in solar cells caused voltage losses of 0.8–1.0 eV due to the formation of charge transfer states and non-radiative relaxation processes [88]. One of the key benefits of organic solar cells is that they can be applied in situations where traditional solar panels made of rigid inorganic materials are not suitable such as in the form of semitransparent PVs for smart windows or as flexible solar cells for textile, electronic skin, food packaging and other applications [89].

Organic PVs are also appealing for aesthetic reasons because it is possible to shape them as desired, e.g., by creating devices in the shape of a plant leaf [90]. Organic solar cells have an advantage over traditional inorganic PV modules in that they can efficiently capture low-level, scattered light and achieve PCEs over 20% when exposed to fluorescent or LED white light. This makes them ideal for a variety of indoor settings. As a result, organic PVs are becoming increasingly important in shaping our future and will continue to have a growing impact in the coming years [91].

2.3.7 NANOROD-BASED DSSC

A PCE of 13% has been reported for an N719-loaded DSSC having mesoporous anatase. But they suffer from slow electron transport speed, which is a major shortcoming of these solar cells [92,93]. One potential solution to this challenge is to employ 1D nanostructured metal oxides such as nanorods, NWs and nanotubes. Nanotubes or nanorods can enhance the lifetime of films by up to three times that of nanoparticle-based films [94]. Numerous methodologies, i.e, hydrothermal synthesis, chemical vapor deposition, direct oxidation of Ti substrate and template assistant approach have been utilized for the synthesis of various TiO_2 nanostructures. To date, a wide variety of anatase morphologies have been produced using hydrothermal methods, including nanostars, nanoflowers, nanorods, nanoparticles and others. However, these morphology adjustments were performed under different experimental conditions. Effective dye loading and electron transportation through TiO_2 film are crucial parameters that depend upon the porosity of the photoanodes, grain boundaries between two nanostructures, surface area and the photoanode. Therefore, it is important to be careful about the nanomorphology of TiO_2 when designing the photoanode for DSSCs. It is well established that by architecting nanoparticles, particle grain boundaries could be tailored. Architecting nanoparticles involves the design and synthesis of nanoscale particles with specific physical and chemical characteristics. An increase in the number of grain boundaries can impede the rapid movement of electrons, thereby reducing efficiency [65].

Economical production of solar cells is being extensively researched. Modifying the metal oxides for nano-morphologies through eco-friendly means is a challenge for energy conversion and storage [92]. Using PVs to harness solar power can be a solution to the global problem of finding sustainable energy sources.

2.3.8 PEROVSKITE SOLAR CELLS

This type of solar cell incorporates the perovskite-structured compounds as an active layer to harvest solar energy. Hybrid organic-inorganic materials based on lead or tin halides, known as perovskite structure, typically have the chemical composition AMX3, where A represents a cation, X is a halogen and M is a metal. The stability of the crystal structure is affected by the size of the cation and its interactions with the corner-sharing cation (MX_6^{4-} octahedral) in the MX6 octahedral structure [95].

PSCs are generally composed of five layers (Figure 2.2). These include a photoanode material like FTO (fluorine-doped tin oxide) or indium tin-doped oxide (ITO) followed by the hole transport layer (HTL), photoactive perovskite layer, electron transport layer and metal contact. As the sunlight irradiates the photoactive perovskite layer, photons with the energy more than the forbidden band width are absorbed and then excite the electrons around the nucleus and generate excitons (electron–hole pairs). The photoactive layer of perovskite has a very low binding energy for electrons, which causes the electron–hole pairs to become free charge carriers. The holes move into the HTL and the electrons move into the electron transport layer, allowing them to reach their corresponding electrodes. This results in an electric current flowing across the electrodes [96].

In 2009, the halide perovskite was used as a light sensitizer in dye-sensitized liquid junction solar cells and the PCE of 3% to 4% was attained [97]. Halide perovskites are a popular choice for solar applications because of their exceptional optoelectronic qualities [98]. Kim et al. in 2012 reported a long-lasting and highly reliable PSC with an improved PEC of 9.7% [99]. Since then, the use of halide perovskites in optoelectronics and an examination of the anomalous electronic characteristics of material have been the subjects of extensive research resulting in the achievement of 23.3% PCE [100]. Significant advancements in PSCs can be

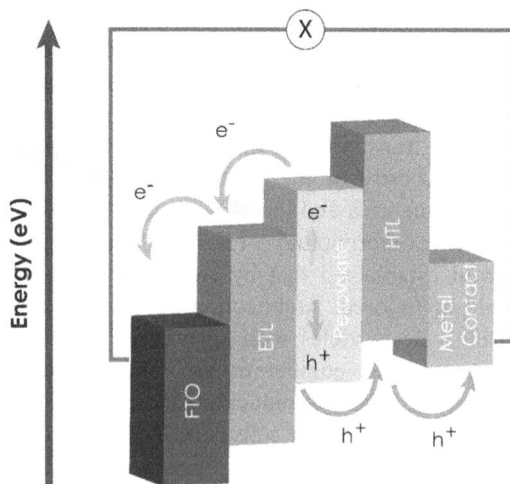

FIGURE 2.2 Working principle of perovskite solar cells.

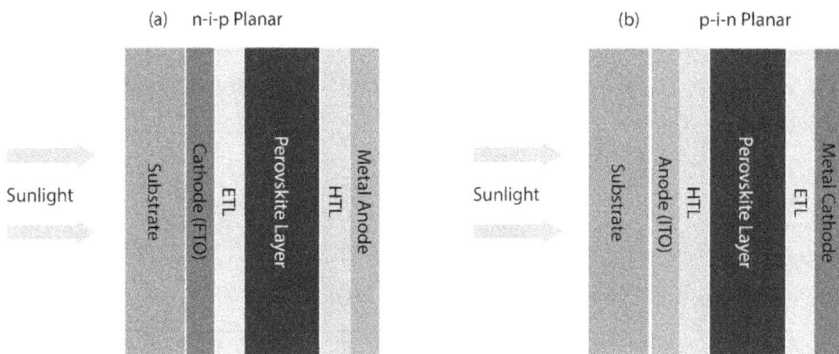

(a) n-i-p Planar

(b) p-i-n Planar

Sunlight — Substrate | Cathode (FTO) | ETL | Perovskite Layer | HTL | Metal Anode

Sunlight — Substrate | Anode (ITO) | HTL | Perovskite Layer | ETL | Metal Cathode

FIGURE 2.3 PCS classification: (a) n-i-p and (b) p-i-n.

associated with the tunable bandgap of perovskites, long charge carrier diffusion length and improvements in perovskite film deposition [101,102]. The main barrier to commercialization has been the stability of halide perovskites. The key to long-term stability in PSCs may lie in understanding the device architecture and PSC degradation mechanisms.

Despite being first investigated a decade ago [103], halide perovskites have only recently received significant attention after their effective use in PV devices. The PV material absorbs light which initiates the functioning of a solar cell. Perovskite material has several benefits over traditional solar materials, including a high absorption efficiency and easy optical bandgap tuning. Halide perovskites have significant absorption due to their direct bandgap materials which make them useful for PV technology [99,104].

Different deposition techniques can be used to create the perovskite layer. Perovskite films produced by various deposition techniques have varying grain sizes, electron and hole mobilities and concentrations of defects in the bulk. The deposition techniques can broadly be divided into vacuum deposition and solution-processed deposition [96].

PSCs are classified as "p-i-n" structured and "n-i-p" structured (Figure 2.3), based on the direction of electron flow as shown in the figure. In "n-i-p" structured PSCs, light-emitting electrons travel from the perovskite layer to underneath the electron selective contact and are gathered at the contacts. The perovskite layer of "p-i-n" structured PSCs allows electrons to flow to the top electron selective contact where they are captured at the metal contact [105].

2.4 CARBON AND SILICON-BASED NANO-ENGINEERING IN THIRD-GENERATION SOLAR CELLS

2.4.1 CARBON ALLOTROPES IN DSSCs

Carbon-based nanomaterials have become the subject of research in recent years due to the advancement of third-generation solar PVs [106,107]. Carbon materials are essential components in many applications, notably catalysis, super capacitors

TABLE 2.4

PCEs of DSSCs Based on Carbon Nanocomposites with Anatase TiO$_2$

Photoanodes	PCE	Year	References
MW CNTs	7.37	2009	[112]
Nanoporous carbon	3.40	2012	[113]
0.3% MWCNTs	9.05	2013	[114]
Graphene	4.03	2015	[115]
Nitrogen-doped graphene@NiO$_2$	9.75	2017	[116]

and electrode materials [108]. This is due to the advantages of carbon materials such as their abundance as a resource, ease of functionality and quick charge transfer rate [109]. In this regard, for portable energy conversion devices, graphene-based PVs with CNTs acting as transporting layers or electrodes have drawn significant interest [110].

Structurally, graphene is a single sheet of graphite that is one carbon atom thick. A CNT, on the other hand, is a sheet of graphene that has been rolled into a hollow cylindrical shape. The weak interlayer forces of graphite can be easily overcome to create a graphene sheet. The resulting graphene is made up of a hexagonal arrangement of carbon atoms connected by sp [1] bonds, which can extend indefinitely in a 2D space. Due to their exceptional chemical, physical, electrical and optical properties, graphene and CNTs have emerged as leading contenders for use in a wide range of commercial and academic endeavors. The high efficiency and affordable manufacturing cost lead to graphene and CNTs increasingly being used in the PV industry [111]. A summary of the impact of combining carbon nanomaterials with anatase TiO$_2$ on the efficiency of DSSCs is given in Table 2.4.

Recently, Casaluci et al. described the fabrication of larger DSSC modules using a practical spray coating approach with graphene ink as a counter electrode. The TCO substrate was treated with a spray of graphene-based ink that was produced via liquid phase exfoliation of graphite [117].

2.4.2 CARBON-BASED ALLOTROPES IN ORGANIC SOLAR CELLS

In polymer solar cells, Park et al. attained power conversion efficiency of 6.1% and 7.1% by using flexible anode and cathode made of graphene-based materials. CNTs are used along with polymers in PV solar cells as electron acceptors [118,119]. Graphene and CNTs because of their high electrical conductivity, high carrier mobility and transparency have the potential to be used as transparent electrode materials for solar cells. By incorporating graphene-based materials into the device structure, it is possible to enhance the performance and stability of solar cells, resulting in an increased PCE [120,121]. BHJ polymer solar PVs have grown in popularity due to their lower manufacturing costs, and they are relatively simple to produce on flexible substrates using roll-to-roll solution techniques [122].

TABLE 2.5
Summary of Perovskite Solar Cell Employing Carbon Nanomaterials in an Interfacial Layer, Electron Transport Layer and HTL

Use	Materials	PCE	Short-Circuit Current Density	Open Circuit Potential	Fill Factor	Year	References
Interfacial layer	GO	16.11	22.06	1.03	71	2016	[126]
	CQD	16.4	22.64	1.019	71.6	2017	[127]
	EVA-CNT	17.1	22	1.07	72	2019	[128]
Electron transport layer	PCBM	3.9	10.32	0.6	61	2013	[129]
	Fullerene	20.14	23.95	1.08	77	2018	[125]
	GQD	17.8	21.92	1.12	76	2017	[130]
Hole transport layer	MWCNT	15.5	20.3	1.1	70	2016	[131]
	SM-CNT	18.3	22.7	1.12	73.8	2020	[132]
	CNT@G	19.56	23.31	1.08	77	2018	[133]

2.4.3 CARBON NANOMATERIALS IN PEROVSKITE SOLAR CELLS

Graphene and CNTs are effective HTL for solar cells due to their property to inhibit electron–hole recombination on the electrode. Recently, graphene oxide (GO) has been used as an HTL to enhance the efficiency of PSCs [123]. They are also utilized as an interface layer among the perovskite layer and HTL in order to reduce the surface traps, which ultimately promotes rapid hole extraction across different function layers. This method enhanced the efficiency of perovskite-based solar cells up to 16% [124]. Additionally, the use of carbon allotropes in the electron transport layer remarkably enhanced the efficiency to 20.14% [125]. Table 2.5 provides an overview of the implementation of carbon nanomaterials in various layers such as the interfacial layer, electron transport layer and HTL.

2.4.4 SILICON NANOWIRE-BASED SOLAR CELLS

The global energy shortage crisis can be addressed by utilizing the potential of PVs. Silicon (Si) is the most durable, non-toxic and abundant substance and possess brilliant electronic characteristics. Despite this, the cost of PV technology has been unable to remain competitive in large-scale industrial production. The solar cell market primarily utilizes crystalline silicon wafers that are typically between 0.2 and 0.3 mm thick. Silicon materials attribute significantly to the overall cost of PV cells [134]. TF-based silicon solar PVs have been a new research interest and appear to be a suitable alternative [135,136]. Inexpensive substrates, i.e., glass, stainless steel or plastic are used by such devices. Reducing the thickness of the silicon layers can lead to inadequate light absorption, and low carrier collection can be caused by the non-optimized crystal structure. These factors will eventually result in low energy conversion efficiency. Light-trapping technology can be used to improve the performance of TF solar cells by providing effective light absorption and broad-band enhancement. Light-trapping structures can also be designed into nano-holes, nano-cones and NWs to boost the overall performance of cells. Scientists are focusing on silicon nanowires (SiNWs) because of their wide potential applications, e.g., chemical sensors, field-effect transistors, solar cells and field emitters [137]. The challenge is to make NW synthesis controllable, economical and simple. Two methodologies can be utilized for preparing SiNWs:

1. Bottom-up approach: Vapor deposition techniques (both CVD and PVD) and laser ablation.
2. Top-down approach: Etching and lithography.

The light absorption efficiently increases by using NWs in a radial configuration. Research in SiNWs is directed to achieve high optical carrier collection efficiency and ultra-low reflectance in the radial direction. The use of nanostructures in solar cells can result in shorter carrier collection paths and a reduction in carrier recombination rate. Additionally, solar cells with a high surface-to-volume ratio have both positive and negative effects on the overall characteristics of the cell. Structural properties enhance the light absorption and junction area, but the carriers have shorter lifetimes due to the high rate of recombination brought on by surface and interfacial defects, which lowers efficiency. Multiple research reports are made on surface passivation. Investigations are made into passivation layers of SiO_2, Al_2O_3, SiNx and hydrogenated amorphous Si. The nanostructure can be subjected to surface treatment procedures to minimize surface and interfacial flaws. 1D SiNWs are becoming an area of interest as an encouraging material for emerging solar PVs. Multiple methodologies can be implemented for the preparation of SiNWs. The vapor-liquid-solid growth method is a highly effective method for growing 1D SiNWs [138,139].

Other 1D nanostructure preparation strategies include top-down methods of reactive ion etching (RIE) and metal-catalyzed electroless etching (MCEE). In RIE, the Bosche process and nanoimprinting along with a Si substrate are used to fabricate SiNW arrays. A 30-nm thick TF of chromium is patterned on silicon using UV-imprint lithography. Pressure and heat are usually applied during the imprinting procedure. For MCEE, a simple and economical solution etching process is utilized to prepare 1D SiNWs. Electroless etching can be performed by dipping Si wafers in a solution of $AgNO_3$ and hydrofluoric acid for the creation of wafer-scale SiNW arrays. The MCEE process can result in highly oriented Si nanostructures and SiNW arrays. This results in SiNWs with an aspect ratio of 25 to 100 and random distribution [140]. SiNWs are used to substitute the traditional texture with anti-reflection characteristics. The light undergoes many internal reflections, causing an extended optical path prior to absorption. Since the prepared SiNW samples were black, the material was named "black silicon" [141]. For planar-junction solar cells, electron-and-hole is created when Si absorbs a photon of light. The generated charge is far from the electrode and there is a greater probability for recombining. Because of this, producing electricity requires a large charge carrier diffusion distance. On the other hand, solar cells with nanostructures absorb light from many directions and thus are more effective.

Silicon nanostructure is an efficient method for creating economical and high-performing solar cells due to its superior properties for light absorption, carrier transport and charge separation. An approach for creating economical Si-based solar cells is the use of radial p-n junction SiNW arrays that can suppress reflectance over a broad range of spectral wavelengths and improve light trapping and reduce carrier recombination. SiNW-based solar cells hold a lot of potential but further research is needed to address the challenge of recombination of carriers [65].

2.4.5 MULTI-JUNCTION SOLAR CELLS

In 1961, Shockley and Queisser using the first-principle calculations demonstrated that the PCE of single-junction solar cells cannot exceed 33% of the total sunlight it receives on the surface [142]. It is due to the fact that single-junction solar cells are designed with single bandgap. If the bandgap is kept low, then only the solar radiations with low energy will take part in PV process and the radiations with high energy (greater than the bandgap) will be wasted. Similarly, if the bandgap is kept high, the radiations with low energy will be wasted. So, there was a need of solar cells that can utilize most of the solar spectrum.

MJSCs were introduced to surpass the Shockley–Queisser limit. These cells are designed with different bandgaps by employing different semiconductor junctions. These multiple p-n junctions are connected in series to absorb different wavelength ranges of sunlight, and it allows the PCE to surpass the theoretical Shockley–Queisser limit.

The working mechanism of a MJSC is displayed in Figure 2.4. When the sunlight penetrates through the top contact layer of a multi-junction cell with the highest bandgap, energy will collect light (photons) from the highly energized solar radiations. Only the high-energy radiations compatible with the high bandgap will take part in the PV mechanism and the low-energy radiations will pass through to the next layer, which has a lower bandgap. Each semiconductor absorbs the photons with energy between its energy bandgaps. Hence, a large part of solar radiations

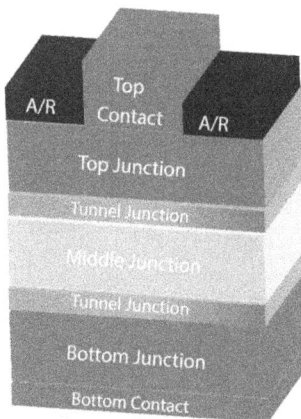

FIGURE 2.4   Schematic of a multi-junction solar cell.

(both low and high energy) will be effectively used, and it will ultimately increase the PCE of solar cells [143].

In 1988, double heterostructure gallium arsenide (GaAs) tunneling junction approach was used for the first time and a PCE of 20% was achieved [144]. After double heterostructure, three-layer approach was used and the PCE was enhanced further. Over the next decades, more junctions were stacked with a five-junction solar cells having an efficiency up to 35%. Recently, the efficiency of 39.2% was achieved under 1 sun and 47% under 143 suns using the six-junction approach [145].

2.5 EFFICIENCY OF SOLAR CELLS

In conclusion, the development of solar PVs over the past 10 years has resulted in a significant reduction in the cost per kWh of solar energy. However, making this technology accessible to third-world countries remains a challenge. The first generation of solar cells, based on conventional silicon, have efficiencies ranging from

TABLE 2.6
Efficiency Table

Type of Solar Cell	Structure	PCE (%)	V_{oc} (V)	J (mA/cm^2)	FF	Year	References
Si (crystalline cell)	AR/Front passivation/n:c-Si/ i:a-Si/p:a-Si, n:a-Si/ Rear electrodes	26.7	0.738	42.65	84.9	2017	[146]
GaAs	AR/AlAs/GaAs/ contact	27.6	1.1272	29.78	86.7	2018	[147]
InP (crystalline cell)	Front surface CL/LnP EL/LnP BL/ backsurface CL	24.2	0.939	31.15	82.6	2013	[148]
CIGS	AR/BZO/BL/CIGS/ MoSe/Mo/SLG	23.35	0.734	39.58	80.4	2018	[149]
Si (amorphous cell)		10.2	0.896	16.36	69.8	2014	[150]
Perovskite (cell)	Glass/ITO/PolyTPD/ PFN/perovskite/LiF/ BCP/Ag	23.7	1.213	24.99	78.3	2022	[151]
Dye (cell)		13%	0.91	18.1	0.78	2013	[152]
Organic (cell)		15.2	0.8467	24.24	74.3	2020	[153]
GaAs/CIGS multi-junction solar cells	InGaP/TJ/GaAs/ Ln$_2$O$_3$:Ce/CdS/ CIGS/Mo/SLG	29.3	2.97	12.41	0.80	2022	[154]
GaInAs/GaAsP triple junction	AR/GaLnP/GaInAs/ GaAsP/GaLnP CGB/Au	39.5				2022	[155]

13% to 24% but are relatively expensive to produce. The second generation of solar cells, based on TFs such as CdTe and CIGS, offer a more economical option with efficiencies ranging from 13% to 23.3%. The limitations of these cells have led to the development of third-generation solar cells, based on nanotechnology, which offer more efficient and cost-effective options such as perovskite and organic solar cells with efficiencies ranging from 10% to 22%. Table 2.6 provides a summary of the efficiencies of various solar cells.

2.6 SUMMARY

Development of new materials is essential now to address a number of challenges including lifetime, efficiency, technology readiness, raw material cost and functionality. Nanomaterials come out as a vital contender for large-scale and low-cost manufacturing by using processes based on chemical synthesis. These manufacturing processes are economical because they are carried out at low temperatures. Even though the conversion efficiencies of silicon-based solar PVs are remarkably higher than nanomaterial-based solar PVs, the latter remain attractive due to their low production cost.

REFERENCES

1. de Amorim, W.S., et al., *The nexus between water, energy, and food in the context of the global risks: An analysis of the interactions between food, water, and energy security*. Environmental Impact Assessment Review, 2018. **72**: p. 1–11.
2. Ahmadi, M.H., et al., *Solar power technology for electricity generation: A critical review*. Energy Science & Engineering, 2018. **6**(5): p. 340–361.
3. Hosker, E.E., *World Energy Outlook 2022*, International Energy Agency, OECD (2022).
4. Ghasemzadeh, F., & Shayan, M.E., *Nanotechnology in the service of solar energy systems*. Nanotechnology and the Environment, 2020. 59. DOI: 10.5772/intechopen.93014.
5. Abdelkareem, M.A., et al., *Recent progress in the use of renewable energy sources to power water desalination plants*. Desalination, 2018. **435**: p. 97–113.
6. Kopp, G., & Lean, J.L., *A new, lower value of total solar irradiance: Evidence and climate significance*. Geophysical Research Letters, 2011. **38**(1): L01706, p. 1–7.
7. Raj, C.C., & Prasanth, R., *A critical review of recent developments in nanomaterials for photoelectrodes in dye sensitized solar cells*. Journal of Power Sources, 2016. **317**: p. 120–132.
8. Dupont, E., Koppelaar, R., & Jeanmart, H., *Global available solar energy under physical and energy return on investment constraints*. Applied Energy, 2020. **257**: p. 113968.
9. Wilberforce, T., et al., *Prospects and challenges of concentrated solar photovoltaics and enhanced geothermal energy technologies*. Science of The Total Environment, 2019. **659**: p. 851–861.
10. Rezk, H., Abdelkareem, M.A., & Ghenai, C., *Performance evaluation and optimal design of stand-alone solar PV-battery system for irrigation in isolated regions: A case study in Al Minya (Egypt)*. Sustainable Energy Technologies and Assessments, 2019. **36**: p. 100556.

11. Rabaia, M.K.H., et al., *Environmental impacts of solar energy systems: A review.* Science of The Total Environment, 2021. **754**: p. 141989.
12. Sharma, S., Jain, K.K., Sharma, A., *Solar cells: In research and applications—a review.* Materials Sciences and Applications, 2015. **06**(12): p. 1145–1155.
13. Kalogirou, S.A., *Photovoltaic systems*, in Solar Energy Engineering, Elsevier, (2009) 469–519.
14. Liu, W., et al., *Highly efficient CIGS solar cells based on a new CIGS bandgap gradient design characterized by numerical simulation.* Solar Energy, 2022. **233**: p. 337–344.
15. Bosio, A., Pasini, S., & Romeo, N., *The history of photovoltaics with emphasis on CdTe solar cells and modules.* Coatings, 2020. **10**(4): p. 344.
16. Mufti, N., et al., *Review of CIGS-based solar cells manufacturing by structural engineering*, Solar Energy, 2020. **207**: p. 1146–1157.
17. Asim, N., Mohammad, M., & Badiei, M., *Novel nanomaterials for solar cell devices*, in Nanomaterials for Green Energy, Edited by Bharat A. Bhanvase, Vijay B. Pawade, Sanjay J. Dhoble, Shirish H. Sonawane, Muthupandian Ashokkumar, Elsevier, Netherlands (2018) 227–277.
18. Das, N., *Nanostructured solar cells*. BoD–Books on Demand (2017). Published by Intech open publisher.
19. Almosni, S., et al., *Material challenges for solar cells in the twenty-first century: Directions in emerging technologies*. Science and Technology of advanced Materials, 2018. **19**(1): p. 336–369.
20. Inganäs, O., & Sundström, V., *Solar energy for electricity and fuels*. Ambio, 2016. **45**: p. 15–23.
21. Yu, Q., et al., *High-efficiency dye-sensitized solar cells: The influence of lithium ions on exciton dissociation, charge recombination, and surface states*. ACS Nano, 2010. **4**(10): p. 6032–6038.
22. Abate, A., et al., *Lithium salts as "redox active" p-type dopants for organic semi-conductors and their impact in solid-state dye-sensitized solar cells*. Physical Chemistry Chemical Physics, 2013. **15**(7): p. 2572–2579.
23. Fraas, L.M., *History of solar cell development*, in Low-Cost Solar Electric Power, Springer, (2014) 1–12.
24. Ernst, M., et al., *Characterization of recombination properties and contact resistivity of laser-processed localized contacts from doped silicon nanoparticle ink and spin-on dopants*. IEEE Journal of Photovoltaics, 2017. **7**(2): p. 471–478.
25. Sites, J.R., *Quantification of losses in thin-film polycrystalline solar cells*. Solar Energy Materials and Solar Cells. 2003. **75**(1-2): p. 243–251.
26. Kephart, J.M., et al., *Reduction of window layer optical losses in CdS/CdTe solar cells using a float-line manufacturable HRT layer*. in 2013 IEEE 39th Photovoltaic Specialists Conference (PVSC), IEEE (2013).
27. McCandless, B.E. & Dobson, K.D., *Processing options for CdTe thin film solar cells*, Solar Energy, 2004. **77**(6): p. 839–856.
28. Sites, J., & Pan, J., *Strategies to increase CdTe solar-cell voltage*. Thin Solid Films, 2007. **515**(15): p. 6099–6102.
29. Ghamsari-Yazdel, F., Gharibshahian, I., & Sharbati, S., *Thin oxide buffer layers for avoiding leaks in CIGS solar cells: A theoretical analysis*. Journal of Materials Science: Materials in Electronics, 2021. **32**(6): p. 7598–7608.
30. Jackson, P., et al., *Effects of heavy alkali elements in Cu (In, Ga) Se2 solar cells with efficiencies up to 22.6%*. Physica Status Solidi (RRL)–Rapid Research Letters, 2016. **10**(8): p. 583–586.

31. Dimitrov, D.Z., & Du, C.H., *Crystalline silicon solar cells with micro/nano texture.* Applied Surface Science, 2013. **266**: p. 1–4.
32. Liu, Y., Xiu, Y., & Wong.C., *Micro/nano structure size effect on superhydrophobicity and anti reflection of single crystalline Si solar cells.* in 2010 Proceedings 60th Electronic Components and Technology Conference (ECTC), IEEE (2010).
33. Mutlak, F.A., et al., *Improvement of absorption light of laser texturing on silicon surface for optoelectronic application.* Optik, 2021. **237**: p. 166755.
34. O'Regan, B., & Grätzel, M., *A low-cost, high-efficiency solar cell based on dye-sensitized colloidal TiO2 films.* Nature, 1991. **353**(6346): p. 737–740.
35. Fujishima, A., & Honda, K., *Electrochemical photolysis of water at a semiconductor electrode.* Nature, 1972. **238**(5358): p. 37–38.
36. Park, H., et al., *Surface modification of TiO2 photocatalyst for environmental applications.* Journal of Photochemistry and Photobiology C: Photochemistry Reviews, 2013. **15**: p. 1–20.
37. Vlachopoulos, N., et al., *Very efficient visible light energy harvesting and conversion by spectral sensitization of high surface area polycrystalline titanium dioxide films.* Journal of the American Chemical Society, 1988. **110**(4): p. 1216–1220.
38. Hoye, R.L.Z., Musselman, K.P., MacManus-Driscoll, J.L., *Doping ZnO and TiO2 for solar cells.* APL Materials, 2013. **1**(6): p. 060701.
39. Sharma, K., Sharma, V., Sharma, S.S., *Dye-sensitized solar cells: Fundamentals and current status.* Nanoscale Research Letters, 2018. **13**(1): p. 381–381.
40. Lee, K., et al., *Incorporating carbon nanotube in a low-temperature fabrication process for dye-sensitized TiO$_2$ solar cells.* Solar Energy Materials and Solar Cells, 2008. **92**(12): p. 1628–1633.
41. Sun, S., Gao, L., & Liu, Y., *Enhanced dye-sensitized solar cell using graphene-TiO2 photoanode prepared by heterogeneous coagulation.* Applied Physics Letters, 2010. **96**(8): p. 083113.
42. Ilahi, N.A., et al. *Electrospinning titanium dioxide (TiO$_2$) nanofiber for dye sensitized solar cells based on Bryophyta as a sensitizer,* J. Phys.: Conf. Ser., 2017. **795**, p. 012033.
43. Chuangchote, S., Sagawa, T., & Yoshikawa, S., *Efficient dye-sensitized solar cells using electrospun TiO2 nanofibers as a light harvesting layer.* Applied Physics Letters, 2008. **93**(3): p. 033310.
44. Liu, Y.Y., et al., *Ultra-long hierarchical bud-like branched TiO2 nanowire arrays for dye-sensitized solar cells.* Thin Solid Films, 2017. **640**: p. 14–19.
45. Koo, H.J., et al., *Nano-embossed hollow spherical TiO2 as bifunctional material for high-efficiency dye-sensitized solar cells.* Advanced Materials, 2008. **20**(1): p. 195–199.
46. Bwana, N.N., *Improved short-circuit photocurrent densities in dye-sensitized solar cells based on ordered arrays of titania nanotubule electrodes.* Current Applied Physics, 2009. **9**(1): p. 104–107.
47. Nikolay, T., et al., *Electronic structure study of lightly Nb-doped TiO2 electrode for dye-sensitized solar cells.* Energy & Environmental Science, 2011. **4**(4): p. 1480.
48. Zhang, J., et al., *Effect of cerium doping in the TiO2 photoanode on the electron transport of dye-sensitized solar cells.* The Journal of Physical Chemistry C, 2012. **116**(36): p. 19182–19190.
49. Latini, A., et al., *Efficiency Improvement of DSSC Photoanode by Scandium Doping of Mesoporous Titania Beads.* The Journal of Physical Chemistry C, 2013. **117**(48): p. 25276–25289.
50. Chou, C.C., et al., *Highly efficient dye-sensitized solar cells based on panchromatic ruthenium sensitizers with quinolinylbipyridine anchors.* Angewandte Chemie International Edition, 2014. **53**(1): p. 178–183.

51. Dissanayake, M.A.K.L., et al., *A novel multilayered photoelectrode with nitrogen doped TiO₂ for efficiency enhancement in dye sensitized solar cells.* Journal of Photochemistry and Photobiology A: Chemistry, 2017. **349**: p. 63–72.

52. Mahmoud, M.S., et al., *Demonstrated photons to electron activity of S-doped TiO2 nanofibers as photoanode in the DSSC.* Materials Letters, 2018. **225**: p. 77–81.

53. Minoura, H., & Yoshida, T., *Electrodeposition of ZnO/dye hybrid thin films for dye-sensitized solar cells.* Electrochemistry, 2008. **76**(2): p. 109–117.

54. Chen, Y. C., Li, Y.J., & Hsu, Y.K., *Enhanced performance of ZnO-based dye-sensitized solar cells by glucose treatment.* Journal of Alloys and Compounds, 2018. **748**: p. 382–389.

55. Basu, K., et al., *Enhanced photovoltaic properties in dye sensitized solar cells by surface treatment of SnO₂ photoanodes.* Scientific reports, 2016. **6**(1): p. 1–10.

56. Liu, X., et al., *Niobium pentoxide nanotube powder for efficient dye-sensitized solar cells.* New Journal of Chemistry, 2016. **40**(7): p. 6276–6280.

57. Kakiage, K., et al., *Highly-efficient dye-sensitized solar cells with collaborative sensitization by silyl-anchor and carboxy-anchor dyes.* Chemical Communications, 2015. **51**(88): p. 15894–15897.

58. Guo, D., et al., *Hierarchical TiO2 submicrorods improve the photovoltaic performance of dye-sensitized solar cells.* ACS Sustainable Chemistry & Engineering, 2017. **5**(2): p. 1315–1321.

59. Feldt, S.M., et al., *Design of organic dyes and cobalt polypyridine redox mediators for high-efficiency dye-sensitized solar cells.* Journal of the American Chemical Society, 2010. **132**(46): p. 16714–16724.

60. Horiuchi, T., et al., *High efficiency of dye-sensitized solar cells based on metal-free indoline dyes.* Journal of the American Chemical Society, 2004. **126**(39): p. 12218–12219.

61. Chandiran, A.K., et al., *Analysis of electron transfer properties of ZnO and TiO2 photoanodes for dye-sensitized solar cells.* ACS Nano, 2014. **8**(3): p. 2261–2268.

62. Quintana, M., et al., *Comparison of dye-sensitized ZnO and TiO2 solar cells: Studies of charge transport and carrier lifetime.* The Journal of Physical Chemistry C, 2007. **111**(2): p. 1035–1041.

63. Gong, J., J. Liang, and K. Sumathy, *Review on dye-sensitized solar cells (DSSCs): Fundamental concepts and novel materials.* Renewable and Sustainable Energy Reviews, 2012. **16**(8): p. 5848–5860.

64. Gong, J., et al., *Review on dye-sensitized solar cells (DSSCs): Advanced techniques and research trends.* Renewable and Sustainable Energy Reviews, 2017. **68**: p. 234–246.

65. Thomas, S., et al., *Nanomaterials for Solar Cell Applications.* 2019: Elsevier.

66. Vittal, R. and K.-C. Ho, *Zinc oxide based dye-sensitized solar cells: A review.* Renewable and Sustainable Energy Reviews, 2017. **70**: p. 920–935.

67. Fakharuddin, A., et al., *A perspective on the production of dye-sensitized solar modules.* Energy & Environmental Science, 2014. **7**(12): p. 3952–3981.

68. Keis, K., et al., *Studies of the adsorption process of Ru complexes in nanoporous ZnO electrodes.* Langmuir, 2000. **16**(10): p. 4688–4694.

69. Jose, R., V. Thavasi, and S. Ramakrishna, *Metal oxides for dye-sensitized solar cells.* Journal of the American Ceramic Society, 2009. **92**(2): p. 289–301.

70. Godinho, K.G., A. Walsh, and G.W. Watson, *Energetic and electronic structure analysis of intrinsic defects in SnO2.* The Journal of Physical Chemistry C, 2009. **113**(1): p. 439–448.

71. Kim, M.-H. and Y.-U. Kwon, *Semiconducting divalent metal oxides as blocking layer material for SnO2-based dye-sensitized solar cells.* The Journal of Physical Chemistry C, 2011. **115**(46): p. 23120–23125.

72. Sayama, K., H. Sugihara, and H. Arakawa, *Photoelectrochemical properties of a porous Nb2O5 electrode sensitized by a ruthenium dye*. Chemistry of Materials, 1998. **10**(12): p. 3825–3832.

73. Rani, R.A., et al., *Highly ordered anodized Nb2O5 nanochannels for dye-sensitized solar cells*. Electrochemistry Communications, 2014. **40**: p. 20–23.

74. Suresh, S., et al., *The role of crystallinity of the Nb2O5 blocking layer on the performance of dye-sensitized solar cells*. New Journal of Chemistry, 2016. **40**(7): p. 6228–6237.

75. S. Albrecht, C. Cymorek, and J. Eckert, *Niobium and Niobium Compounds*. **Vol. 24**. 2011, Weinheim: Wiley-VCH.

76. Rani, R.A., et al., *Thin films and nanostructures of niobium pentoxide: Fundamental properties, synthesis methods and applications*. Journal of Materials Chemistry A, 2014. **2**(38): p. 15683–15703.

77. Reznichenko, L., et al., *Invar effect in n-Nb2O5, αht-Nb2O5, and L-Nb2O5*. Crystallography Reports, 2009. **54**(3): p. 483–491.

78. Ai, X., et al., *Ultrafast electron transfer from Ru polypyridyl complexes to Nb2O5 nanoporous thin films*. The Journal of Physical Chemistry B, 2004. **108**(34): p. 12795–12803.

79. Suresh, S., et al., *Phase modification and morphological evolution in Nb2O5 thin films and its influence in dye-sensitized solar cells*. Applied Surface Science, 2017. **419**: p. 720–732.

80. Sampaio, R.N., et al., *Correlation between charge recombination and lateral hole-hopping kinetics in a series of cis-Ru (phen')(dcb)(NCS) 2 dye-sensitized solar cells*. ACS Applied Materials & Interfaces, 2017. **9**(39): p. 33446–33454.

81. Sayama, K. and H. Arakawa, *Effect of Na2CO3 addition on photocatalytic decomposition of liquid water over various semiconductor catalysis*. Journal of Photochemistry and Photobiology A: Chemistry, 1994. **77**(2-3): p. 243–247.

82. Ou, J.Z., et al., *Elevated temperature anodized Nb2O5: A photoanode material with exceptionally large photoconversion efficiencies*. ACS Nano, 2012. **6**(5): p. 4045–4053.

83. Sahu, A., A. Garg, and A. Dixit, *A review on quantum dot sensitized solar cells: Past, present and future towards carrier multiplication with a possibility for higher efficiency*. Solar Energy, 2020. **203**: p. 210–239.

84. Hao, M., et al., *Ligand-assisted cation-exchange engineering for high-efficiency colloidal Cs1− xFAxPbI3 quantum dot solar cells with reduced phase segregation*. Nature Energy, 2020. **5**(1): p. 79–88.

85. Hu, L., et al., *Flexible and efficient perovskite quantum dot solar cells via hybrid interfacial architecture*. Nature Communications, 2021. **12**(1): p. 1–9.

86. Green, M.A., et al., *Solar cell efficiency tables (version 51)*. Progress in Photovoltaics: Research and Applications, 2017. **26**(1): p. 3–12.

87. Tang, C.W., *Two-layer organic photovoltaic cell*. Applied Physics Letters, 1986. **48**(2): p. 183–185.

88. Benduhn, J., et al., *Intrinsic non-radiative voltage losses in fullerene-based organic solar cells*. Nature Energy, 2017. **2**(6): p. 1–6.

89. Costa, J.C., et al., *Flexible sensors—from materials to applications*. Technologies, 2019. **7**(2): p. 35.

90. Välimäki, M., et al., *Custom-shaped organic photovoltaic modules—freedom of design by printing*. Nanoscale Research Letters, 2017. **12**(1): p. 1–7.

91. Yin, H., et al., *Designing a ternary photovoltaic cell for indoor light harvesting with a power conversion efficiency exceeding 20%*. Journal of Materials Chemistry A, 2018. **6**(18): p. 8579–8585.

92. Mathew, S., et al., *Dye-sensitized solar cells with 13% efficiency achieved through the molecular engineering of porphyrin sensitizers.* Nature Chemistry, 2014. **6**(3): p. 242–247.

93. Yum, J.-H., et al., *A cobalt complex redox shuttle for dye-sensitized solar cells with high open-circuit potentials.* Nature Communications, 2012. **3**: p. 631.

94. Melcarne, G., et al., *Surfactant-free synthesis of pure anatase TiO2 nanorods suitable for dye-sensitized solar cells.* Journal of Materials Chemistry, 2010. **20**(34): p. 7248.

95. Park, N.-G., *Perovskite solar cells: An emerging photovoltaic technology.* Materials Today, 2015. **18**(2): p. 65–72.

96. Bhattarai, S., et al., *A detailed review of perovskite solar cells: Introduction, working principle, modelling, fabrication techniques, future challenges.* Micro and Nanostructures, 2022. p. 207450.

97. Kojima, A., et al., *Organometal halide perovskites as visible-light sensitizers for photovoltaic cells.* Journal of the American Chemical Society, 2009. **131**(17): p. 6050–6051.

98. Park, N.-G., T. Miyasaka, and M. Grätzel, *Organic-Inorganic Halide Perovskite Photovoltaics.* Cham, Switzerland: Springer, 2016.

99. Kim, H.-S., et al., *Lead iodide perovskite sensitized all-solid-state submicron thin film mesoscopic solar cell with efficiency exceeding 9%.* Scientific Reports, 2012. **2**(1): p. 1–7.

100. Green, M., et al., *Solar cell efficiency tables (version 57).* Progress in Photovoltaics: Research and Applications, 2021. **29**(1): p. 3–15.

101. Shockley, W. and H. Queisser, *Detailed balance limit of efficiency of p–n junction solar cells*, in Renewable Energy. 2018, Routledge. p. 35–54.

102. Xing, G., et al., *Long-range balanced electron-and hole-transport lengths in organic-inorganic CH3NH3PbI3.* Science, 2013. **342**(6156): p. 344–347.

103. Møller, C.K., *Crystal structure and photoconductivity of caesium plumbohalides.* Nature, 1958. **182**(4647): p. 1436–1436.

104. Lee, M.M., et al., *Efficient hybrid solar cells based on meso-superstructured organometal halide perovskites.* Science, 2012. **338**(6107): p. 643–647.

105. Lemercier, T., et al., *A comparison of the structure and properties of opaque and semi-transparent NIP/PIN-type scalable perovskite solar cells.* Energies, 2020. **13**(15): p. 3794.

106. Ago, H., et al., *Composites of carbon nanotubes and conjugated polymers for photovoltaic devices.* Advanced Materials, 1999. **11**(15): p. 1281–1285.

107. Suzuki, K., et al., *Application of carbon nanotubes to counter electrodes of dye-sensitized solar cells.* Chemistry Letters, 2003. **32**(1): p. 28–29.

108. Lin, X.-X., et al., *Ionothermal synthesis of microporous and mesoporous carbon aerogels from fructose as electrode materials for supercapacitors.* Journal of Materials Chemistry A, 2016. **4**(12): p. 4497–4505.

109. Gao, P., et al., *Molecular engineering of functional materials for energy and opto-electronic applications.* CHIMIA, 2015. **69**(5): p. 253.

110. Kuila, T., et al., *Recent advances in the efficient reduction of graphene oxide and its application as energy storage electrode materials.* Nanoscale, 2013. **5**(1): p. 52–71.

111. Allen, M.J., V.C. Tung, and R.B. Kaner, *Honeycomb carbon: A review of graphene.* Chemical Reviews, 2009. **110**(1): p. 132–145.

112. Muduli, S., et al., *Enhanced conversion efficiency in dye-sensitized solar cells based on hydrothermally synthesized TiO2–MWCNT nanocomposites.* ACS Applied Materials & Interfaces, 2009. **1**(9): p. 2030–2035.

113. Kim, J.-M. and S.-W. Rhee, *Electrochemical properties of porous carbon black layer as an electron injector into iodide redox couple.* Electrochimica Acta, 2012. **83**: p. 264–270.

114. Chan, Y.F., et al., *Dye-sensitized TiO2 solar cells based on nanocomposite photo-anode containing plasma-modified multi-walled carbon nanotubes.* Progress in Photovoltaics: Research and Applications, 2013. **21**(1): p. 47–57.

115. Mehmood, U., et al., *Improving the efficiency of dye sensitized solar cells by TiO2-graphene nanocomposite photoanode.* Photonics and Nanostructures – Fundamentals and Applications, 2015. **16**: p. 34–42.

116. Ranganathan, P., et al., *Enhanced photovoltaic performance of dye-sensitized solar cells based on nickel oxide supported on nitrogen-doped graphene nanocomposite as a photoanode.* Journal of Colloid and Interface Science, 2017. **504**: p. 570–578.

117. Casaluci, S., et al., *Graphene-based large area dye-sensitized solar cell modules.* Nanoscale, 2016. **8**(9): p. 5368–5378.

118. Park, H., et al., *Flexible graphene electrode-based organic photovoltaics with record-high efficiency.* Nano Letters, 2014. **14**(9): p. 5148–5154.

119. Kymakis, E. and G.A. Amaratunga, *Carbon nanotubes as electron acceptors in polymeric photovoltaics.* Reviews on Advanced Materials Science, 2005. **10**(4): p. 300–305.

120. Acik, M. and S.B. Darling, *Graphene in perovskite solar cells: Device design, characterization and implementation.* Journal of Materials Chemistry A, 2016. **4**(17): p. 6185–6235.

121. Agresti, A., et al., *Graphene-perovskite solar cells exceed 18% efficiency: A stability study.* ChemSusChem, 2016. **9**(18): p. 2609–2619.

122. Rafique, S., et al., *Fundamentals of bulk heterojunction organic solar cells: An overview of stability/degradation issues and strategies for improvement.* Renewable and Sustainable Energy Reviews, 2018. **84**: p. 43–53.

123. Wang, Y., et al., *Carbon nanotube bridging method for hole transport layer-free paintable carbon-based perovskite solar cells.* ACS Applied Materials & Interfaces, 2018. **11**(1): p. 916–923.

124. Park, M.-A., et al., *Bifunctional graphene oxide hole-transporting and barrier layers for transparent bifacial flexible perovskite solar cells.* ACS Applied Energy Materials, 2021. **4**(9): p. 8824–8831.

125. Ye, Q.-Q., et al., *N-Type doping of fullerenes for planar perovskite solar cells.* ACS Energy Letters, 2018. **3**(4): p. 875–882.

126. Feng, S., et al., *High-performance perovskite solar cells engineered by an ammonia modified graphene oxide interfacial layer.* ACS Applied Materials & Interfaces, 2016. **8**(23): p. 14503–14512.

127. Jin, J., et al., *Enhanced performance and photostability of perovskite solar cells by introduction of fluorescent carbon dots.* ACS Applied Materials & Interfaces, 2017. **9**(16): p. 14518–14524.

128. Mazzotta, G., et al., *Solubilization of carbon nanotubes with ethylene-vinyl acetate for solution-processed conductive films and charge extraction layers in perovskite solar cells.* ACS Applied Materials & Interfaces, 2018. **11**(1): p. 1185–1191.

129. Jeng, J.-Y., et al., *CH3NH3PbI3 perovskite/fullerene planar-heterojunction hybrid solar cells.* Advanced Materials, 2013. **25**(27): p. 3727–3732.

130. Ryu, J., et al., *Size effects of a graphene quantum dot modified-blocking TiO2 layer for efficient planar perovskite solar cells.* Journal of Materials Chemistry A, 2017. **5**(32): p. 16834–16842.

131. Aitola, K., et al., *Carbon nanotube-based hybrid hole-transporting material and selective contact for high efficiency perovskite solar cells.* Energy & Environmental Science, 2016. **9**(2): p. 461–466.

132. Jeon, I., et al., *Carbon nanotubes to outperform metal electrodes in perovskite solar cells via dopant engineerin g and hole-selectivity enhancement*. Journal of Materials Chemistry A, 2020. **8**(22): p. 11141–11147.

133. Li, X., et al., *Perovskite Solar Cells: Unique Seamlessly Bonded CNT@Graphene Hybrid Nanostructure Introduced in an Interlayer for Efficient and Stable Perovskite Solar Cells (Adv. Funct. Mater. 32/2018)*. Advanced Functional Materials, 2018. **28**(32): p. 1870225.

134. Brendel, R., *Thin-Film Crystalline Silicon Solar Cells*. 2003: Wiley.

135. Marrocco, V., et al., *Efficient plasmonic nanostructures for thin film solar cells*, in *SPIE Proceedings*. 2010, SPIE.

136. McEvoy, A., Markvart, T. and Castaner, L., *Practical Handbook of Photovoltaics Fundamentals and Applications*, 1st Edition, 2003, Published by Elsevier Publisher, Netherlands.

137. Shao, M., D.D.D. Ma, and S.T. Lee, *Silicon nanowires – synthesis, properties, and applications*. European Journal of Inorganic Chemistry, 2010. **2010**(27): p. 4264–4278.

138. Peng, K.-Q., et al., *Platinum nanoparticle decorated silicon nanowires for efficient solar energy conversion*. Nano Letters, 2009. **9**(11): p. 3704–3709.

139. Shao, M.-W., et al., *Fabrication and application of long strands of silicon nanowires as sensors for bovine serum albumin detection*. Applied Physics Letters, 2005. **87**(18): p. 183106.

140. Huang, Y.-F., et al., *Improved broadband and quasi-omnidirectional anti-reflection properties with biomimetic silicon nanostructures*. Nature Nanotechnology, 2007. **2**(12): p. 770–774.

141. Syu, H.-J., et al., *Influences of silicon nanowire morphology on its electro-optical properties and applications for hybrid solar cells*. Progress in Photovoltaics: Research and Applications, 2013. **21**(6): p. 1400–1410.

142. Rühle, S., *Tabulated values of the Shockley–Queisser limit for single junction solar cells*. Solar Energy, 2016. **130**: p. 139–147.

143. Cotal, H., et al., *III–V multijunction solar cells for concentrating photovoltaics*. Energy & Environmental Science, 2009. **2**(2): p. 174–192.

144. Sugiura, H., et al., *Double heterostructure GaAs tunnel junction for a AlGaAs/GaAs tandem solar cell*. Japanese Journal of Applied Physics, 1988. **27**(2R): p. 269.

145. Geisz, J.F., et al., *Six-junction III–V solar cells with 47.1% conversion efficiency under 143 Suns concentration*. Nature Energy, 2020. **5**(4): p. 326–335.

146. Yoshikawa, K., et al., *Silicon heterojunction solar cell with interdigitated back contacts for a photoconversion efficiency over 26%*. Nature Energy, 2017. **2**(5): Article number: 17032.

147. Kayes, B.M., et al., *27.6% Conversion efficiency, a new record for single-junction solar cells under 1 sun illumination*, in *2011 37th IEEE Photovoltaic Specialists Conference*. 2011, IEEE.

148. Wanlass, M., *Systems and methods for advanced ultra-high-performance InP solar cells*. 2017, Google Patents.

149. Nakamura, M., et al., *Cd-free Cu(In,Ga)(Se,S)2 thin-film solar cell with record efficiency of 23.35%*. IEEE Journal of Photovoltaics, 2019. **9**(6): p. 1863–1867.

150. Matsui, T., et al., *High-efficiency amorphous silicon solar cells: Impact of deposition rate on metastability*. Applied Physics Letters, 2015. **106**(5): p. 053901.

151. Xu, J., et al., *Triple-halide wide-band gap perovskites with suppressed phase segregation for efficient tandems*. Science, 2020. **367**(6482): p. 1097–1104.

152. Mathew, S., et al., *Dye-sensitized solar cells with 13% efficiency achieved through the molecular engineering of porphyrin sensitizers*. Nature Chemistry, 2014. **6**(3): p. 242–247.

153. Würfel, U., et al., *A 1 cm2 organic solar cell with 15.2% certified efficiency: Detailed characterization and identification of optimization potential.* Solar RRL, 2021. **5**(4): p. 2000802.

154. Makita, K., et al., *Mechanical stacked GaAs//CuIn1−yGaySe2 three-junction solar cells with 30% efficiency via an improved bonding interface and area current-matching technique.* Progress in Photovoltaics: Research and Applications, 2023. **31**(1): p. 71–84.

155. France, R.M., et al., *Triple-junction solar cells with 39.5% terrestrial and 34.2% space efficiency enabled by thick quantum well superlattices.* Joule, 2022. **6**(5): p. 1121–1135.

3 Two-Dimensional and Three-Dimensional (3D) Perovskite Structures for Robust and Efficient Solar Cell Fabrication

Pendyala Naresh Kumar and Lioz Etgar
The Hebrew University of Jerusalem, Israel

3.1 INTRODUCTION

Organic-inorganic hybrid perovskite materials have tremendous potential in photovoltaics. The hybrid perovskite has a ABX_3 chemical formula which provides a 3D structure. The structural constituents are alkylammonium (A+, methylammonium/formamidinium or inorganic Cs+), Pb^{2+}/Sn^{2+} (B^{2+}, divalent inorganic cation), and $I^-/Br^-/Cl^-$ (X^-, a halide anion), respectively. Change of constituents or composition tunes the valance band (VB)/conduction band (CB) position, optoelectronic, and structural properties [1–3]. Owing to the properties such as high absorption coefficients, long-range carrier diffusion lengths, defect tolerance, and longer lifetime values, the power conversion efficiency (PCE) increases rapidly to be more than 25% [4–7]. Even though a lot of work has been evolved in the fabrication of perovskite solar cells (PSCs), there are stability concerns.

The most common organic cation, methylammonium, could easily dissolute when exposed to humidity, and the perovskite lose crystallinity and turns into yellow color due to the increased amount of PbI_2 in perovskite films [8–10]. There are several reports of PSCs fabricated in ambient conditions, compromised with the PCEs [11,12].

Closely related solids exist, based on the same ionic species, which are arranged in 2-dimensional n-lead halide-thick layers separated by sheets of organic cations (barriers) as shown in Figure 3.1a [13,14]. The 2D-perovskite structure has a quantum well (QW) structure with different dielectric constants of the organic part (barrier) and the inorganic part, ~2 and ~6, respectively [15,16]. There are three main 2D perovskite phases Ruddlesden-Popper (RP), Dion-Jacobson (DJ), and alternating cations in the interlayer space (ACI) [17–19]. The layer thickness is dictated by preparation, stoichiometry, and by augmenting the small ammonium cation with a larger alkyl ammonium cation. This yields 2D analogues with different layer (n) values (i.e., n=1,

DOI: 10.1201/9781003364825-3

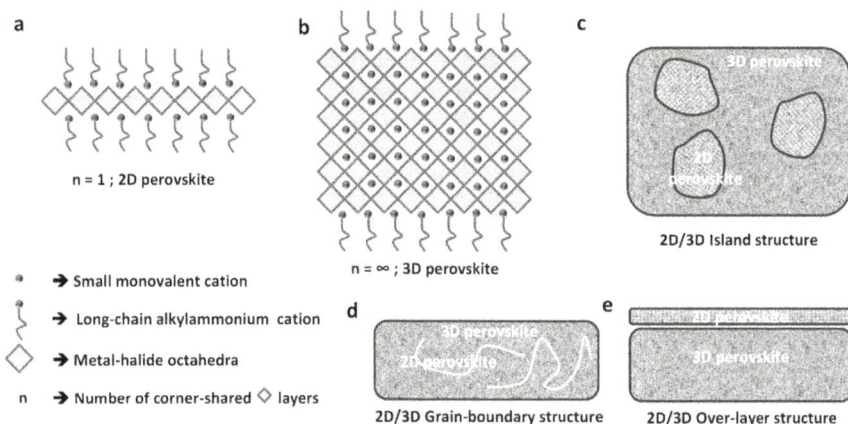

FIGURE 3.1 Broad classification of perovskite structures: 2D perovskite (a), 3D perovskite (b), and 2D-3D combined perovskite structures (c-e).

n=2, ..., n = ∞), where n = ∞ is a cubic 3D perovskite such as MAPbI$_3$, whereas the other n values describe 2D (n=1) or a quasi-2D (n>1) perovskite structure (Figure 3.1a-c). The additional long organic cation, which is added to the perovskite structure (in the 2D structure), is expected to provide hydrophobicity, enhancing the resistivity of the perovskite to humidity. On the other hand, the electronically insulating long organic A' moieties prevent the charge transport between the layered 2D-perovskite with the long-chain barrier molecules for solar cell performance. Recently, the length of the barrier molecules was changed and functionalities were modified for improving charge carrier properties, PCEs, and stability [20–22].

However, when introducing these low-dimensional perovskites into the solar cell, there are several possible structures that can be used with each structure having its own effect on the PV performance and stability.

Figure 3.1 presents several common cases using low-dimensional perovksites in the solar cell: (i) 2D perovskite lower than n=5; (ii) hybrid 2D/3D perovskite where the 2D perovskite are islands inside the 3D perovskite; (iii) 2D perovskite passivate the grain boundaries; (vi) 2D perovskite as an overlayer on top of the 3D perovskite. This perspective will review these cases with an outlook about the use of low-dimensional perovskite inside the solar cell.

3.2   2D PEROVSKITE SOLAR CELLS

There are several reports on 2D PSCs that began in 2014 [23–26]. Several concepts including hot-casting deposition [27], additive engineering [28,29], barrier molecule design [30,31], compositional and solvent engineering [32], vacuum poling [33,34], and interface tuning [35] were implemented to improve the PCEs for 2D PSCs [36].

To achieve a pure 2D-perovskite phase is challenging, and to fabricate high-efficiency solar cell based on this 2D perovskite is even more challenging due to the barrier molecules which act as insulator for charge transport. 2D-perovskite and the phase purity are analyzed using spectroscopy and crystal structure measurements. The concept of pure phase (i.e., distribution of single well width, QWs) and intermediate phase (i.e., distribution of multiple well widths, MQWs) based 2D-perovskite with smaller n values (n = 1, 2, 3, 4, and 5) was studied by Wei Huang et al.

The n-butylamine iodide in the 3D-perovskite precursor is responsible for the non-homogeneous well distribution in the 2D-perovskite structure. But the addition of molten-state n-butylacetate (BAAC) formed a uniformly distributed intermediate gel. Thus, the author obtained a high-phase pure 2D-perovskite. The high-quality QW structures are attributed to the strong-ionic coordination between the n-butylamine acetate and the perovskite framework. The strong ionic interaction between acetate and Pb^{2+} was confirmed from Fourier-transform infrared (FTIR) spectra. At the same time, a distinguishable difference of I^{2+} coordination environment with both the n-butylamines was confirmed using X-ray absorption spectroscopy. The concentration of the Bragg spots obtained for both QW and MQWs is seen in the GIWAXS images. Also, (0 k 0) spots such as (060), (080), and (0100) diffractions were found only for QWs, showing the superior crystallinity and preferred orientation as well as the difference between pure phase 2D and intermediate phase 2D-perovskite of small well distances. Therefore, the obtained pure phase of the 2D perovskite ($BA_2MA_{n-1}Pb_nI_{3n+1}$: n=4) attributed for the high PCE of 16.2% [34]. Moreover, the pure phase of 2D perovskite results in enhanced stability of solar cells.

It was reported that the 2D perovskite phase will tend to grow near the substrate surface, whereas the 3D-perovskite phase tends to expand on the top surface of the perovskite film. Considering this, Dongqin Bi et al. implemented multifunctional interface engineering on top of the $GA_2MA_4Pb_5I_{16}$ perovskite using the same GABr (guanidinium bromide) barrier molecule treatment to tune the secondary crystallization of the perovskite film. The 2D perovskite film was grown in a multi-step approach by spin-coating $MAPbI_3$, GA_2PbI_4, and then GABr at the end. The GABr effect of secondary crystallization was studied by three different concentrations (0, 5, 7.5, 10, 12.5, and 15 mM) and a prominent trend in the PCE and its parameters such as J_{SC}, V_{OC}, and FF was observed (Figure 3.2a). The optimized GABr concentration (10 mM) reported the highest PCE for $GA_2MA_4Pb_5I_{16}$-based 2D-PSC. The author confirmed the reduced defect density of the 2D film for GABr treatment by the blue shift of the PL. At the same time, the slow-decay lifetime value that increased from 148 to 217 ns supported the reduced trap centers for the perovskite film.

In this work, a surface treatment by the barrier molecule corrected the phase distribution; therefore, the PCE enhanced from 15.9% to 19.3%. The corresponding J–V curves, PCE statistics, and the enhanced stability of pure phase 2D perovskite (n = 5) is shown in Figure 3.2b–e [35].

A recent study shows the use of benzyl diammonium iodide (BzDAI) barrier molecule in DJ 2D-perovskite intending to enhance the phase purity of 2D/3D PSCs. In DJ structures, the barrier molecules with both head and tails with ammonium groups tend to attract the perovskite layers with ionic interactions. These are stronger than the van der Waals interactions in the case of RP structure. The BzDAI barrier molecule has some advantages such as short molecular length and the aromatic ring with free π-electrons for the enhanced charge carriers

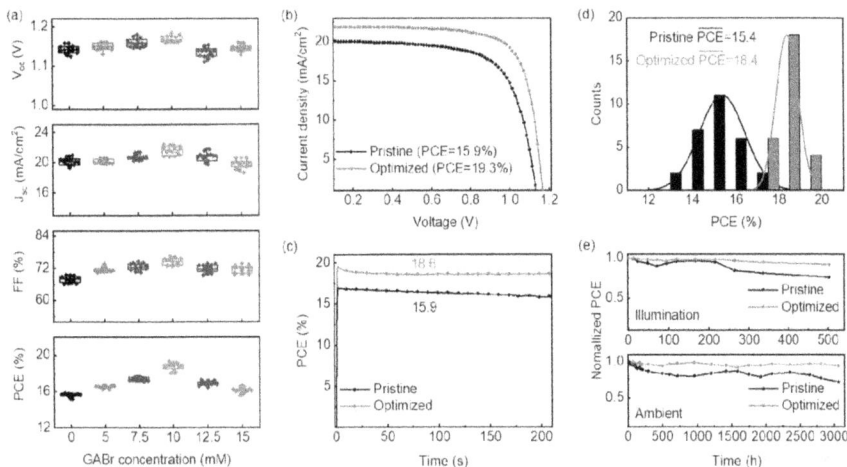

FIGURE 3.2 PCE parameters obtained for different concentrations of GABr (a), J–V curves, (c) steady-state power outputs curves (0.91 and 0.88 V) (b and c), statistics of the PCE (d), and light illumination and ambient stability of pristine and optimized PSCs (e). (Reprinted from the reference [35] with permission).

transport to the selective contacts. The ray patterns recorded for the (BzDA)Pb $(I_{0.93}Br_{0.07})_4$ (n = 1) perovskite found the crystal planes (001) and (110), which were attributed to the zig-zag morphology of the perovskite structure (Figure 3.3a). Hall effect measurements show mobility of 0.75 cm^2/V·s with a p-type characteristic (Figure 3.3c). The final solar cell results with PCE of 15.6%.

This section discusses 2D perovskite with n value lower or equal to five. This is the perovskite structure which is the closest to 2D material. In addition, this is the most challenging perovskite structure to be implemented into the solar cell, due to charge transport difficulties. However, this 2D perovskite shows better stability than the other cases (as presented next); therefore, pushing its PCE is essential as 2D perovskite has the potential to be highly efficient and remain stable at the same time.

3.3 STRUCTURAL DESIGN OF COMBINED 2D AND 3D PEROVSKITE SOLAR CELLS

As mentioned earlier, if the n value is lower than five, the perovskite structure can be considered as 2D; however, when increasing the n value some mixing of n values can be found. Here, we have classified them into three significant kinds of structures as described in Figure 3.1c-e: (1) a hybrid 2D/3D perovskite island structure, where the 2D perovskite grow within the 3D grains. Most of the quasi-2D perovskite structures (n > 5) come into this category with the uneven distribution of different thick 2D perovskites within the resultant perovskite film; (2) grain-boundary 2D/3D perovskite structure if the 2D perovskite tends to grow selectively between the 3D grains; and (3) over layered 2D-3D perovskite structure if the layer of 2D perovskite grows on the surface of the 3D perovskite layer by spin coating or by surface treatment with the barrier molecules.

(a)

(b)

(c)

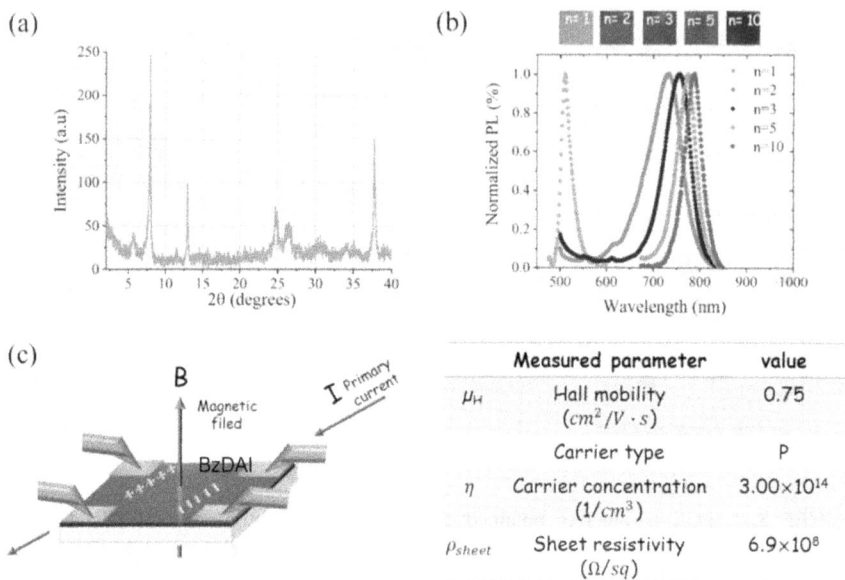

	Measured parameter	value
μ_H	Hall mobility $(cm^2/V \cdot s)$	0.75
	Carrier type	P
η	Carrier concentration $(1/cm^3)$	3.00×10^{14}
ρ_{sheet}	Sheet resistivity (Ω/sq)	6.9×10^8

FIGURE 3.3 The XRD diffraction pattern for pure 2D perovskite (a), the PL and photos of 2D perovskite films (of different n values) (b), the schematic illustration of Hall Effect measurement and the measured parameters of 2D perovskite (n = 10) (c). (Reprinted from reference [1] with permission).

3.3.1 HYBRID 2D/3D PEROVSKITE ISLAND STRUCTURE

When increasing the n value to be more than five, a mixing of 2D with 3D perovskite is created (i.e., quasi 2D perovskite). Studies show that when using quasi-2D perovskite in solar cells, higher V_{oc} and longer electron lifetime can be obtained compared to 3D perovskite [37,38]. Vertically ordered small-n-value to large-n-value nanoplates between the bottom and top surfaces of the quasi-2D perovskite films are normally self-assembled based on unequal growth rates of different-n-value nanoplates. However, Bin Hu et al. introduced the vacuum poling approach to uniformly distribute the different n-value nanoplates (PEA$_2$MA$_{n-1}$Pb$_n$I$_{3n+1}$), which is different from the vertical alignment of different thick QWs of non-uniform distribution of nanoplates, thus obtaining a high PCE of 18.04% [33].

Another study shows different PCEs of 12.51%, 15.78%, and 17.9% for different n values 10, 50, and 90, respectively, for dipropylammonium iodide (DipI)-based Dip$_2$MA$_{n-1}$Pb$_n$I$_{3n+1}$ perovskite. In this case, used the infrared (IR) spectroscopy to follow the perovskite degradation through the diffusion of ammonia/amines to atmosphere and determined that the concentration of perovskite material after 240 min at 100°C is up to 575% greater in 2D/3D perovskite films (n = 10) than that observed in 3D perovskite films. The material stability also improved the thermal stability of photovoltaic devices, offering an efficiency drop of just 4% for n = 50 and n = 10 after thermal annealing, while the performance drop for reference 3D samples under the same conditions was greater than 80% [39] (Figure 3.4).

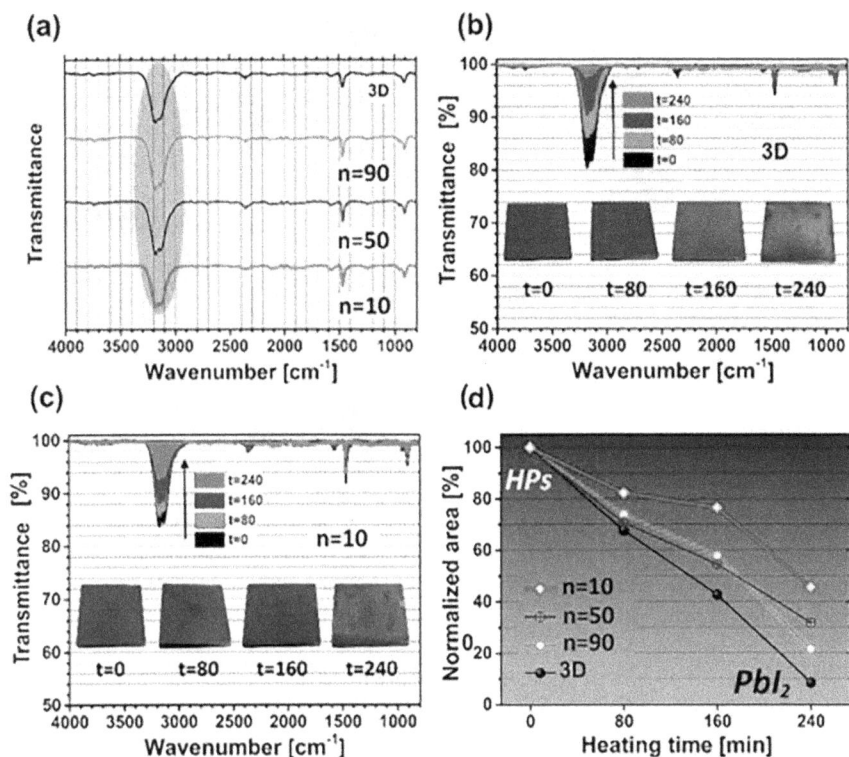

FIGURE 3.4 Thermal stability evaluated by IR spectroscopy T = 100°C; t = 80 min (three times) in air, RH 50–60%. (a) Highlighted N-H stretching IR stretching (b and c). Evolution of N-H peak as a function of heating time (d). Qualitative determination of the concentration of ammonium salts (and therefore perovskite material) by the integration of the N-H peak. (Reprinted from the reference [39] with permission).

Yongsheng Liu's group has worked on a class of aromatic-formamidiniums (PhFACl, pHOPhFACl, and p-FPhFACl) barrier molecules to construct hybrid 2D/3D PSCs. The 2D/3D perovskite synthesis followed a two-step deposition method. First, PbI$_2$ was deposited and then the electrode was dipped in FAI/MAI/MACl and ArFA in isopropanol (IPA) solution. It was observed that the average grain size of hybrid 2D/3D perovskite has increased (~1.4 μm) compared to the control perovskite film due to the addition of ArFA barrier molecules. The authors argue that ArFa barrier molecules passivate the defects which formed NH----I strong hydrogen bonds with the corner-sharing [PbI$_6$]$^{4-}$ octahedrons confirmed from the ^{1}H NMR spectral measurement. High-resolution transmission electron microscope (HRTEM) images of the mixed 2D/3D perovskite confirmed the crystalline phases (Figure 3.5a-c). The orientation of the perovskite crystals was analyzed using grazing incidence wide-angle X-ray scattering (GIWAXS) measurements. The 3D perovskite showed isotropic crystal orientation, where for 2D/3D structure the ring patterns were turned into Bragg spots due to partial crystal orientation (Figure 3.5d-g).

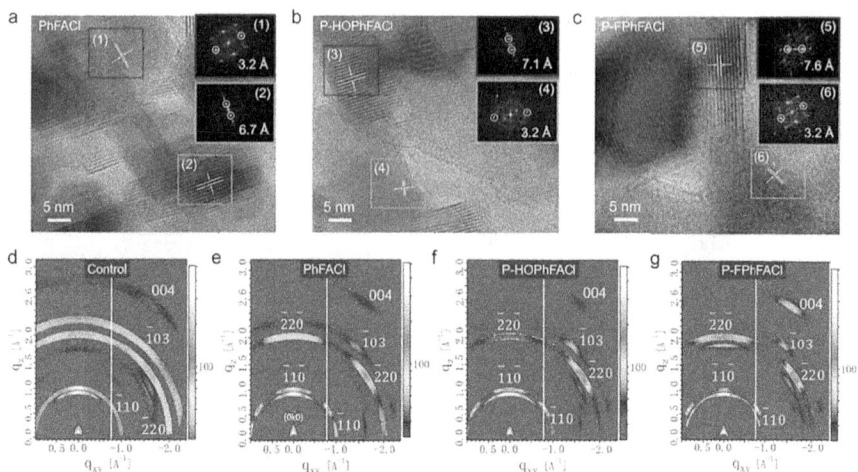

FIGURE 3.5 HRTEM of the 2D/3D hybrid perovskite films based on PhFACl (a), p-HOPhFACl (b), and p-FPhFACl (c). (d–g) GIWAXS patterns of the control (d), PhFACl (e), p-HOPhFACl (f), and pFPhFACl (g)-based perovskite films. (Reprinted from the reference [22] with permission).

Among the three spacer molecules, when PhFACl was incorporated in 2D/3D PSC, the best PCE of 23.36% was achieved. These 2D/3D perovskites showed enhanced stability for 1480 h at 80°C, which retained 87% of its initial PCE, whereas the 3D PSC dropped to 66% of its initial value [22].

2-thiophenemethylammonium (ThMA) was used as a barrier molecule in 2D/3D perovskite structure. As previously reported, the GIWAXS measurements showed some orientation in the case of the 2D/3D structure. However, the stability enhancement was very clear in this case where the cell retained 99% of its initial PCE for hybrid 2D/3D structure, whereas the control device reduced to 80% of its initial value after 1680 h. The stability was tracked using XRD as shown in Figure 3.6a-b. Additional stability measurements was done under illumination for 500 h with a white LED (100 mWcm^{-2}); the PCE for the control device reduced to 0%, but the 2D/3D PSC retained 94% of its initial value. This work indicates that even when adding a small (8 wt%) amount of ThMA as the barrier molecule the influence on the stability is drastic [40].

3.3.2 2D/3D PEROVSKITE GRAIN-BOUNDARY STRUCTURE

This section assumes that the 2D perovskite forms on the grain boundaries of 3D perovskite, which passivates the grains and contributes to the PCE and stability. Wang et al. reported on the addition of butyl ammonium (BA) to mixed cation mixed halide of the composition $(FA_{0.83}Cs_{0.17}Pb(I_{0.6}Br_{0.4})_3)$. The BA to FA/Cs mixing ratio was changed and the influence of this structure on the PV performances and stability was studied. When recorded, the high-resolution scanning electron microscopy (HR-SEM) images for the optimized 2D/3D perovskite

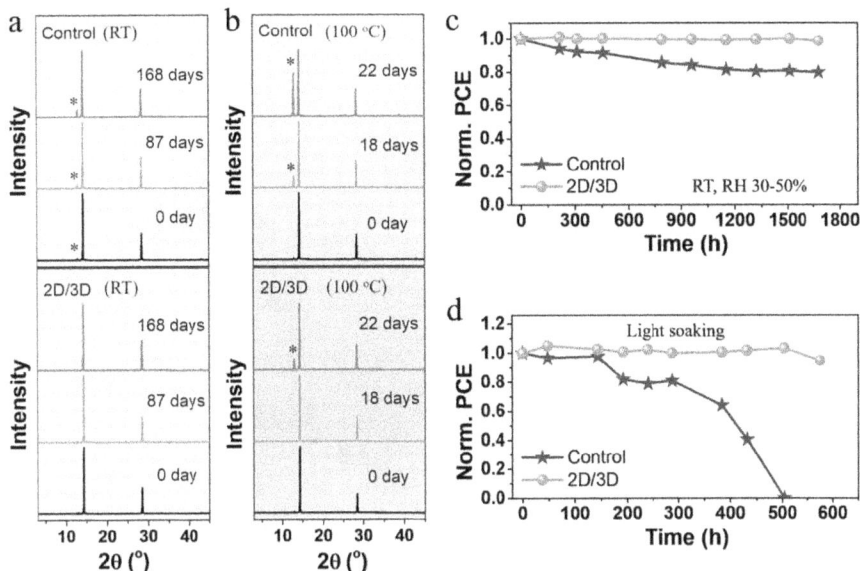

FIGURE 3.6 XRD patterns obtained for control and 2D/3D perovskite films (air, RH = 30–50%) (a), XRD patterns of control and 2D/3D perovskite films (air, at 100°C) (b), stability test of encapsulated devices stored in the dark (c), and un-encapsulated devices exposed to continuous light soaking in N_2 (d). (Reprinted from the reference [40] with permission).

(surface-view), the vertically grown 2D perovskite was observed between the 3D grains. It was expected due to the expelled BA cations out of the 3D perovskite domains during crystallization, simultaneously involved in the growth of 2D perovskite perpendicular to the (010) plane of the substrate. In support of this, the author found a higher fluorescence lifetime for the 2D-3D structure compared to the BA cation-free perovskite (control) device, attributed to the reduced trap state density, thus reducing the trap-assisted charge recombination. In this report, PCE achieved of the PSC was 20.6% and possessed a 50% longer device lifetime compared to the control PSC structure [41].

In 2020, Hin-Lap Yip et al. studied the MA-free 2D-3D perovskite upon addition of β-guanidinopropionic acid (β-GUA) into the 3D perovskite precursor solution ($FA_{0.95}Cs_{0.05}PbI_3$). The author also studied the 2D/3D perovskite by separately adding guanidinium iodide (GAI) and 5-ammonium valeric acid iodide (5-AVAI) into the perovskite precursor and compared it with β-GUA. In all the samples, the X-ray diffraction (XRD) results (Figure 3.7a–c) appeared with standard perovskite phases (α-phase). Interestingly, the author could not find any shift in the (110) diffraction peak for the β-GUA (0, 2, and 5 mol%) added 2D-3D structure. This phenomenon attributed that the grown 2D perovskite with β-GUA molecule did not enter the lattice of 3D perovskite. From the surface-view SEM images (Figure 3.7d–g) of the pristine and other barrier molecule-added perovskite structures, the author found that the 2D perovskite with β-GUA distributed on the upper layer of the 2D-3D perovskite layer, especially between the grain boundaries. However, the emergence of new phases was

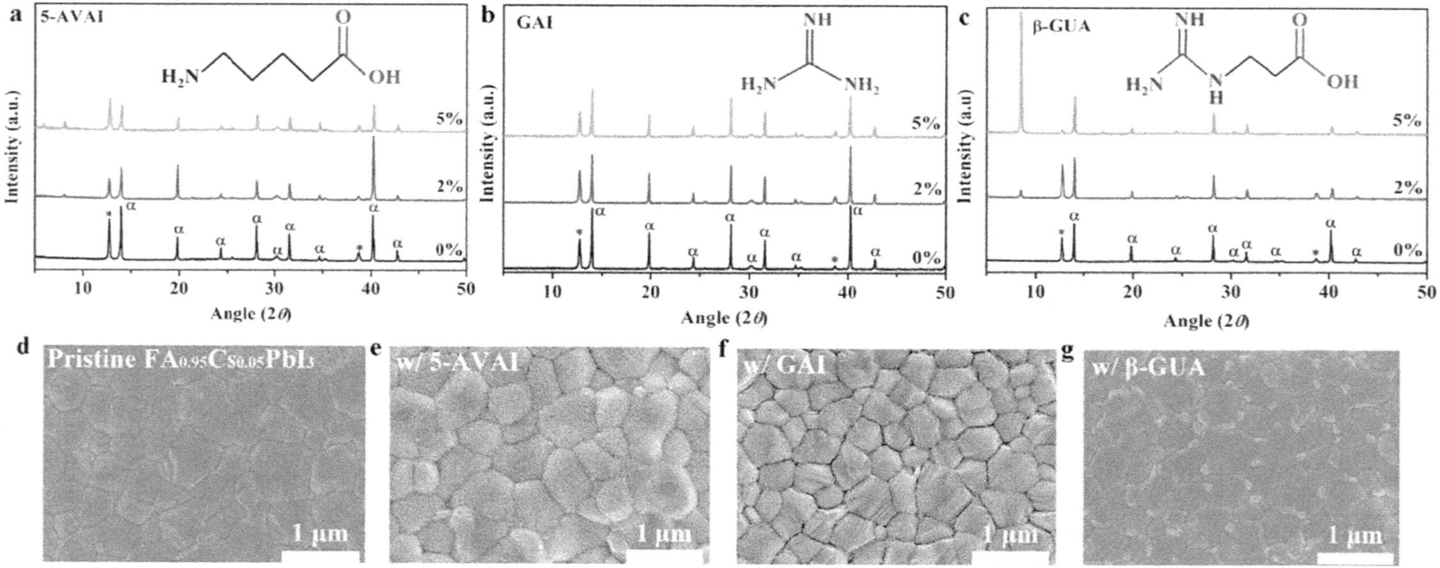

FIGURE 3.7 The perovskite films' XRD results doped with various concentrations of 5-AVAI, GAI, and β-GUA, respectively (a–c). SEM images of pristine (d) and 2 mol% 5-AVAI-doped (e), GAI-doped (f), and β-GUA-doped (g) perovskite films. (Reprinted from the reference [42] with permission).

observed with round shapes at the grain boundaries with the addition of β-GUA (associated with forming 2D perovskite phase). Thus, this work represented the β-GUA molecules that could not enter the 3D perovskite lattice and which were responsible for the growth of grain-boundary-structured 2D perovskite [42].

In another approach, the 2D perovskite is directly added into the precursor solution to obtain the 2D/3D perovskite grain-boundary structure. In 2018, Yang Yang's group added a 1.67 mol% 2D phenylethylammonium lead iodide into the precursor solution and observed the 2D perovskite protected $FAPbI_3$ from moisture and assisted in charge separation/collection. The fabricated PSC has shown the best PCE of 21.06% [43]. In the same year, 2018, Yanlin Song's group studied the grain-boundary structure-based PSCs introducing the $EDBEPbI_4$ microcrystals [2,2-(ethylenedioxy)bis(ethylammonium)) (EDBE)] into the $MAPbI_3$ precursor solution. The obtained $(EDBEPbI_4)_x(MAPbI_3)_{1-x}$ 2D/3D grain-boundary perovskite film was analyzed from XRD and absorption spectroscopy measurements. The perovskite crystal orientations were found in the grain-boundary structure with the reduced arc segments from the GIWAXS measurements compared to the control 3D perovskite film. The SEM confirmed the vertically aligned flack kind of 2D perovskite structures grown between the 3D perovskite gains. The growth of the vertically aligned 2D perovskite flakes between the 3D perovskite grains was attributed to the exposed amine groups present at the two faces of 2D $EDBEPbI_4$, thus finding effective interactions with the 3D perovskite grains. Further, the two-fold increased 3D perovskite grain size supported the author's assumption of amine group interactions of 3D perovskite grains. The author evaluated the photovoltaic performance and stability (Figure 3.8). A large-area solar cell module (active area of 342 cm^2) achieved the PCE of 11.59% with this technique. Moreover, the module

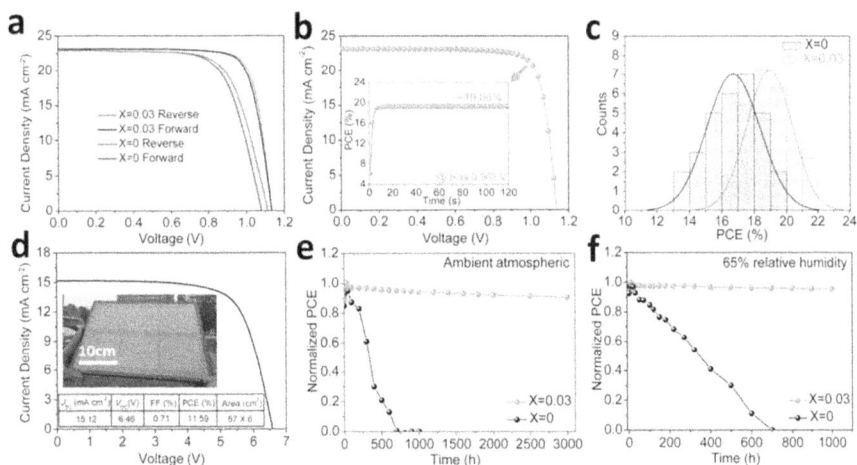

FIGURE 3.8 Hysteresis behavior of the PSCs with varied 2D concentrations: (a) J–V curves obtained for the champion device (x = 0.03), (b) histogram of the number of cells, (c) J–V curve of the module, inset photograph shows the module, (d) and the module stability test (e and f). (Reprinted from the reference [44] with permission).

retained 90% of the initial PCE after 1000 h, whereas the 3D PSC remained with 10% of its initial PCE value after 600 h [44].

Jianxi Yao et al. studied all inorganic 2D/3D perovskite with the $Cs_2PbI_2Cl_2$-$CsPbI_{2.5}Br_{0.5}$ composition. In this case, it was observed that the 2D $Cs_2PbI_2Cl_2$ phase distributed among the grain boundaries of 3D $CsPbI_{3-x}Br_x$ grains. With these 2D/3D grain-boundary structures, the conversion efficiency of the solar cells was 15% and retained 80% of the initial efficiency at a humidity of 70 ± 10% RH under continuous heating at 80°C for 12 h. Considerably, the Voc was found to be 1.21 V, whereas for 3D control device it was found to be 0.81 V (with PCE 7.3%). This obviously improved Voc of 1.21 V was attributed to the configuration of the heterojunction of 2D and 3D crystals that suppressed the nonradiative recombination within the device [45].

Unlike the quasi-2D/3D Island perovskite structures, in the case of 2D/3D grain-boundary structures, the 2D perovskite have insulating barrier molecules between the grain boundaries that is not present within the 3D perovskite. However, it passivates the defects over 3D perovskite, forming the vertical junctions. Thus, the 2D/3D perovskite grain-boundary structure is adequate. However, optimization of precursor concentration and the barrier molecules for the selective grain-boundary growth of 2D perovskite is crucial.

3.3.3 2D/3D Perovskite Over-Layer Structure

2D/3D Perovskite over-layer structure is also known for passivation treatment on top of template (3D) perovskite layers [46–48]. In this case, the 2D perovskite layer tends to grow evenly on top of the 3D perovskite (bulk) from simple spin-coating or immersion in the barrier molecule or 2D perovskite solution. Thus, the 2D perovskite with organic barrier molecules protects the solar cell from ambient conditions but does not disturb the 3D perovskite lattice. At the same time, the top layer of 2D perovskite will concentrate the charge carriers by the formation of the band alignment.

In 2019, Shengzhong Liu et al. reported the effect of barrier molecule's fluorination on constructed 2D-3D perovskite over-layered structure and their PCEs. The author chose the aliphatic molecule BA and its fluorinated counterpart 4,4,4-trifluorobutylammonium (FBA) while aromatic molecule PEA and its fluorinated counterpart 4-fluorophenylethylammonium (FPEA). The fabrication process of 2D-3D heterojunction from spin-coating technique is shown in Figure 3.9. The surface is studied from SEM and AFM images and shows the growth of plate-like 2D perovskite morphology on top of 2D/3D perovskite films. A templated growth of QWs by the underlying 3D phase was found to be a general property within 2D/3D films irrespective of fluorination or compositional engineering. The author has studied the formation of the built-in band alignment and spontaneous charge transfer from QWs to the underlying 3D layer at the 2D/3D heterojunction. The better QW distribution and corresponding faster charge-transfer kinetics enabled the significant optimization of optoelectronic properties such as enhanced carrier mobility, enhanced charge collection, reduced charge nonradiative recombination, and reduced loss in potential of 2D/3D devices. The resulted PSCs were stored under ambient conditions

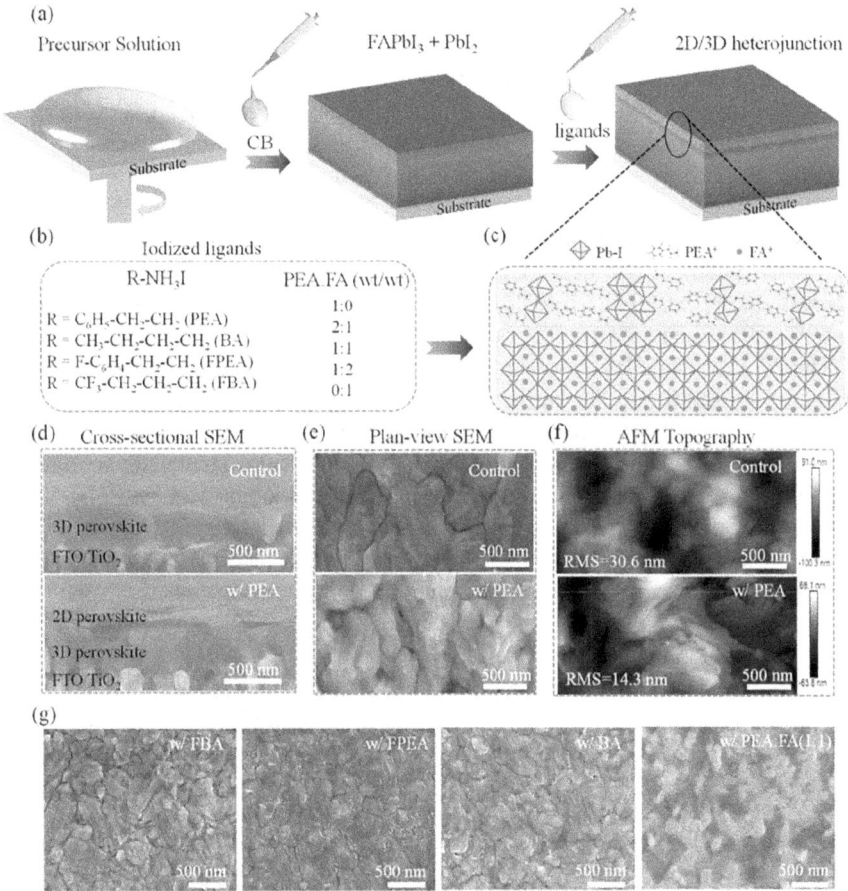

FIGURE 3.9 Scheme represented the 2D/3D PSC fabrication process (a), different barrier molecules with different compositions (b), schematic 2D/3D hierarchical structure (c), cross-sectional (d), surface-view (e) SEM images, AFM images (f), and surface-view of the FBA, FPEA, BA, and PEA: FA (1:1)-based 2D/3D perovskite films (g). (Reprinted from the reference [49] with permission).

(air, RH = 30–40%); the FPEA barrier molecule-based 2D/3D PSC has shown an 84% retention of its initial PCE. In contrast, the control device has dropped to 3% of its initial PCE value [49].

In 2020, Chun-Chao Chen's group worked on the fabrication of 2D/3D so that the 2D perovskite capping layer was given on top of the 3D perovskite layer by the spin-coating isopentylammonium iodide (PNAI) in IPA (20 mM) solution. In this report, the Author has done a systematic evolution on 2D perovskite crystallization and the reduction of residuals PbI_2 in the bottom 3D perovskite layer from XRD measurements performed for different temperatures treated 2D/3D perovskite films. In such a way, the quality of the transformation process of PNAI into 2D perovskite upon annealing at different temperatures were studied. After depositing the PNAI on top of 3D perovskite, the films were annealed at 20°C (PANI-20), 50°C (PANI-50), 70°C (PNAI-70), 90°C (PNAI-90), and 120°C (PNAI-120). When measuring the XRD, a decrease in peak intensities at 5.7 and 12.7 degrees corresponds to PNAI and residual PbI_2 as the annealing temperature increased to 90°C but then retained a weak peak at 5.7 degrees. At the same time, gradually increased peak intensity at 4.3 degrees was attributed to the evolution of the 2D perovskite from PNAI during the annealing treatment process. The transformation of PNAI into 2D perovskite upon increased annealing temperature was further confirmed from the GIWAXS patterns. The best PCE obtained was 20.26% for the control device, whereas the PNAI-90-based 2D-3D device has shown an improved efficiency of 22.62% [50] (Figure 3.10).

The 2D perovskite layer was grown from the other approach after immersing the 3D perovskite film into the barrier molecule solution, as reported below.

To attribute the hydrophobicity character for 2D perovskite top layer in 2D/3D PSCs, in 2019, Yuhang Liu et al. employed the pentafluoro phenylethylammonium (FEA) barrier molecule. Here, the top layer FEA lead iodide [$(FEA)_2PbI_4$], an ultrahydrophobic 2D perovskite material, was developed on 3D perovskite films from the immersion method. The measured water droplet contact angles for 2D/3D film was 96 degrees and remained at the same value after 3 min, whereas for the 3D

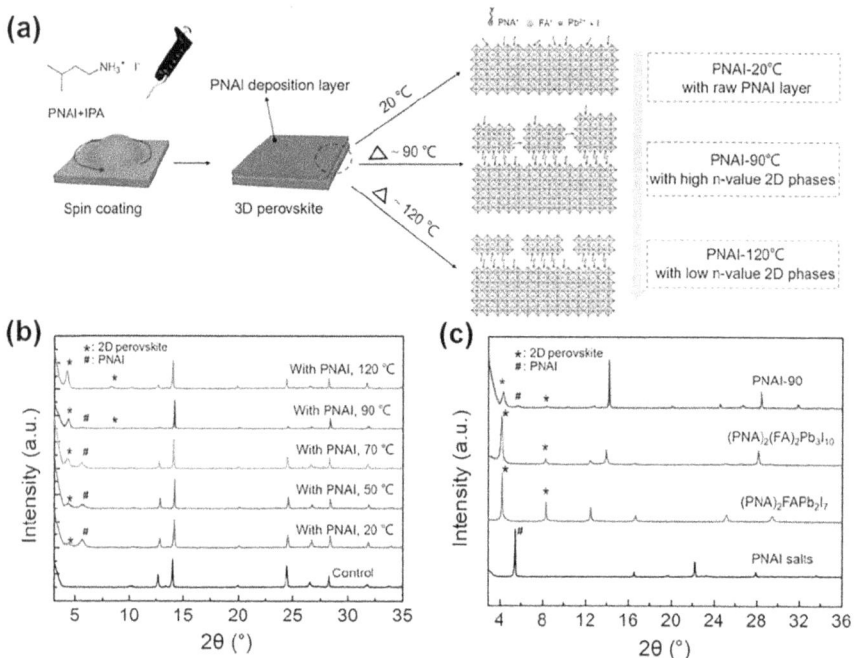

FIGURE 3.10 The schematic represented fabrication of perovskite film and annealing procedure (a). XRD patterns were obtained for perovskite films (b). (c) XRD patterns of PNAI-90 treated perovskite film, different 2D perovskite phases, and PNAI salts. (Reprinted from the reference [50] with permission).

perovskite film, it was 44, and after three min, it reduced to 38 degrees. Thus, this work has proven the tunable properties for 2D/3D PSCs by functional modification of barrier molecules. When measuring the photoluminescence (PL) spectra, the 2D/3D perovskite over-layer film showed a five-folded intensity compared to the 3D perovskite film. At the same time, the TRPL measurement showed a two-fold increased charge carrier lifetime. When evaluating the stability of the un-encapsulated PSCs, the 2D-3D PSCs retained 90% of the initial PCEs when stored under ambient conditions (RH = 40%), whereas the 3D PSC retained only 43% [51].

Differing from solution approaches, Mansoo Choi et al., in 2021, reported the solid-state in-plane growth (SIG) for growing 2D perovskite on top of 3D perovskite layer for the benefit of intact properties by applying pressure and heat so as to avoid solvent use over the 3D perovskite during the 2D perovskite layer. During this process, the authors found the growing of 2D seeds in-plane direction initially and grown gradually. After a specific time, the 2D film was detached, and a new 2D-3D bilayer was observed. However, maintaining a particular temperature played a vital role in forming 2D seeds in the SIG process. In this way, the PCE of 24.59% was obtained for the 2D-3D PSC, whereas PCE of 22.39% was obtained for the control device. When measuring the stability by storing the PSCs at RH = 85%, RT, the 2D-3D PSC degraded only a 2.7% of its initial value after 1083 h. In contrast, the control device showed 41.4% degradation from its initial PCE after 400 h [52]. Thus, growing the 2D/3D perovskite over-layer is an efficient way to passivate the 3D perovskite surface. Still, the optimization of thickness, barrier molecule, and treating temperatures play an essential role in the fabrication of an efficient PSC.

In the case of all the above 2D/3D PSCs, the stability of the solar cells was found to be more significant compared to their counterpart 3D PSCs. The perovskite composition, type of perovskite structure, PCE parameters, and stability of the device are provided in Table 3.1.

3.4 SUMMARY AND OUTLOOK

It summarises that the addition of long-chain aliphatic or aromatic barrier molecules making 3D/bulk into a 2D or quasi 2D perovskite and classified it from 3D. The low-dimensional 2D perovskite-based solar cells are reported with moderate PCEs, and they need to be improved. Whereas 2D-3D PSCs are reported with high PCEs and improved stability towards the humid-air conditions than the control (3D perovskite-based) devices. Though it is a good sign for the PSCs, for future directions, there is a need for probing into more for 2D-perovskite layer passivation, carrier mobilization, and charge injection behavior when present together with the 3D. The 2D-3D structures could be either on the surface, vertically between grains, or covering specific trap-state locations in bulk. The high efficiency and improved stability are found for 2D PSCs in recent reports, especially for phase pure perovskites than that for intermediate phase 2D perovskite structures. However, the 2D-3D perovskite solar cells are efficient than the 2D perovskite, as reported. Therefore, it is the right way to focus on phase pure 2D perovskite passivation to fabricate 2D-3D PSCs. There is also a need for concentrating on controlled crystal growth for 2D/3D PSCs.

TABLE 3.1

Some Recent (since 2017/2018) 2D with PCE >14%, Quasi-2D and 2D/3D Perovskite-Based Solar Cell Parameters with High PCE >20% and Stability Parameters

Perovskite/Precursor	Perovskite Structure	J_{SC} (mAcm^{-2})	V_{OC} (V)	FF (%)	PCE (%)	Stability (Retained η/Time)	Year Ref	References
2D (n≤5) perovskite structure								
$(PEA)_2(MA)_4Pb_5I_{16}$	2D (n=5)	15.8	1.19	75	14.1	90%/1080 h	2018	[53]
$BA_2MA_4Pb5I_{16}$	2D (n=5)	19.0	1.11	64	14.3	98%/2400 h	2020	[54]
$(GA)(MA)_nPb_nI_{3n+1}$	2D (n =3)	18.8	1.15	68	14.7	88%/5760 h	2019	[55]
$(PPA)_2(MA)_2Pb_3I_{10}$	2D (n =3)	17.2	1.20	72	14.9	60%/600 h	2020	[24]
$(ThMA)_2(MA)_{n-1}Pb_nI_{3n+1}$	2D (n=3)	18.9	1.07	76	15.4	90%/1000 h	2018	[56]
$(ThDMA)(MA)_4Pb_5I_{16}$	2D (n=5)	19.5	1.07	75	15.7	95%/1655 h	2020	[57]
$(BA,GA)_2MA_4Pb_5I_{16}$	2D (n=5)	16.9	1.17	81	15.9	93%/1200 h	2019	[58]
$BA_2MA_3Pb_4I_{13}$	2D (n=4)	16.7	1.31	74	16.2	>90%/4680 h	2021	[34]
$(GA)_2MA_4Pb_5I_{16}$	2D (n=5)	21.9	1.17	75	19.3	94%/3000 h	2021	[35]
Hybrid 2D-3D perovskite structures								
$[(NH_4)_{2.4}(FA)_{n-1}Pb_nI_{3n+1.4}I_{0.85}(MAPbBr_3)_{0.15}$	Island	22.8	1.09	73	18.2	90%/1000 h	2018	[59]
$3BBAI{:}MACl{:}PbI_2$ (molar ratios 1.8:2.5:3)	Island	18.2	1.23	81	18.2	82%/2400 h	2018	[60]
$ThMA /(MA_xFA_{1-x})_{n-1}Pb_nI_{3n+1}^{2-}$	Island	22.8	1.16	81	21.5	99%/1680 h	2019	[40]
$PEA_2MA_{n-1}PbnI_{3n+1}$	Island	17.9	1.22	82	18.0	96%/5760 h	2019	[33]
$BA_{0.05}(FA_{0.83}Cs_{0.17})_{0.91}Pb(I_{0.8}Br_{0.2})_3$	grain-boundary	22.7	1.14	80	20.6	80%/4000 h	2017	[41]
$(PEA)_2PbI_4/FACsPbI_3$	grain-boundary	25.3	1.05	72	20.2	>98%/400 s	2019	[61]
$(PEA_2PbI_4)/ FAPbI_3$	grain-boundary	24.4	1.13	76	21.1	98%/1392 h	2018	[43]

Composition	Type					Stability	Year	Ref.
$(EDBE)PbI_4/Cs_{0.05}(FA_{0.83}MA_{0.17})_{0.95}Pb(I_{0.83}Br_{0.17})_3$	grain-boundary	23.5	1.13	79	21.1	90%/3000 h	2018	[44]
$\beta\text{-}GUA / FA_{0.95}Cs_{0.05}PbI_3$	grain-boundary	24.4	1.14	80	22.2	88%/400 h	2020	[42]
$Cs_2PbI_2Cl_2/CsPbI_{2.5}Br_{0.5}$	grain-boundary	16.6	1.21	75	15.1	95%/1440	2020	[45]
$FAI/MAI/MACl/ArFACl$	grain-boundary	24.7	1.16	81	23.4	99%/2400 h	2021	[22]
$CEA/(Cs_{0.1}FA_{0.9})Pb(I_{0.9}Br_{0.1})_3$	over-layer	22.7	1.10	79	20.1	92%/2400 h	2019	[62]
$(PEAI)/(FAPbI_3)_{0.85}(MAPbBr_3)_{0.15}[10\ V\% CsPbI_3]$	over-layer	23.4	1.15	78	21.0	–	2019	[63]
$VBABr/(MAPbBr_3)_{0.15}(FAPbI_3)_{0.85}\ (5\% Cs)$	over-layer	21.5	1.20	79	20.4	90%/2300 h	2019	[64]
$FAI:PbI_2(1:1.08) /FAPbI_3$	over-layer	24.2	1.14	77	21.1	84%/1440 h	2019	[49]
$FEA/ FAPbI_3$	over-layer	25.8	1.09	78	22.2	90%/1000 h	2019	[51]
$(3AMP)(MA_{0.5}FA_{0.5})_3(Pb_{0.5}Sn_{0.5})_4I_{13}/ MA_{0.5}FA_{0.5}Pb_{0.5}Sn_{0.5}I_3$	over-layer	28.6	0.88	80	20.1	99%/100 h	2020	[65]
$NMABr/ (FA)_{0.95-x}Cs_{0.05}MA_xPbI_{3-x}Cl_x$	over-layer	23.6	1.13	79	21.0	80%/350 h	2020	[66]
$ODAI_2 /(FA_{0.85}MA_{0.15})Pb(I_{0.85}Br_{0.15})_3$	over-layer	24.2	1.10	81	21.6	97%/640 h	2020	[67]
$VBABr/(FAPbI_3)_{0.95}(MAPbBr_3)_{0.05}$	over-layer	23.8	1.14	80	21.7	>98%/60 s	2020	[68]
$PNAI/(CsPbI_3)_x(FAPbI_3)_y(MAPbBr_3)_{1-x-y}$	over-layer	23.8	1.16	82	22.6	89%/1000 h	2020	[50]
$(BA)_2PbI_4/(FAPbI_3)_{0.95}(MAPbBr_3)_{0.05}$	over-layer	24.7	1.18	84	24.6	>98%/7000 h	2021	[52]

ACKNOWLEDGMENTS

Author would like to acknowledge Israel Ministry of Energy.

REFERENCES

1. Cohen, B., El, Li Y., Meng, Q. & Etgar, L. Dion-Jacobson two-dimensional perovskite solar cells based on benzene dimethanammonium cation. *Nano Lett.* **19**, 2588–2597 (2019).
2. McMeekin, D. P. *et al.* A mixed-cation lead mixed-halide perovskite absorber for tandem solar cells. *Science (80-.).* **351**, 151–155 (2016).
3. Hassan, Y. *et al.* Ligand-engineered bandgap stability in mixed-halide perovskite LEDs. *Nature* **591**, 72–77 (2021).
4. Min, H. *et al.* Efficient, stable solar cells by using inherent bandgap of α-phase formamidinium lead iodide. *Science (80-.).* **366**, 749–753 (2019).
5. Zhu, H. *et al.* Tailored Amphiphilic Molecular Mitigators for Stable Perovskite Solar Cells with 23.5% Efficiency. *Adv. Mater.* **32**, 1–8 (2020).
6. Jiang, Q. *et al.* Surface passivation of perovskite film for efficient solar cells. *Nat. Photonics* **13**, 460–466 (2019).
7. Yoo, J. J. *et al.* Efficient perovskite solar cells via improved carrier management. *Nature* **590**, 587–593 (2021).
8. Alberti, A. *et al.* Pb clustering and PbI2 nanofragmentation during methylammonium lead iodide perovskite degradation. *Nat. Commun.* **10**, 2196, 1–11 (2019).
9. Dhamaniya, B. P., Chhillar, P., Roose, B., Dutta, V. & Pathak, S. K. Unraveling the effect of crystal structure on degradation of methylammonium lead halide perovskite. *ACS Appl. Mater. Interfaces* **11**, 22228–22239 (2019).
10. Wang, Q. *et al.* Scaling behavior of moisture-induced grain degradation in poly-crystalline hybrid perovskite thin films. *Energy Environ. Sci.* **10**, 516–522 (2017).
11. Rai, M. *et al.* Hot dipping post treatment for improved efficiency in micro patterned semi-transparent perovskite solar cells. *J. Mater. Chem. A* **6**, 23787–23796 (2018).
12. Seetharaman, M. S. *et al.* Efficient organic-inorganic hybrid perovskite solar cells processed in air. *Phys. Chem. Chem. Phys.* **16**, 24691–24696 (2014).
13. Safdari, M. *et al.* Layered 2D alkyldiammonium lead iodide perovskites: Synthesis, characterization, and use in solar cells. *J. Mater. Chem. A* **4**, 15638–15646 (2016).
14. Li, Y., Zheng, G., Lin, C. & Lin, J. Synthesis, structure and optical properties of different dimensional organic-inorganic perovskites. *Solid State Sci.* **9**, 855–861 (2007).
15. Gao, Y. *et al.* Molecular engineering of organic–inorganic hybrid perovskites quantum wells. *Nat. Chem.* **11**, 1151–1157 (2019).
16. Kepenekian, M. *et al.* Concept of lattice mismatch and emergence of surface states in two-dimensional hybrid perovskite quantum wells. *Nano Lett.* **18**, 5603–5609 (2018).
17. Grancini, G. *et al.* One-Year stable perovskite solar cells by 2D/3D interface engineering. *Nat. Commun.* **8**, 1–8 (2017).
18. Lin, Y. *et al.* Enhanced thermal stability in perovskite solar cells by assembling 2D/3D stacking structures. *J. Phys. Chem. Lett.* **9**, 654–658 (2018).
19. Lekina, Y. & Shen, Z. X. Excitonic states and structural stability in two-dimensional hybrid organic-inorganic perovskites. *J. Sci. Adv. Mater. Devices* **4**, 189–200 (2019).
20. Panuganti, S. *et al.* Distance dependence of förster resonance energy transfer rates in 2D perovskite quantum wells via control of organic spacer length. *J. Am. Chem. Soc.* **143**, 4244–4252 (2021).
21. Hu, J. *et al.* Synthetic control over orientational degeneracy of spacer cations enhances solar cell efficiency in two-dimensional perovskites. *Nat. Commun.* **10**, 1276, 1–11 (2019).

22. Liu, T. *et al.* Spacer engineering using aromatic formamidinium in 2D/3D hybrid perovskites for highly efficient solar cells. *ACS Nano* **15**, 7811–7820 (2021).

23. Cao, D. H., Stoumpos, C. C., Farha, O. K., Hupp, J. T. & Kanatzidis, M. G. 2D homologous perovskites as light-absorbing materials for solar cell applications. *J. Am. Chem. Soc.* **137**, 7843–7850 (2015).

24. Zhu, T. *et al.* Novel quasi-2D perovskites for stable and efficient perovskite solar cells. *ACS Appl. Mater. Interfaces* **12**, 51744–51755 (2020).

25. Quan, L. N. *et al.* Ligand-stabilized reduced-dimensionality perovskites. *J. Am. Chem. Soc.* **138**, 2649–2655 (2016).

26. Tsai, H. *et al.* High-efficiency two-dimensional Ruddlesden-Popper perovskite solar cells. *Nature* **536**, 312–317 (2016).

27. Liao, K. *et al.* Hot-casting large-grain perovskite film for efficient solar cells: Film formation and device performance. *Nano-Micro Lett.* **12**, 156, 1–22 (2020).

28. Sun, J. *et al.* Enhancement of 3D/2D perovskite solar cells using an F4TCNQ molecular additive. *ACS Appl. Energy Mater.* **3**, 8205–8215 (2020).

29. Liu, Z. *et al.* Controllable two-dimensional perovskite crystallization via water additive for high-performance solar cells. *Nanoscale Res. Lett.* **15**, 108, 1–7 (2020).

30. Cohen, B. El, Wierzbowska, M. & Etgar, L. High efficiency quasi 2D lead bromide perovskite solar cells using various barrier molecules. *Sustain. Energy Fuels* **1**, 1935–1943 (2017).

31. Li, Z. *et al.* A new organic interlayer spacer for stable and efficient 2D Ruddlesden–Popper perovskite solar cells. *Nano Lett.* **19**, 5237–5245 (2019).

32. Zheng, H. *et al.* Synergistic effect of additives on 2D perovskite film towards efficient and stable solar cell. *Chem. Eng. J.* **389**, 124266, 1–9 (2020).

33. Zhang, J. *et al.* Uniform permutation of quasi-2D perovskites by vacuum poling for efficient, high-fill-factor solar cells. *Joule* **3**, 3061–3071 (2019).

34. Liang, C. *et al.* Two-dimensional Ruddlesden–Popper layered perovskite solar cells based on phase-pure thin films. *Nat. Energy* **6**, 38–45 (2021).

35. Huang, Y. *et al.* Stable layered 2D perovskite solar cells with an efficiency of over 19% via multifunctional interfacial engineering. *J. Am. Chem. Soc.* **143**, 3911–3917 (2021).

36. Lan, C., Zhou, Z., Wei, R. & Ho, J. C. Two-dimensional perovskite materials: From synthesis to energy-related applications. *Mater. Today Energy* **11**, 61–82 (2019).

37. Cohen, B. El, Wierzbowska, M. & Etgar, L. High efficiency and high open circuit voltage in quasi 2D perovskite based solar cells. *Adv. Funct. Mater.* **27**, 1604733, 1–7 (2017).

38. Gharibzadeh, S. *et al.* Record open-circuit voltage wide-bandgap perovskite solar cells utilizing 2D/3D perovskite heterostructure. *Adv. Energy Mater.* **9**, 1803699, 1–10 (2019).

39. Rodríguez-Romero, J. *et al.* Widening the 2D/3D perovskite family for efficient and thermal-resistant solar cells by the use of secondary ammonium cations. *ACS Energy Lett.* **5**, 1013–1021 (2020).

40. Zhou, T. *et al.* Highly efficient and stable solar cells based on crystalline oriented 2D/3D hybrid perovskite. *Adv. Mater.* **31**, 1901242, 1–9 (2019).

41. Wang, Z. *et al.* Efficient ambient-air-stable solar cells with 2D-3D heterostructured butylammonium-caesium-formamidinium lead halide perovskites. *Nat. Energy* **2**, 17135, 1–10 (2017).

42. Yao, Q. *et al.* Graded 2D/3D perovskite heterostructure for efficient and operationally stable MA-free perovskite solar cells. *Adv. Mater.* **32**, 2000571, 1–10 (2020).

43. Lee, J. W. *et al.* 2D perovskite stabilized phase-pure formamidinium perovskite solar cells. *Nat. Commun.* **9**, 1–10 (2018).

44. Li, P. *et al.* Phase pure 2D perovskite for high-performance 2D–3D heterostructured perovskite solar cells. *Adv. Mater.* **30**, 1805323, 1–8 (2018).
45. Li, Z. *et al.* 2D-3D Cs2PbI2Cl2-CsPbI2.5Br0.5mixed-dimensional films for all-inorganic perovskite solar cells with enhanced efficiency and stability. *J. Phys. Chem. Lett.* **11**, 4138–4146 (2020).
46. Rahmany, S. & Etgar, L. Two-dimensional or passivation treatment: The effect of hexylammonium post deposition treatment on 3D halide perovskite-based solar cells. *Mater. Adv.* **2**, 2617–2625 (2021).
47. Madhavan, V. E. *et al.* CuSCN as hole transport material with 3D/2D perovskite solar cells. *ACS Appl. Energy Mater.* **3**, 114–121 (2020).
48. Chen, J., Seo, J. Y. & Park, N. G. Simultaneous improvement of photovoltaic performance and stability by in situ formation of 2D perovskite at (FAPbI3) 0.88(CsPbBr3)0.12/CuSCN interface. *Adv. Energy Mater.* **8**, 1702714, 1–15 (2018).
49. Niu, T. *et al.* Interfacial engineering at the 2D/3D heterojunction for high-performance perovskite solar cells. *Nano Lett.* **19**, 7181–7190 (2019).
50. He, M. *et al.* Compositional optimization of a 2D-3D heterojunction interface for 22.6% efficient and stable planar perovskite solar cells. *J. Mater. Chem. A* **8**, 25831–25841 (2020).
51. Liu, Y. *et al.* Ultrahydrophobic 3D/2D fluoroarene bilayer-based water-resistant perovskite solar cells with efficiencies exceeding 22%. *Sci. Adv.* **5**, 1–9 (2019).
52. Jang, Y. W. *et al.* Intact 2D/3D halide junction perovskite solar cells via solid-phase in-plane growth. *Nat. Energy* **6**, 63–71 (2021).
53. Fu, W. *et al.* Two-dimensional perovskite solar cells with 14.1% power conversion efficiency and 0.68% external radiative efficiency. *ACS Energy Lett.* **3**, 2086–2093 (2018).
54. Zheng, F. *et al.* Revealing the role of methylammonium chloride for improving the performance of 2D perovskite solar cells. *ACS Appl. Mater. Interfaces* **12**, 25980–25990 (2020).
55. Zhang, Y. *et al.* Dynamical transformation of two-dimensional perovskites with alternating cations in the interlayer space for high-performance photovoltaics. *J. Am. Chem. Soc.* **141**, 2684–2694 (2019).
56. Lai, H. *et al.* Two-dimensional Ruddlesden-Popper perovskite with nanorod-like morphology for solar cells with efficiency exceeding 15%. *J. Am. Chem. Soc.* **140**, 11639–11646 (2018).
57. Lu, D. *et al.* Thiophene-based two-dimensional Dion–Jacobson perovskite solar cells with over 15% efficiency. *J. Am. Chem. Soc.* **142**, 11114–11122 (2020).
58. Lian, X. *et al.* Two-dimensional inverted planar perovskite solar cells with efficiency over 15%: Via solvent and interface engineering. *J. Mater. Chem. A* **7**, 18980–18986 (2019).
59. Zheng, H. *et al.* New-type highly stable 2D/3D perovskite materials: The effect of introducing ammonium cation on performance of perovskite solar cells. *Sci. China Mater.* **62**, 508–518 (2019).
60. Yang, R. *et al.* Oriented quasi-2D perovskites for high performance optoelectronic devices. *Adv. Mater.* **30**, 1804771, 1–8 (2018).
61. Thote, A. *et al.* Stable and reproducible 2D/3D formamidinium-lead-iodide perovskite solar cells. *ACS Appl. Energy Mater.* **2**, 2486–2493 (2019).
62. Liu, G. *et al.* Introduction of hydrophobic ammonium salts with halogen functional groups for high-efficiency and stable 2D/3D perovskite solar cells. *Adv. Funct. Mater.* **29**, 1807565, 1–9 (2019).
63. Heo, S. *et al.* Dimensionally engineered perovskite heterostructure for photovoltaic and optoelectronic applications. *Adv. Energy Mater.* **9**, 1902470, 1–8 (2019).

64. Proppe, A. H. *et al.* Photochemically cross-linked quantum well ligands for 2D/3D perovskite photovoltaics with improved photovoltage and stability. *J. Am. Chem. Soc.* **141**, 14180–14189 (2019).
65. Ke, W. *et al.* Narrow-bandgap mixed lead/tin-based 2D Dion-Jacobson perovskites boost the performance of solar cells. *J. Am. Chem. Soc.* **142**, 15049–15057 (2020).
66. Zhao, S. *et al.* Cascade Type-II 2D/3D perovskite heterojunctions for enhanced stability and photovoltaic efficiency. *Sol. RRL* **4**, 2000282, 1–9 (2020).
67. Jiang, X. *et al.* Dion-Jacobson 2D-3D perovskite solar cells with improved efficiency and stability. *Nano Energy* **75**, 104892 (2020).
68. Teale, S. *et al.* Dimensional mixing increases the efficiency of 2D/3D perovskite solar cells. *J. Phys. Chem. Lett.* **11**, 5115–5119 (2020).

4 Strategies and Progress in Fabrication of Metal Halide Perovskite Nanocrystal Solar Cells

Ananthakumar Soosaimanickam
R&D Division, Intercomet S.L, Madrid, Spain

4.1 METAL HALIDE PEROVSKITE NANOCRYSTALS AND THEIR SURFACE PROPERTIES – A NUT SHELL

Recently emerged metal halide perovskite nanocrystals (PNCs) have attracted much attention among researchers to utilize them for several kinds of potential applications. As an exceptional material, PNCs are showing outstanding optical properties that encompass them for several kinds of optoelectronic devices including solar cells, sensors, photodetectors, lasers and light-emitting diodes (LEDs) [1–5]. With the general formula of AMX_3 (A= methylammonium, MA^+ or $CH_3NH_3^+$, formamidinium $CH(NH_2)^+$ or FA^+, or cesium Cs^+, etc., B=Pb^{2+}, Ge^{2+}, Sn^{2+}, etc., X=Cl, Br and I), these compound NCs are showing exemplary optical properties with interesting structural features [6,7]. The structure of these materials comprises PbX_6^{4-} ocatahedral unit where the A^+ cations are filled with the voids created by the four neighbouring group PbX_6^{4-} octahedra. In the case of perovskite derivative structures such as Cs_4PbX_6 and $CsPb_2X_5$ they have decoupled and alternative layers of corner-sharing $[PbX_6]^{4-}$ octahedra structure [8].

These materials possess exciting properties such as high absorption co-efficient, long carrier-lifetime and higher carrier mobility. Especially, their higher photoluminescent quantum yield (PLQY), narrow photoluminescence (PL) spectra and large exciton binding energy values are helping them to surpass and replace the traditionally used semiconductor NCs to fabricate energy-efficient devices [9,10]. The exciting binding energy of the hybrid PNCs is at least five times higher than their bulk counterpart [11] and thus there are large number of free carriers possible under this circumstance. Moreover, the large Stroke shift in the PL spectra of the PNCs significantly reduces the self-absorption effect which helps to convert the photon energy. Also, these materials are found to be highly defect tolerance which results in exceptional performance in device fabrication. Most importantly, these NCs are having ionic surface that distinguishes them from the other compound NC-based colloids under the atmosphere of different kinds of ligands and solvents [12].

DOI: 10.1201/9781003364825-4

Because of the excellent structural and optical properties, the charge transfer between these NCs in solid state is effectively engineered through different aspects for solar cell applications. It is expected that because of their outstanding structural and optical properties, PNC-based solar cells are assumed to overcome Shockley–Queisser (SQ) limit. Use of smaller PNCs, i.e. perovskite quantum dots (PQDs) in the bulk perovskite solar cells has demonstrated that PQDs are having potential of delivering high efficiency [13–15]. Therefore, fabrication of PNC-based solar cells have recently attracted and efficiency of over 13% has been demonstrated [16].

Literature sources indicate that PNCs are generally prepared using two well-known synthetic approaches, namely ligand-assisted reprecipitation (LARP) and hot-injection method through which we can obtain colloidally prepared functionalized NCs with desired dimension [17]. Although several assumptions are postulated regarding their growth mechanism, the exact steps associated with the PNCs still remain as obscure [18]. Also, these methods are generally assisted with the highly unsaturated, higher boiling point solvents and ligands such as oleylamine (OAm), oleic acid (OA) and 1-octadecene (1-ODE). These solvents and ligands are generally covered on the PNCs or PQDs surface and retard the charge transport of NCs or QD films.

Since organic ligands play a key role in the fabrication of PNC-based optoelectronic devices, several criteria are postulated for an ideal choice. This includes: (a) should have a large dipole moment in order to dissolve in the organic solvents; (b) should have a small barrier for the charge conduction across the ligand; (c) should have strong binding ability to passivate the PNCs; (d) should minimize the moisture penetration; and (e) should be commercially available for mass production [19]. Despite this appropriate ligand protection, these NCs often undergo stability issues owing to the highly ionic surface which renders them to carry out further processing. However, experimental studies have demonstrated that it is possible to enhance the properties of these NCs by applying suitable organic and inorganic compounds on their surface [20].

These so-called post-modification or post-treatment processes are highly beneficial for the fabrication of optoelectronic devices, especially for solar cells. There are different perspectives that have postulated regarding the structural and stability issues related with the PNCs and facts related with these concepts are varied with time to time. Deposition of PNCs using different kinds of solvents to fabricate films conveys that these NCs could be processed through approaches such as spin coating, dip coating and spray deposition. When these PNCs are combined or incorporated with bulk perovskite layer, they are found to be useful in improving the performance of the solar cell [21–23].

A numerical analysis predicts that the bilayer of $CsPbI_3$ QDs/$FAPbI_3$ QDs with the device structure FTO/TiO_2/$CsPbI_3$ QDs/$FAPbI_3$ QDs/PTAA/MoO_3/Ag can deliver 18.55%, which reveals the potential pathways to improve the PNC-based solar cell [24]. The incorporation of PNCs or PQDs into the organic bulk-heterojunction solar cells generally play several roles including (a) mediator in the acceptor phase; (b) could promote molecular ordering in the active layers; (c) their high dielectric constant and strong fluorescence could suppress the charge

High-performance PQDSCs

Compositional Engineering
- A-site doping
- B-site doping
- X-site exchange
- Lead-free

Reduce defect density
Enhance carrier lifetime
Passivate surface defects
Promote radiative recombination

Surface Chemistry
- Surface modification
- Ligand regulation
- Post-treatment

Reduce surface vacancies
Facilitate carrier transport
Surface passivation
Enhance coupling between dots

Device Operation
- Device architecture
- PQD layer engineering
- Charge transport layer

Energy level alignment
Efficient charge extraction
Improve charge transport rate

Stability
- Hydrothermal stability
- Storage stability
- Illumination stability

Form the ordered structure
Long-term operation

Electrode
PQDs
Glass/ITO(FTO)

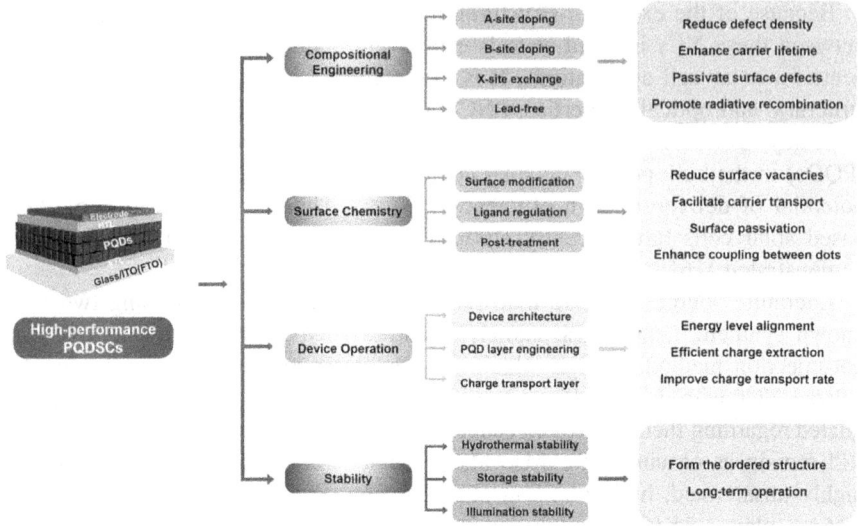

FIGURE 4.1 Different aspects associated with the synthesis, processing and fabrication of a PQD-based solar cell. Reprinted with permission from [29].

recombination; and (d) high extinction coefficient and light scattering property could improve the light absorbing property of the active layers [25].

Small-sized PNCs, which are described as PQDs, show interesting performance in solar cells, owing to the so-called quantum-confinement effect [25–29]. Because of the low voltage losses, slow cooling of photogenerated hot carriers, tunable bandgap and compositional tunability, PQDs are considered as building blocks for the modern flexible optoelectronic devices. Although these small QDs are having a large number of surface defects, advanced surface-treatment approaches and related methodologies have solved the existing setbacks in this area in order to reach their superior optical performance. Also, their active layer fabrication process is similar to the already demonstrated quantum-dot sensitized solar cells (QDSSCs). In this aspect, several efforts have been carried out to fabricate highly efficient perovskite QD-based solar cells and more than 15% efficiency has been achieved [25]. Thus, it is necessary to understand the current trend in this topic in order to address several issues associated with this field. This chapter discusses about the recent innovations in the fabrication of PNC-based solar cells. The overall directions associated with the fabrication of PQDs or PNC-related solar cells are represented in Figure 4.1 [29].

4.2 GENERAL STRATEGIES INVOLVING IN THE FABRICATION OF PEROVSKITE NANOCRYSTAL-BASED SOLAR CELLS

To fabricate a QD or NC-based film assembly, it is important to prepare the nano-materials with narrow dispersion. Also, fabrication of array of QDs and NCs is assisted with the fact that how the solvent is useful for generating such assembly through cost-effective approaches. Along with this, conducting ability of ligands, surface defects, halide vacancies and ion migration are considerably affecting the performance of PNC-based solar cells. Fabrication of PNC-based solar cells generally requires a surface treatment using an organic or inorganic compound, which improves the carrier transport. This is because the insulating property of the native ligands forces us to replace them to improve the carrier transport. Also, multiple purification of PNCs often creates large number of surface traps that quench the photovoltaic performance.

The intergap states, midgap states, undercoordinated Pb^{2+} ions and halide vacancies existing in PNCs are fulfilled by the post-surface modification treatment which helps to fabricate highly efficient, structurally ordered thin films. In order to improve the film quality, this surface treatment is carried out during layer-by-layer deposition of the NCs or as post-surface modification after synthesizing the NCs. These two approaches are found to be beneficial in improving the electronic properties of the NCs film which actually assist in the improvement of charge carrier transport. After this treatment, the NCs surface is defect-free and also it influences on the grain growth. It is demonstrated that the traditionally used ligands OAm and OA significantly influence on the stability and phase transformation of PNCs [30]. This is because the oleylammonium ion (OAm^+) on the PNCs surface makes a dynamic equilibrium with oleate ion (OA^-) and also influences the structural reorganization.

Along with ligands, surface treatment through purifying solvents also greatly influences the performance of the solar cells. Purifying solvents will not only eliminate the excessive ligands present on the surface but also influence the device performance. However, excessive purification steps remove the protecting ligands from the PNCs surface, thereby decreasing their optical properties. According to literature, generally medium polarity solvents are preferred to purify the PNCs owing to the ionic surface. Specifically, it is found that ethyl acetate (EtoAc) and methyl acetate (MeOAc) treatment of PNCs significantly influences the solar cell performance [31]. Compared with EtOAc, the use of MeOAc is found to be efficient, owing to its smaller molecular structure. Also, the defect sites of the PQDs are fulfilled with the MeOAc washing, which improves the charge transport [32]. Furthermore, the use of MeOAc is found to be stabilizing the PLQY of $FAPbI_3$ QDs compared with the use of EtOAc [33].

Also, the ligand management is excellently manipulated by the polarity of the solvents which additionally benefits the fabrication of high-efficiency solar cells. In this treatment, the fabricated NCs film is dipped for few seconds in the solvent which may consist the additives, and then deposited by another layer of NCs. For example, along with MeOAc washing, incorporation of $Pb(NO_3)_2$ in the same is found to be improving the carrier transport properties through passivation [34]. Here, the native ligands are washed by the solvent MeOAc whose places are consequently refilled by the nitrate salt. Also, the metal ions in the solvent are penetrating into the depth of the fabricated QD layers which essentially help to improve the photocurrent. This is actually bringing the NCs quite closer together thereby filling the voids of the NCs surface which is beneficial for the NCs coupling. Such QDs-QDs coupling effect in films significantly influence the external quantum efficiency (EQE) value of the QDs films compared with the QDs in solution [33]. This process, which is also called as ligand-exchange process of the PNCs, is highly efficient for the fabrication of solar cells with excellent performance.

However, a systematic study on the effect of MeOAc treatment with $CsPbI_3$ QDs reveals that excessive purification cycles could induce the phase transformation along with the degradation of the fabricated films [35]. Furthermore, the use of MeOAc to purify $CsSnI_3$ QDs is found to be not suitable and results in degradation [36]. Therefore, important care should be undertaken for the solvent-based ligand-exchange process. This ligand-exchange process is very familiar in the case of QDSSCs which generally alter the energy level, passivation of midgap trap states and increased electronic coupling of QDs. A combined use of MeOAc and EtOAc with different kinds of salts in the same investigation for the surface treatment of PNCs is also reported [37].

Cho et al. carried out a solid-state ligand exchange on $CsPbBr_3$ NC films using NaOAC dissolved caroboxylate ester solutions such as MeOAc, EtOAc, propyl acetate (PrOAc) and butyl acetate (BuOAc) [38]. The authors have found that the mixture of EtOAc/BuOAc was highly efficient in stripping the native ligands over others. This is due to the fact that the native ligands are highly soluble in the longer chain alkyl acetates compared with shorter chain, which makes the ligand-exchange highly feasible. The schematic diagram of this ligand-exchange process is shown in Figure 4.2.

Therefore, this layer-by-layer (LBL) deposition approach is highly efficient with the assistance of surface-treatment process strategies. With this approach, NC films with the thickness of several hundred nanometers are prepared and the device performance is finally measured. For each layer, the treatment is carried out in the presence of appropriate solvent to replace the native insulating ligands and the inclusion of additives in the solvent is useful to repair the defect sites. The efficacy of this approach generally relies on several factors including the amount of additives, polarity and composition of the solvent, treatment time, influence of temperature and pH during the post-treatment.

For the post-treatment of the $CsPbX_3$ QDs film to fabricate the high-efficiency solar cells, most of the investigations are dealing with the use of formamidinium (FA^+) cations to passivate the NCs surface. Because of this, the mobility of the PNCs layer significantly improves thereby delivering higher efficiency. Although it is generally postulated as surface defects that are passivated by the additive treatment, there are other speculations too. For example, on the treatment of FAI on $CsPbI_3$ QDs, it is speculated that the treatment (a) may refill the voids of the QDs, (b) could make a partial ligand-exchange, (c) could form a perovskite/QDs shell and (d) could induce a grain growth on the QDs film [34]. The use of FAI for the surface treatment is a generally accepted approach to improve the solar cell performance. It is demonstrated that this FAI treatment is quite effective to eliminate the PNCs surface-bound OAm^+ ions [37].

The influence of post-treatment on the optical properties is also important and along with the increase in the intensity of the PL spectra, the shift in the UV-visible and PL spectra strongly conveys how this post-treatment is structurally influenced with different additives. In general, the post-treatment of the QDs film enhances the electronic coupling, charge transport and reduced recombination loss for the

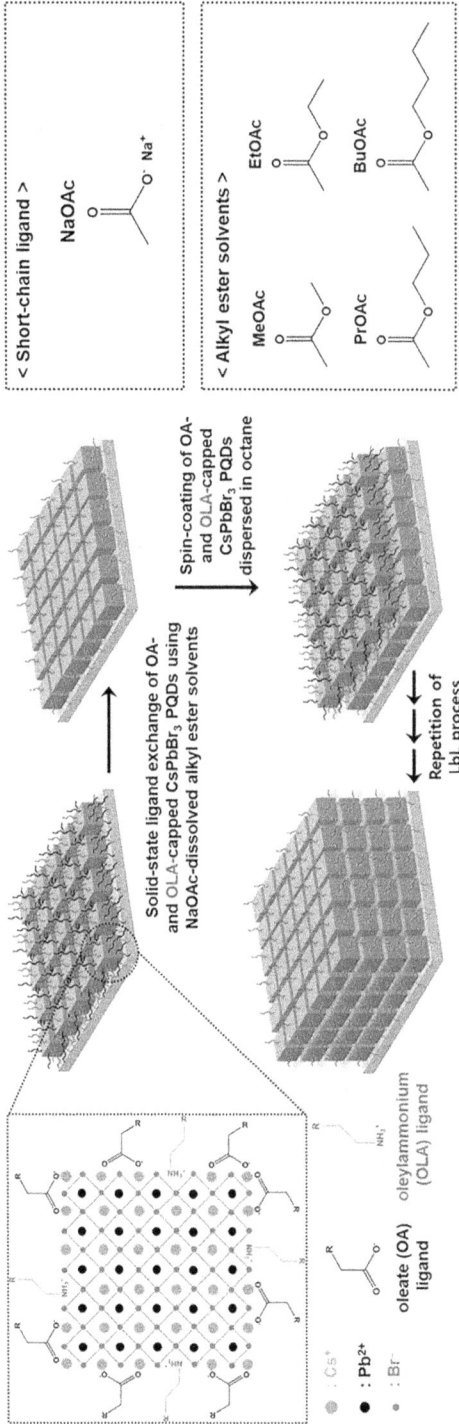

FIGURE 4.2 Schematic representation of the solid-state ligand exchange process using MeOAc, EtOAc, PrOAc, BuOAc dissolved in NaOAc. Reprinted with permission from Ref. [38].

improvement of photocurrent. An ideal post-treatment should not shift the optical spectrum in either way and it should only improve the PL properties through surface passivation.

For the hole-transporting purposes, organic semiconducting polymers such as 2,2,7,7-tetrakis-(N, N-di-p-methoxyphenylamine)-9,9-bifluorene (spiro-OMeTAD) and poly-triarylamine (PTAA) are used which efficiently extract the photogenerated holes. However, inorganic hole-transporters such as Cu_2O, CuI, NiO, MoS_2 and CuSCN have also been found as useful for this purpose to fabricate PNC-based solar cells with good stability [39].

4.3 APPLICATION OF METAL HALIDE PEROVSKITE NANOCRYSTALS FOR SOLAR CELLS

4.3.1 ORGANIC-INORGANIC HYBRID LEAD HALIDE PEROVSKITE NANOCRYSTALS FOR SOLAR CELLS

In recent years, organic-inorganic hybrid lead halide perovskites are highly exploited for photovoltaic devices and impressive efficiency performance has been realized. Therefore, nano dimension of these materials is also getting interest to fabricate future electronic devices with enhanced stability and performance. Xue et al. followed an interesting polarity-dependent solvent-washing strategy on $FAPbI_3$ QDs films [40]. Here, first the authors used 2-pentanol to wash the as-synthesized $FAPbI_3$ QDs and second the QDs were washed using mixed solvents acetonitrile/toluene solvent mixture. In every cycle, the native ligand density was found to be decreasing and the authors were able to reduce the ligand density from 3.6 to 1.2. Also, the fabricated solar cell with the device structure $ITO/SnO_2/FAPbI_3$ QDs/Spiro-oMeTAD/Au showed efficiency of 8.38% with high stability over bulk $FAPbI_3$-based devices. Although 2-pentanol is used in this case, in an another study it is observed that the use of 2-pentanol reduced the PLQY of $FAPbI_3$ QDs [41].

Following this, Ding et al. fabricated heterojunction of $FAPbI_3$ QDs (size: ~7–12 nm)/NiO_x structure and achieved 9.4% efficiency with J_{sc}=13.3 mA/cm^2 [42]. Here, the photoexcited carrier injection rate is favourable to achieve such higher efficiency, which imparts the deposition of additional layers with the perovskite QDs system. Additives like benazamidine hydrochloride that are added with the purifying solvent MeOAc rectify the vacancies and improve the performance of $FAPbI_3$ QD-based solar cell [43].

Xu et al. used different kinds of acetate salts of organic cations such as FA$^+$, guanidinium (GA$^+$) and PEA$^+$ to exchange the ligands from $FAPbI_3$ QDs [44]. The authors postulated here the complete removal of OAm and OA ligands by FA$^+$, GA$^+$, PEA$^+$ and OAc$^-$ ions which also fulfills the surface defects. With this, the authors achieved 10.13% efficiency in the case of FAOAc-treated QD-based solar cell. Also, the FAOAc-treated QDs exhibited 4.58 ns, which was the longest carrier lifetime compared with other samples. This emphasized the fact of PNCs surface repairing using potential compounds and molecules. As a champion record, Hao et al. fabricated a solar cell using $Cs_{1-x}FA_xPbI_3$ QDs in which the controlled OA-ligand management was able to deliver 16.6% with V_{oc}=1.17 V [45]. Here, the OA-rich environment reduced the surface defects that improve the PLQY and so a higher efficiency was achieved. Along with higher V_{oc}, the fabricated device structure ITO/SnO$_2$/Cs$_{1-x}$FA$_x$PbI$_3$ QDs/Spiro-MeOTAD/Au showed J_{sc}=18.3 mA/cm^2 and FF=78.3.

It is observed that regulation of ligands will not only affect the morphology and optical properties but also efficiency of the fabricated device. Instead of the use of OAm as ligand and solvent, the use of oleylammonium iodide (OAmI) entirely prohibits the proton exchange between OAm and OA and this has an effect on the efficiency. Using this approach, Ding et al. achieved 13.8% efficiency for the fabricated device structure FTO/SnO$_2$/FAPbI$_3$ QDs/Spiro-OMeTAD/Au [41]. In this case, lead acetate (Pb(CH$_3$COO)$_2$) and oleylammonium iodide (OAmI) are used as the sources of Pb^{2+} and OAm$^+$ which significantly help to improve the passivation without the proton exchange process between OAm and OA.

It is evidently demonstrated that there is a dynamic equilibrium existing between OAm and OA which is present on the NCs surface and this is avoided using a specific source. Because of this, the PLQY of the prepared FAPbI$_3$ QDs reached 88.79% which improved the device efficiency. Using surface reconfiguration approach in the presence of an ionic liquid formamidine thiocyanate (FASCN), Zhang et al. achieved ~15% efficiency which also showed excellent stability [46]. These results gave a big hope of making highly efficient hybrid PNC-based solar cells. However, because of less stability and structural complications associated with the substitution of methyl cation in the lattice, hybrid PNC-based solar cells are given less attention compared to the pure inorganic PNC-based solar cells.

4.3.2 ALL-INORGANIC LEAD HALIDE PEROVSKITE NANOCRYSTALS FOR SOLAR CELLS

Generally, PNCs undergo different kinds of structural and morphological modifications under the influence of ligands and additives. Especially, organic and inorganic ligands are critically influencing the surface defects of these NCs, which improves their carrier properties. Hence, the ligand selection for the passivation of PNCs is an important area of research and considerable developments have been achieved on this part [47]. It is visible that the fabrication of NC or QD-based solar cells normally involve surface treatment which evidently improves the carrier properties. Because of their smaller atomic radii, alkali metal ions (e.g., Na$^+$, K$^+$) significantly influence the perovskite solar cells. By adopting this concept, surface treatment using different kinds of alkali halides, acetates, nitrates with PNCs is demonstrated by different research groups [48–50].

Compared with other halide PNCs, CsPbI$_3$ NC or QD-based solar cells are gaining much attention, owing to their light harvesting property in the broad spectrum of the visible region [51,52]. However, because of their structural properties, CsPbI$_3$ exists in the orthorhombic phase, which is not suitable for device applications. Thus, retaining CsPbI$_3$ in the cubic phase is challenging but it is achieved by synthetic methods through several ways. The important consideration here is that the cubic phase of bulk CsPbI$_3$ is not stable at room temperature;

however, Swarnkar et al. have proved that α-CsPbI$_3$ QDs prepared by the hot-injection method is highly stable and the fabricated QDs film-based solar cell could deliver 10.77% efficiency [53]. This is said to be the first report on PQD-based solar cell. Despite this, the stability of the CsPbI$_3$ QDs is much lower than the other halide counterparts, which imparts the necessity of surface treatment to retain the properties.

When these α-CsPbI$_3$ QDs are coupled with FAPbI$_3$ QDs layer, such bilayer stacked structures impressively deliver 15.6% with V_{oc}=1.22 V [54]. Besides, this bilayer configuration additionally helps to avoid the phase transformation of α-CsPbI$_3$ QDs, which is a typical issue faced in this compound NCs. Here, the thermal annealing of these stacked layers results in a cation exchange at the junction which in turn results in the formation of Cs$_x$FA$_{1-x}$PbI$_3$ QDs that helps to achieve higher efficiency with improved stability. It is found that TiO$_2$ can promote the phase transformation of α-CsPbI$_3$ QDs and by choosing SnO$_2$ as the electron transport layer, this issue can be resolved. In this modification, the less hydrophilic, chloride-modified SnO$_2$ protects the phase transformation, which results in 14.5% efficiency [55].

Phase transformation of α-CsPbI$_3$ QDs can also be prohibited by using octylamine (OTAm) together with OAm, which provides a tight binding on the surface [56]. Besides, it is achieved by doping rare-earth elements, for example, ytterbium doping [57], transition metal ions (Zn^{2+}, Mn^{2+}) doping [58], engineering electron transport layer (ETL) [59] and passivating a shell of lead chalcogenide NCs [60]. These investigations indicate that by protecting α-CsPbI$_3$ QDs using an appropriate method, the efficiency as well as stability of the solar cell can be improved. Kim et al. used phenylethylammonium ion (PEA$^+$) post-treatment on the MeOAc-treated CsPbI$_3$ QDs film [61]. This treatment removes the insulating OAm/OA ligands from the QDs and improves the efficiency up to 14.1% through efficient coupling between the QDs. Earlier, using a benzene group for phenyltrimethyl ammonium bromide treatment, 11.2% efficiency was demonstrated [62].

In an another investigation, 3,3-diphenylpropylammonium iodide (DPAI) was used to develop CsPbI$_3$ PNCs ink for the general coating formulation to fabricate LED and solar cells [63]. Here, the CsPbI$_3$ NCs were ligand-exchanged by DPA$^+$ ions through solution phase and fabricated via a single-step approach. Because of the efficient passivation, the fabricated solar cell delivered impressive 14.92% efficiency with V_{oc}=1.21 V and J_{sc}=16.64 mA/cm^2. A pseudo-solution-phase-ligand exchange approach of CsPbI$_3$ QDs in the presence of the combined use of PEA$^+$ and 2-(4-fluorophenyl) ethyl ammonium (FPEA$^+$) is also beneficial to improve the efficiency above 14% [64]. Mostly, these additives are dissolved in the mild polarity solvent such as EtOAC.

Sanehira et al. used FAI in EtOAc to treat the surface of CsPbI$_3$ QDs and found that this treatment incredibly helps achieve 13.4% efficiency with J_{sc}=14.37 mA/cm^2 [34]. Moreover, this treatment additionally improves the lifetime and mobility of the fabricated QDs film, which helps to fabricate other kinds of optoelectronic devices. The MeOAc treatment is also found to be effective with amines and for example, together with di-n-propylamine, this greatly improves the surface passivation through acylation of OA and amine which leads to efficiency of nearly 15% [65]. When PNCs are treated with alkali metal-based salts, the PL properties are significantly enhanced [66]. Adopting this concept, Kim et al. used sodium acetate (NaOAC) dissolved in MeOAc to treat $CsPbI_3$ QDs film and this treatment has led to the efficiency of the fabricated solar cells to 12.4% for the device structure FTO/TiO_2/$CsPbI_3$ QDs/Spiro-oMeTAD/Au [67]. Interestingly, this efficiency has been further improved to 13.4% by treating with FAI in EtOAc. Here, the native OAm/OA ligands are effectively replaced by the incoming acetate ions which ultimately improves the carrier transport. However, not all acetate compounds are found to be useful for this purpose and use of caesium acetate (CsOAc) is found to be ineffective and decrease the performance of the fabricated solar cells. This kind of acetate salt dissolved in ethyl or methyl acetate solvent mixture-based treatment is also realized in few other cases, suggesting the reproducibility and effectivity of this approach.

On the contrary, Ling et al. observed that the treatment of $CsPbI_3$ QDs film using CsOAc dissolved in EtOAc results in the passivation of Cs^+ vacancies and improves the efficiency of the fabricated solar cells [68]. Here, the authors investigated with the effect of other cesium salts such as Cs_2CO_3, CsI and $CsNO_3$ but only the use of CsOAc provides a higher efficiency of 14% with J_{sc}=14.9 mA/cm^2 for the device structure FTO/TiO_2 (40 nm)/$CsPbI_3$ QDs (300 nm)/PTAA (80 nm)/MoO_x (8 nm)/ Ag (120 nm). Interestingly, here the use of FAI treatment instead of CsOAc results in similar enhancement of mobility product, which shows the critical role of CsOAc on the QDs film treatment. The removal of native ligands from the PNCs surface in the presence of alkyl acetate takes place through a demonstrated mechanism.

As an observed fact, the purification of PNCs involves moderate polarity solvents such as methyl acetate, ethyl acetate and tertiary butyl alcohol. The replacement of oleate molecules from the PQDs surface through MeOAc takes place through a hydrolysis process. Here, air MeOAc is hydrolysed through an *in-situ* process that produces acetic acid and methanol [69]. Then, this acetic acid replaces the oleate molecules from the PQDs surface and exists as acetate ion. It should be noted that this MeOAc is not capable of removing OAm^+ ions, probably owing to their strong binding ability compared with the oleate ions. These studies imply a fact that depends on the conditions and other factors, the surface treatment is affected. Purification of $CsPbI_3$ QDs is important to achieve narrow size dispersion and this significantly influences in the formation of QDs array as well as device efficiency [70]. In this study, through MeOAc-assisted purification and consequent gel permeation chromatography technique (GPC), it was able to achieve QDs with size ~10.6 ± 1.2 nm along with 70% PLQY. These QDs were used to fabricate FTO/TiO_2/$CsPbI_3$ QDs/ Spiro-OMeTAD/MoO_x/Ag solar cells that impressively delivered 15.3% efficiency. The schematic diagram of the purification process and structural and electrical characteristics of $CsPbI_3$ QDs and fabricated devices are shown in Figure 4.3.

Although FAI treatment enhances the efficiency of PNC-based solar cells, it is quite challenging to maintain stability. Also, most of the FAI-based treatment results in cation exchange reaction (i.e. formation of $Cs_xFA_{1-x}CsPbI_3$ QDs) and this often results in issues with absorption properties. To solve this, Kim et al. used phenylethylammonium cations to treat the $CsPbI_3$ QDs surface and incredibly this treatment delivered 14.1% efficiency with excellent stability for 15 days [71]. In most of the cases, around 300–800 nm thickness is generally followed to fabricate PQD-based solar cells and for a case, it is demonstrated that over 12% efficiency is

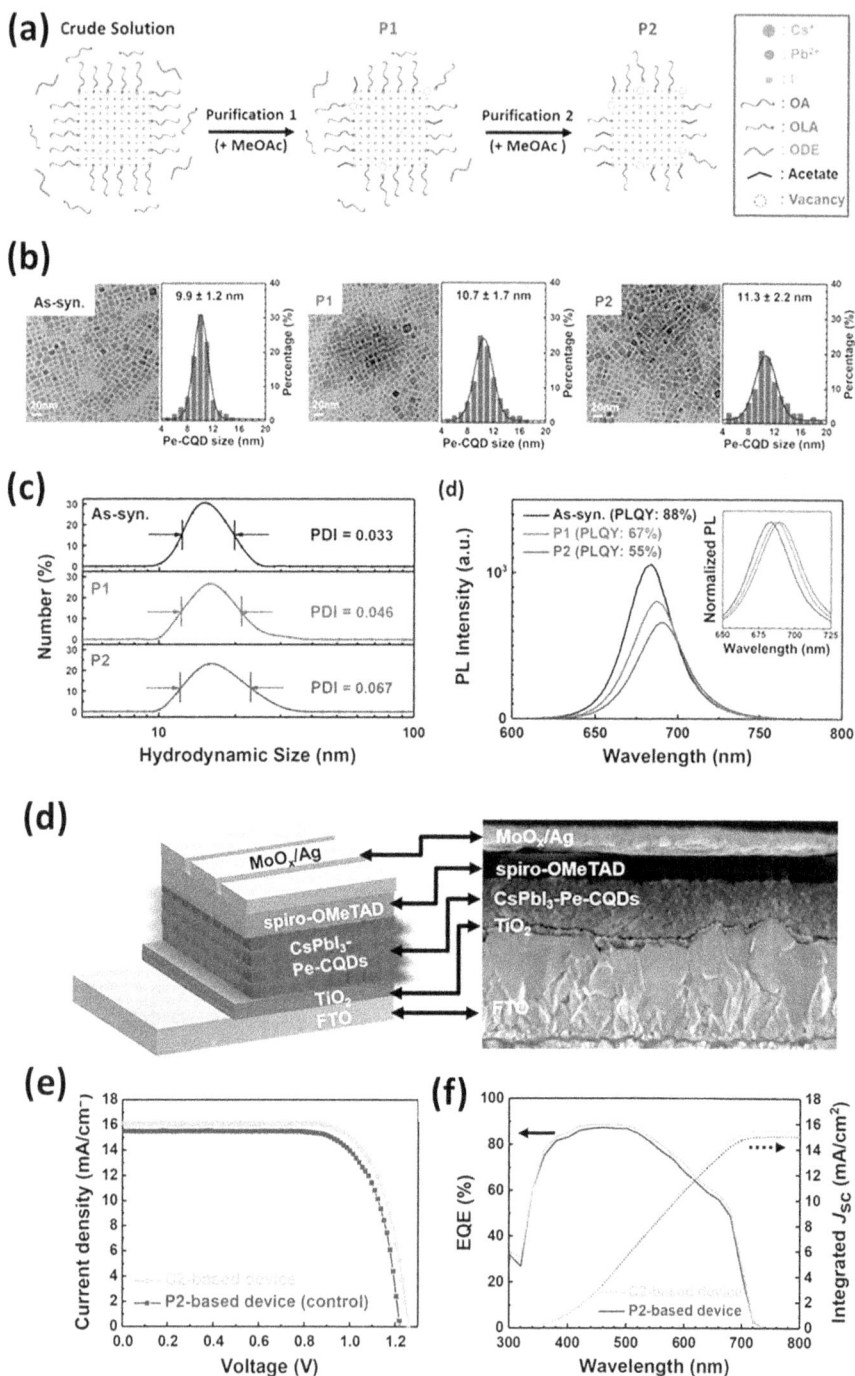

FIGURE 4.3 (a) Schematic representation of purification of CsPbI$_3$ QDs using MeOAc; (b) high-resolution transmission electron microscopy (HR-TEM) images of CsPbI$_3$ QDs with

possible with the CsPbI$_3$ QDs array with 1.2 μm thickness [72]. Similar to this a lot of evidence provided the fact that surface treatment using suitable organic or inorganic compound is an integrated way to fabricate highly efficient solar cells [42].

Because of their short chain and flexibility with the surface of NCs, thiols are playing significant role in modifying the structural and optical properties of PNCs. Thiols, a sulphur-containing compound, can generate sulphide (S^{2-}) that assist in the synthesis and modification of the functional properties of traditional semiconductor NCs. The interaction of S^{2-} with the PNCs surface often modifies the phase and also improves the optical properties [73–75]. Hochan Song et al. observed that with the treatment of CsPbI$_3$ PNCs using L-cysteine, surface defects could be controlled excellently and the efficiency can be improved to a maximum of 15.5% [76]. Similarly, incorporation of L-phenylalanine makes similar kinds of effect on the photovoltaic properties through passivation of the surface defects, thereby reaching nearly 15% efficiency [77].

Compounds such as benzoic acid [78], p-mercaptopyridine [79], ethylene diamine tetracetic acid (EDTA) [80] and ionic liquid [81] are found to be efficient candidates in improving the surface of the CsPbI$_3$ PNCs which results in higher efficiency. In this, recently Huang et al. attempted the surface treatment of 2,5-thiophenedicarboxylic acid through a solution-phase anchoring strategy which resulted in an impressive 16.14% efficiency [82]. These results clearly predict the potentiality of the PNCs in solar cells with respect to the functional group. The major concerns about the improving performance of PNC-based solar cells are improving V$_{oc}$. The loss in V$_{oc}$ is associated with different parameters including (a) the defects present on PNCs due to ligands and synthesis conditions; (b) intrinsic voltage loss due to ligands and mobile ions of PNCs; (c) band misalignment; (d) non-radiative recombination that occurs at the perovskite and transport layer interface; and (e) interstitial defects present in QDs [25,83]. The V$_{oc}$ loss is also accompanied with the presence of too many grain boundaries in the NCs layer and this could be overcome through forming different bandgap QDs which result in a kind of gradient alignment [84].

Although most of the CsPbI$_3$ QD-related studies describe about the surface modification and additive-related treatment, it is possible to improve the efficiency through modifications in the structure. For example, through prevention of the octahedral tilting in CsPbI$_3$ QDs by means of trimethylsulfonium iodide (TMSI) ligand, an impressive record of 16.64% efficiency was achieved [85]. Use of ligands such as trioctylphosphine (TOP) and tertiary-butyl iodide (TBI) could solve the iodide vacancies on CsPbI$_3$ QDs and the reduced charge-carrier recombination in this case results in the improved performance upto 16.21% [86]. Other

---

two cycles of purification; (c) dynamic light scattering (DLS) and photoluminescence (PL) and absolute PLQY of the as-synthesized and two-cycles purified CsPbI$_3$ QDs; (d) schematic representation of the device structure with cross-sectional scanning-electron microscopy (SEM) image; (e) current–voltage curve (I–V) of the best-performing device; and (f) external quantum efficiency (EQE) and integrated J$_{sc}$ curve of the best performing solar cell. Reprinted with permission from Ref. [70].

modifications with CsPbI$_3$ QDs are also getting interest to generate high photo-current. Zhang et al. doped an organic compound F6TCNNQ with the as-prepared CsPbI$_3$ QDs in order to convert them to p-type [72]. These QDs were deposited with the undoped CsPbI$_3$ QDs layer (n-type) to fabricate a p-n homojunction. The fabricated p-n homo-junction with the structure FTO/TiO$_2$/n-type CsPbI$_3$ QDs/p-type CsPbI$_3$ QDs/PTAA/MoO$_3$/Ag delivered 15.29% efficiency for the active layer thickness ~600 nm which is higher than the device with CsPbI$_3$ QDs only (13.50%). Here, when the fabrication is achieved using higher thickness of the homo-junction, i.e. with 1.2 µm, resulted in 12.28% efficiency. Devices with such higher thickness imply that it is possible to construct highly dense photoactive layers for the large area coating. It is also possible to modify the interface between QDs/hole-transporter in order to improve the performance.

Ji et al. inserted an organic polymer PBDB-T between QDs/PTAA heterojunc-tion and due to the modification in energy level alignment, the fabricated devices based on CsPbI$_3$ and FAPbI$_3$ QDs delivered 13.8% and 13.2% efficiency, respec-tively [87]. Here, the modification of energy level leads to the improvement in J$_{sc}$, which results in higher efficiency. Since bi-precursor route probably results in the formation of other perovskite derivatives such as Cs$_4$PbI$_6$, a tri-precursor meth-odology of synthesizing CsPbI$_3$ QDs under optimized Cs/Pb ratio results in high production yield which could improve the efficiency upto 14.24% [87]. As PNCs are susceptible with the doping of transition metal ions, such effort could modify the optical properties and improve device performance.

Compared with other elements, the role of zinc (Zn^{2+}) in improving the optical properties, especially PLQY, is studied by many research groups. It is well known that the incorporation of Zn^{2+} with the traditional semiconductor NCs enhanced the optical properties incredibly. When Zn^{2+} is incorporated on the PNCs surface, the defect states are extremely minimized which result in higher PL intensity. Through synchrotron-based extended X-ray absorption fine structure (EXFAS) analysis, Bi et al. found that the incorporation of Zn^{2+} ions with CsPbI$_3$ QDs improved the local ordering of the lattice and reduced the octahedral distortion [88]. In this case, the prepared CsPb$_{0.93}$Zn$_{0.07}$I$_3$ QDs delivered 95% PLQY with longer lifetime (22.2 secs) due to the ultralow trap density. Moreover, the fabricated highly stable solar cell using these QDs with the structure glass/FTO/TiO$_2$(50 nm)/CsPb$_{0.93}$Zn$_{0.07}$I$_3$ QDs (200 nm)/Spiro-OMeTAD (120 nm)/Ag (100 nm) delivered 14.8% with V$_{oc}$=1.19 V, J$_{sc}$=16.4 mA/cm^2, FF=0.76. Here, the CsPbI$_3$ QD-based solar cells delivered only 9.7% effi-ciency, which clearly envisages the doping effect of Zn^{2+}. This is because addition of Zn^{2+} significantly enhances the Goldschmidt factor and also formation energy, which ultimately improves the stability [25]. When iodide of zinc salts is used as the source for doping, for example, ZnI$_2$, the iodide ions additionally enrich the CsPbI$_3$ QDs surface and enhance the efficiency of solar cells above 16% [89]. Other than cationic doping, the spectrum of the PNCs could be enlarged through modifying halide composition. For example, a compositionally tuned CsPbI$_3$ QDs with Br$^-$ result in CsPb(I$_{1-x}$Br$_x$)$_3$ QDs and this improve the phase stability and also helping to achieve efficiency of 12.31% [90]. Here, due to the incorporation of Br-, the fabricated solar cells are able to retain their PCE stability upto 15 days, which shows the importance of compositional engineering for the fabrication of highly stable PNCs solar cells.

To fabricate the electron transport TiO$_2$ layer, a potential method is the deposition of two different layers of TiO$_2$ that consists of compact and mesoporous architecture. This assembly actually facilitates the fast electron transfer from the active layer in order to extract it to harvest large amount of photons. Interestingly, when Cs$^+$ ions are used for this TiO$_2$ layer, this improves the charge transfer between TiO$_2$/perovskite QDs interface and the electron injection rate is enhanced. Specifically, Chen et al. observed that compared with Cs-treated-compact-TiO$_2$ layer, the e$^-$ injection rate is found to be higher in the case of Cs-treated-mesoporous-TiO$_2$ layer, which is reflected in the increase of J$_{sc}$ [91]. The schematic diagram of the fabricated device and its electrical characteristics, as well as the electrical characteristics of the Cs-treated and untreated CsPbI$_3$ QDs film is shown in Figure 4.4.

Here, using CsOAc/MeOAc treatment, the authors achieved 14.32% efficiency for fabricated solar cells. Because of the mesoporous layer, it is possible that CsPbI$_3$ QDs could be embedded with the m-TiO$_2$ layer and improve the charge transport, as observed experimentally [90]. Other than electron transport layer and organic hole transport layer, it is possible to couple PNCs with inorganic NC-based hole-transporter. Liu et al. used ligand-exchanged Cu$_{12}$Sb$_4$S$_{13}$ QDs as hole-transporter to fabricate CsPbI$_3$ QD-based solar cell [92]. Here, the insulating OAm ligands are replaced with highly conductive 3-mercaptopropionic acid (MPA) in order to improve dot-to-dot coupling. Using these ligand-exchanged C$_{12}$Sb$_4$S$_{13}$ QDs, the fabricated device assembly FTO/c-TiO$_2$/m-TiO$_2$/CsPbI$_3$ QDs/C$_{12}$Sb$_4$S$_{13}$ QDs/Au is able to deliver higher J$_{sc}$=18.28 mA/cm^2 with 10% efficiency. Also, the fabrication of nanophotonic structure on the hole-transporting polymer (HTL) of CsPbI$_3$ QD-based solar cells has been observed as improving the performance considerably [93].

There are considerable developments ongoing in recent years related to CsPbI$_3$ QD-based solar cells. For example, in order to minimize the use of toxic Pb^{2+} compound, the source materials are derived from the waste of Pb battery and this has delivered impressive performance of ~14% efficiency [94]. Also, the fabrication of semi-transparent solar cells using CsPbI$_3$ QDs is an emerging promising concept and such effort with PTAA hole-transporter resulted in 9.6% efficiency using silver electrode [95]. The fabrication of flexible PQDs solar cells [96,97], synthesizing PQDs with iodine-rich surface [98], dual passivation or dual post-treatment approach [99,100], fabrication of polymer (or) organic molecules/PQDs composite-based solar cells [101,102] are showing promising outcome for the further improvement of this technology. These analyses indicate that by engineering in different directions on the layers of the PNC-based solar cells, it is possible to improve the performance together with stability.

Despite many of the studies devoted for the exploration of use of CsPbI$_3$ QDs for PV applications, the bromide counterpart of all-inorganic PNCs has also been studied. Especially, surface-engineered CsPbBr$_3$ QDs are showing interesting performance in solar cells. Also, their alloyed composition with iodide delivers interesting results for device performance. Ghosh et al. prepared CsPbBr$_{1.5}$I$_{1.5}$ QDs (approx. 10 nm size) using the hot-injection method and the fabricated solar cell structure FTO/TiO$_2$/CsPbX$_3$/Spiro-MeOTAD/MoO$_3$/Ag delivered ~8% efficiency with excellent stability [103]. Here, the authors were able to tune the absorption

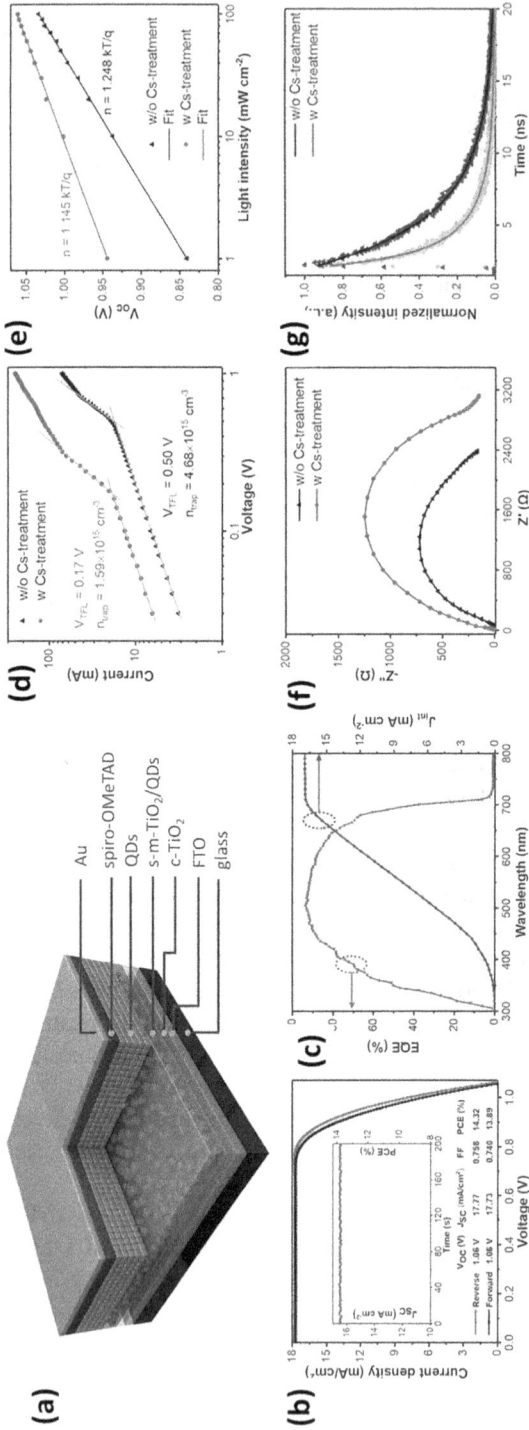

FIGURE 4.4 (a) Schematic representation of the fabricated solar cell; (b) I–V curve of the best-performing device; (c) EQE curve of the best-performing device; (d) space-charge limited current vs voltage of the Cs-treated and untreated CsPbI₃ QDs film; (e) light intensity vs V_oc curve of the Cs-treated and untreated CsPbI₃ QDs film; (f) Nyquist plots; and (g) time-resolved photoluminescence (TRPL) of the Cs-treated and untreated CsPbI₃ QDs film. Reprinted with permission from Ref. [91].

spectra in the range of 500–720 nm. Later, with the Ag^+ doping in the same composition, the authors were able to achieve 9.67% efficiency [104].

Zhang et al. fabricated a dense film of guanidinium thiocyanate/acetone-treated $CsPbBr_3$ QDs [105]. This treatment modifies the fermi level as well as improves the electrical coupling between the QDs. The fabricated semi-transparent solar cells with these QDs ($FTO/TiO_2/CsPbBr_3$ QDs/PTAA/MoO$_3$/Au) delivered 5.01% efficiency. Besides, using PTB7 as hole-transporter, the authors were able to achieve V_{oc}=1.65 V, which is the highest record in this category. Because of their bright luminescent property, $CsPbBr_3$ QDs are much concentrated to fabricate LEDs. The reasons for the low efficiency of these QD-based devices is postulated with different opinions. With the help of Kelvin probe force microscopy (KPFM), it was experimentally observed that the charge transport between ETL/$CsPbBr_3$ QDs and HTL/$CsPbBr_3$ QDs is quite different and abnormal to provide higher efficiency [106]. The fabricated device structure and its I–V curve, vacuum level alignment and fermi-level alignment of the device before and after contact, the schematic diagram of the charge separation in solar cells are shown in Figure 4.5.

Cho et al. fabricated a thick, pin-hole free $CsPbBr_3$ NCs film through a solvent-miscibility-assisted ligand-exchange method that produced V_{oc}=1.6 V with 4.23% efficiency [38]. Similar to other kinds of PNCs, $CsPbBr_3$ NCs are highly adoptive for surface treatment process. Although the halide vacancy in $CsPbBr_3$ NCs lies between $CsPbCl_3$ and $CsPbI_3$ NCs, near-unity PLQY is achieved through potential surface modification strategies, which is quite useful to fabricate LEDs and solar cells [107]. These investigations reveal the possibility of extending the efficiency through different strategies. Unlike $CsPbI_3$ QDs, the stability of $CsPbBr_3$ QDs is quite high and this actually supports to synthesize NCs with impressive PL properties.

4.3.3 Lead-Free Metal Halide Perovskite Nanocrystals for Solar Cells

Owing to the toxicity associated with Pb^{2+} in the hybrid as well as all-inorganic PNCs, research has been intensified to find alternative or to replace Pb^{2+} using less-toxic or non-toxic elements in the perovskite structure. In this aspect, several studies describe the use of other potential elements to replace Pb^{2+} in the perovskite structure. Substitution of other elements in the place of Pb^{2+} often induces different kinds of variations in the structure and optical properties. This replacement may be a complete or partial removal and if it is partial, alloying occurs with different compositions. Quite less hybrid perovskite NCs have been studied for solar cell applications due to their poor photovoltaic performance. For example, a solar cell that consists of $MASnBr_{3-x}I_x$ QDs active layers delivers less than 1% efficiency [108]. Therefore, all-inorganic PNC-based solar cells are investigated through different aspects. Use of lead-free PNCs with Pb^{2+}-perovskite-based solar cells is found to be beneficial for the fabrication of highly efficient solar cells. For example, it is found that $CsSnBr_{3-x}I_x$ QDs are efficient in charge extraction when they are deposited on the $CsPbBr_3$ perovskite layer [109].

Liu et al. fabricated a solar cell using $CsSn_{0.6}Pb_{0.4}I_3$ QDs as active layer [110]. This alloyed composition was quite useful in achieving 2.9% efficiency with high photocurrent, owing to the fast electron transfer rate. Interestingly, the prepared

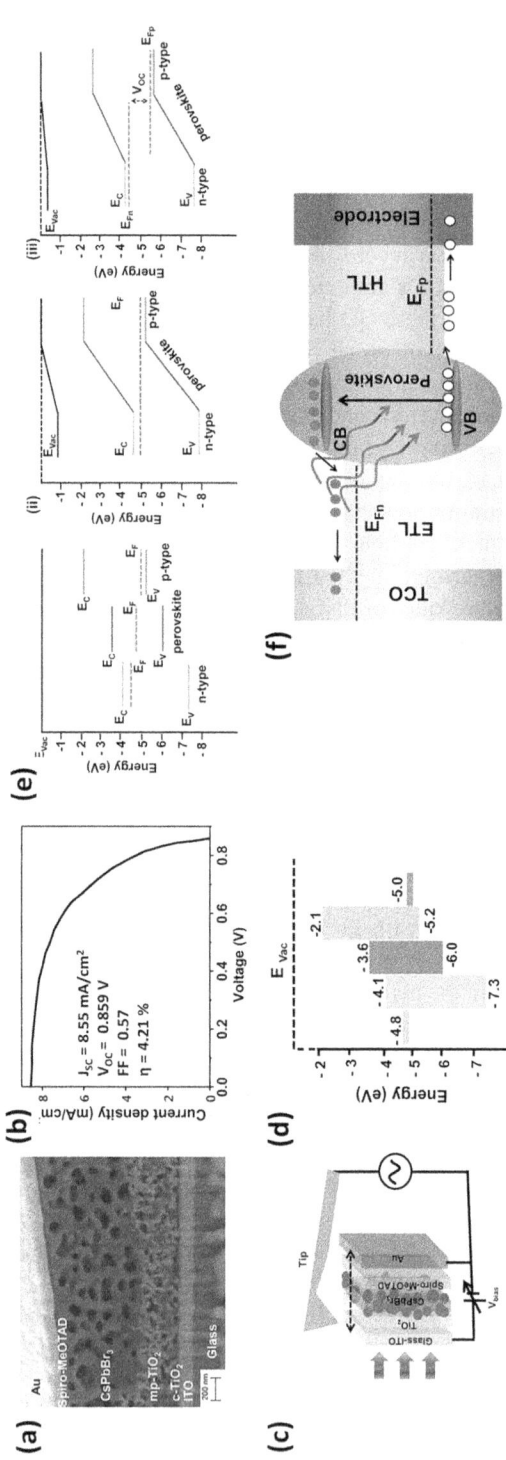

FIGURE 4.5 (a) Cross-sectional SEM image of the fabricated device; (b) I–V curve of the fabricated device; (c) schematic representation of the device; (d) energy-band level representation of the device structure (e) (i) vacuum level alignment along n-type/perovskite/p-type material in solar cell before contact, (ii) fermi level alignment after contact in dark, (iii) separated fermi level of electrons and holes; (f) schematic representation of the charge separation in solar cells. Reprinted with permission from Ref. [106].

$CsSn_{0.6}Pb_{0.4}I_3$ QDs inks are found to be highly stable (over five months) compared with the unstable $CsSnI_3$ QDs. In an another investigation, it is proved that the use of additives such as triphenylphosphite could stable the $CsSnI_3$ QDs and can improve the efficiency over 5% [111]. Chen et al. prepared $CsSnX_3$ (X=Cl, Br and I) nanorods using a solvothermal approach in the presence of diethylenetriamine (DETA) [112]. Here, the authors found that $CsSnI_3$ nanorods with Cs/Sn/I ratio 1/1.2/3.2 showed impressive power-conversion efficiency of ~13%. This result shows that despite the large number of surface defects, enhanced charge separation is possible in lead-free PNCs. Double perovskites, a potential compound in the lead-free perovskite materials, are investigated much in the recent years for several kinds of applications. Caesium silver bismuth bromide, $Cs_2AgBiBr_6$ a double perovskite indirect bandgap (E_g=~2 eV) semiconductor compound, is widely studied in recent years for solar cell fabrication owing to its high stability [113].

Ahmad et al. synthesized $Cs_2AgBiBr_6$ nanocrystals through the hot-injection method and carried out a layer-by-layer solid-state ligand exchange using tetrabutyl ammonium bromide (TBAB) [114]. With this film, the fabricated solar cell with the structure, $FTO/TiO_2/Cs_2AgBiBr_6NCs/Spiro-MeOTAD/Ag$ delivered 0.46% efficiency. Kumar et al. prepared $Cs_2AgBi_{0.6}Sb_{0.4}Br_6$ NCs and using trimethyl silyl iodide (TMSI), the prepared NCs were halide-exchanged to convert it to $Cs_2AgBi_{0.6}Sb_{0.4}(Br_{0.278}I_{0.722})_6$ [115]. These NCs were further dispersed in chlorobenzene and used in the device structure glass/$ITO/NiO_x/Cs_2AgBi_{0.6}Sb_{0.4}Br_6/PCBM/Ag$ which resulted in 0.09% efficiency. Although the performance of these solar cells is quite low compared with the Pb-based PNC-based solar cells, through depositing additional layers and surface engineering strategies, it is possible to achieve higher efficiency. Another bismuth-based lead-free perovskite compound, caesium bismuth iodide ($Cs_3Bi_2I_9$) has also received much attention in solar cell applications. Through the formation of polycrystalline $Cs_3Bi_2I_9$ films, Bai et al. prepared ultrathin nanosheets of the same and applied for solar cells [116]. Here, the nanosheets films were prepared through the dissolution-recrystallization method by means of a spin-coating process in the presence of solvent mixture (DMF and CH_3OH). With the device structure $FTO/c-TiO_2/Cs_3Bi_2I_9/HTM/Au$, authors achieved 3.20% efficiency using CuI as a hole-transporter. Such a high value of bismuth perovskite NCs ensures that further improvement is possible through altering the device structure.

Chen prepared $CsGeX_3$ quantum rods (QRs) through a solvothermal approach to fabricate a solar cell [117]. Through modifying halides, the absorption spectrum of the prepared nanorods was tuned from 565 nm to 655 nm. The fabricated solar cells using $CsGeCl_3$, $CsGeBr_3$ and $CsGeI_3$ nanorods delivered 2.57%, 4.92% and 4.94% efficiency, respectively. All these investigations indicate that lead-free PNCs have a promising avenue for the further improvement of efficiency and stability. Other than these mentioned class of the PNCs, compound NCs such as $Bi_{13}S_{18}I_2$, $KBaTeBiO_6$ NCs and $(NH_4)_3Sb_2I_xBr_{9-x}$ were also utilized for photovoltaic applications; however, all these compound NCs resulted in very low efficiency [118]. Because of their large bandgap and large number of interstitial and surface defects, lead-free PNC-based solar cells suffer poor efficiency. Also, the oxidizing ability of lead-free elements such as Sn^{2+} and Ge^{2+} is higher than that of Pb^{2+} and so there are significant challenges with the fabrication of solar cells with these compounds.

However, novel approaches in the fabrication and synthesis are expected to solve the existing scenario.

4.4 SUMMARY AND PERSPECTIVES

Although PNCs are much studied with different aspects, their use on the flexible substrate is limited and this is one of the major setbacks in developing PNCs for modern generation solar cells. The complication associated with the delamination of QDs film after deposition must be solved for large-scale production. Novel strategies in compositional tuning, doping and insertion of new kinds of additives to prevent phase transformation and corresponding degradation of layer could help solve existing problems to some extent. The ongoing efforts in the development of protection layer and encapsulation strategies will be expected to deliver fruitful results for the fabrication of highly stable, higher efficiency perovskite NC-based solar cell modules. For the synthesis aspects of PNCs, adopting a modern approach such as an automated microfluidic synthesis method could solve the setbacks related to the large-scale production of NCs.

Development of high-charge carrier transport materials with high carrier mobility is also important to enhance the performance from the current level. Also, a common platform related to the different kinds of PNCs and their use in the solar cells can address the complications related to device performance. Incorporating PNCs into the polymer matrix for high stability should be enlarged with novel semiconducting organic polymers in order to fabricate devices with longer lifetime. Besides, retaining the PL properties of QDs films in solid state through post-treatment could establish much development in achieving higher efficiency PV devices. While fabricating PNCs films in a large scale, attention should be paid on the pin-hole free layers to achieve higher performance. For this, it is essential to explore novel binders and additives that does not affect the carrier transport properties.

Large-scale production of PNCs through solution methods often has a lot of challenges that must be solved for the roll-to-roll (R2R) production of solar modules. Also, a customized strategy on the purification and functionalization of PNCs using strongly adherent ligand molecules could minimize the stability issues of the fabricated solar cell devices. Additionally, surface modification and post-treatment of PNCs play an important role in improving the efficiency of solar cells, and so novel approaches in this direction may alleviate the hurdles associated with the surface manipulation of PNCs. Stability of PNC-based solar cells under different ambient conditions is another issue that needs to be solved for the improvement of further aspects. However, the improved stability assisted with lead-free PNCs can solve this, although the performance of such devices needs to be improved much. The current progress in these directions assures that the future generation of solar cell devices with PNCs will have high efficiency along with excellent stability.

ACKNOWLEDGEMENTS

Ananthakumar Soosaimanickam sincerely thanks MINISTERIO DE CIENCIA E INNOVACIÓN, Spain for funding through Torres-Quevedo programme

(PTQ2020–011398). The author also sincerely thanks the excellent support rendered by the management of Intercomet S.L for carrying out this work.

REFERENCES

1. D. Chen, X. Chen, Luminescent perovskite quantum dots: synthesis, microstructures, optical properties and applications, *J. Mater. Chem. C*, 2019, 7, 1413–1446.

2. Y. Fu, H. Zhu, J. Chen, M. P. Hautzinger, X.-Y.Zhu, S.Jin, Metal halide perovskite nanostructures for optoelectronic applications and the study of physical properties, *Nat. Rev. Mater.*, 2019, 4, 169–188.

3. T. Qiao, D. H. Son, Synthesis and properties of strongly quantum-confined cesium lead halide perovskite nanocrystals, *Acc. Chem. Res.*, 2021, 54(6), 1399–1408.

4. H. He, S. Mei, Z. Wen, D. Yang, B. Yang, W. Zhang, F. Xie, G. Xing, R. Guo, Recent advances in blue perovskite quantum dots for light-emitting diodes, *Small*, 2022, 18, 2103527.

5. J. Chen, Y. Zhou, Y. Fu, J. Pan, O. F. Mohammed, O. M. Bakr, Oriented halide perovskite nanostructures and thin films for optoelectronics. *Chem. Rev.* 2021, 121(20), 12112–12180.

6. A. Dey, J. Ye, A. De, E. Debroye, S. K. Ha, E. Bladt, A. S. Kshirsagar, Z. Wang, J. Yin, Y. Wang, L. N. Quan, F. Yan, M. Gao, X. Li, J. Shamsi, T. Debnath, M. Cao, M. A. Scheel, S. K. Jain, J. A. Steele, M. Gerhard, L. Chouhan, K. Xu, X-G. Wu, Y. Li, Y. Zhang, A. Dutta, C. Han, I. Vincon, A. L. Rogach, A. Nag, A. Samanta, B. A. Korgel, C-J. Shih, D. R. Gamelin, D. H. Son, H. Zeng, H. Zhong, H. Sun, H. V. Demir, I. G. Schblykin, I. Mora-Sero, J. K. Stolarczyk, J. Z. Zhang, J. Feldmann, J. Hofkens, J. M. Luther, J. Perez-Prieto, L. Li, L. Manna, M. I. Bodnarchuk, M. V. Kovalenko, M. B. J. Roeffaers, N. Pradhan, O. F. Mohammed, O. M. Bakr, P. Yang, P. Muller-Buschbaum, P. V. Kamat, Q. Bao, Q. Zhang, R. Krahne, R. E. Galian, S. D. Stranks, S. Bals, V. Biju, W. A. Tisdale, Y. Yan, R. L.According to literature, generally medium Z. Hoye, L. Polavarapu, State of the art and prospects for halide perovskite nanocrystals, *ACS Nano*, 2021, 15(7), 10775–10981.

7. K. Wang, D. Yang, C. Wu, M. Sanghadasa, S. Priya, Recent progress in fundamental understanding of halide perovskite semiconductors, *Prog. Nat. Sci.*, 2019, 106, 100580.

8. J. Shamsi, A. S. Urban, M. Imran, L. D. Trizio, L. Manna, Metal halide perovskite nanocrystals: synthesis, post-synthesis modifications, and their optical properties, *Chem. Rev.*, 2019, 119(5), 3296–3348.

9. Y. Mu, Z. He, K. Wang, X. Pi, S. Zhou, Recent progress and future prospects on halide perovskite nanocrystals for optoelectronics and beyond, *iScience*, 2022, 25(11), 105371.

10. A. F. Gualdron-Reyes, S. Masi, I. Mora-Sero, Progress in halide-perovskite nanocrystals with near-unity photoluminescence quantum yield, *Trends in Chemistry*, 2021, 3(6), 499–511.

11. K. Zheng, Q. Zhu, M. Abdellah, M. E. Messing, W. Zhang, A. Generalov, Y. Niu, L. Ribaud, S. E. Canton, T. Pullerits, Exciton binding energy and the nature of emissive states in organometal halide perovskites, *J. Phys. Chem. Lett.*, 2015, 6(15), 2969–2975.

12. N. Fiuza-Maneiro, K. Sun, L. Lopez-Fernandez, S. Gomez-Grana, P. Muller-Buschbaum, L. Polavarapu, Ligand chemistry of inorganic lead halide perovskite nanocrystals, *ACS Energy Lett.*, 2023, 8(2), 1152–1191.

13. Y. Zhou, X. Luo, J. Yang, Q. Qiu, T. Xie, T. Liang, Application of quantum dot interface modification layer in perovskite solar cells: progress and perspectives, *Nanomaterials*, 2022, 12(12), 2102.

14. W. Chi, S. K. Banerjee, Application of perovskite quantum dots as an absorber in perovskite solar cells, *Angew. Chem.*, 2022, 61(9), e202112412.
15. W. Chi, S. K. Banerjee, Performance improvement of perovskite solar cells by interactions between nano-sized quantum dots and perovskite, *Adv. Fun. Mater.*, 2022, 32(28), 2200029.
16. N. S. Peighambardoust, E. Sadeghi, U. Aydemir, Lead halide perovskite quantum dots for photovoltaics and photocatalysis, *ACS Appl. Nano. Mater.*, 2022, 5(10), 14092–14132.
17. M. Liu, H. Zhang, D. Gedamu, P. Fourmont, H. Rekola, A. Hiltunen, S. G. Cloutier, R. Nechache, A. Priimagi, P. Vivo, Halide perovskite nanocrystals for next-generation optoelectronics, *Small*, 2019, 15(28), 1900801.
18. N. Pradhan, Growth of lead halide perovskite nanocrystals: still in mystery, *ACS Phys. Chem Au*, 2022, 2(4), 268–276.
19. H. Song, J. Yang, W. H. Jeong, J. Lee, T. H. Lee, J. W. Yoon, H. Lee, A. J. Ramadan, R. D. J. Oliver, S. C. Cho, S. G. Lim, J. W. Jang, Z. Yu, J. T. Oh, E. D. Jung, M. H. Song, S. H. Park, J. R. Durrant, H. J. Snaith, S. U. Lee, B. R. Lee, H. Choi, A universal perovskite nanocrystal ink for high-performance optoelectronic devices, *Adv. Mater.*, 2022, 2209486.
20. K. Hills-Kimball, H. Yang, T. Cai, J. Wang, O. Chen, Recent advances in ligand design and engineering in lead halide perovskite nanocrystals, *Adv. Sci.*, 2021, 8(12), 2100214.
21. Y. Chen, Y. Zhao, Incorporating quantum dots for high efficiency and stable perovskite photovoltaics, *J. Mater. Chem. A*, 2020, 8, 25017–25027.
22. Y. Chen, N. Wei, Y. Miao, H. Chen, M. Ren, X. Liu, Y. Zhao, Inorganic CsPbBr$_3$ perovskite nanocrystals as interfacial ion reservoirs to stabilize FAPbI$_3$ perovskite for efficient photovoltaics, *Adv. En. Mater.*, 2022, 12(16), 2200203.
23. S. Zhang, Y. J. Yoon, X. Cui, Y. Chang, M. Zhang, S. Liang, C-H. Lu, Z. Lin, Tailoring interfacial carrier dynamics via rationally designed uniform CsPbBr$_x$I$_{3-x}$ quantum dots for high-efficiency perovskite solar cells, *J. Mater. Chem. A*, 2020, 8, 26098–26108.
24. M. Mehrabian, E. N. Afshar, P. Norouzzadeh, Numerical simulation of bilayer perovskite quantum dot solar cell with 18.55% efficiency, *Opt Quant Electron*, 2022, 54, 439.
25. L. Liu, A. Najar, K. Wang, M. Du, S. Liu, Perovskite quantum dots in solar cells, *Adv. Sci.*, 2022, 9(7), 2104577.
26. J. A. Dias, S. H. Santangeli, S. J. L. Ribeiro, Y. Messaddeq, Perovskite quantum dot solar cells: An overview of the current advances and future perspectives, *RRL Solar*, 2021, 5(8), 2100205.
27. X. Ling, J. Yuan, W. Ma, The rise of colloidal lead halide perovskite quantum dot solar cells, Acc. *Mater. Res.* 2022, 3(8), 866–878.
28. A. O. El-Ballouli, O. M. Bakr, O. F. Mohammed, Compositional, processing and interfacial engineering of nanocrystal- and quantum dot-based perovskite solar cells, *Chem. Mater.*, 2019, 31(17), 6387–6411.
29. J. Chen, D. Jia, E. M. J. Johansson, A. Hagfeldt, X. Zhang, Emerging perovskite quantum dot solar cells: feasible approaches to boost performance, *Energy Environ. Sci.*, 2021, 14, 224–261.
30. S. Ananthakumar, J. R. Kumar, S. M. Babu, Cesium lead halide (CsPbX$_3$, X=Cl, Br, I) perovskite quantum dots- synthesis, properties, and applications: a review of their present status, *J. Photonics Energy*, 2016, 6(4), 042001.
31. T. Yang, Z. Zhang, Y. Ding, N. Yin, X. Liu, Nondestructive purification process for inorganic perovskite quantum dot solar cells, *J. Nano. Res.*, 2019, 21, 1–8.

32. C-S. Jo, K. Noh, H. Yoo, Y. Kim, J. Jang, H. H. Lee, Y-J. Jung, J-H. Lee, Y-J. Jung, J-H. Lee, J. Han, J. Lim, S-Y. Cho, Solution-processed fabrication of light-emitting diodes using CsPbBr3 perovskite nanocrystals, *ACS Appl. Nano Mater.*, 2020, 3(12), 11801–11810.

33. C. Ding, F. Liu, Y. Zhang, D. Hirotani, X. Rin, S. Hayase, T. Minemoto, T. Masuda, R. Wang, Q. Shen, Photoexcited hot and cold electron and hole dynamics at $FAPbI_3$ perovskite quantum dots/metal oxide heterojunctions used for stable perovskite quantum dot solar cells, *Nano Energy*, 2020, 67, 104267.

34. E. M. Sanehira, A. R. Marshall, J. A. Christians, S. P. Harvey, P. N. Ciesielski, L. M. Wheller, P. Schulz, L. Y. Lin, M. C. Beard, J. M. Luther, Enhanced mobility $CsPbI_3$ quantum dot arrays for record-efficiency, high-voltage photovoltaic cells, *Sci. Adv.*, 2017, 3(10), 1–8.

35. R. Han, Q. Zhao, J. Su, X. Zhou, X. Ye, X. Liang, J. Li, H. Cai, J. Ni, J. Zhang, Role of methyl acetate in highly reproducible efficient $CsPbI_3$ perovskite quantum dot solar cells, *J. Phys. Chem. C*, 2021, 125(16), 8469–8478.

36. F. Liu, Y. Zhang, C. Ding, S. Kobayashi, T. Izuishi, N. Nakazawa, T. Toyoda, T. Ohta, S. Hayase, T. Minemoto, K. Yoshino, S. Dai, Q. Shen, Highly luminescent phase-stable $CsPbI_3$ perovskite quantum dots achieving near 100% absolute photoluminescence quantum yield, *ACS Nano*, 2017, 11(10), 10373–10383.

37. L. M. Wheeler, E. M. Sanehira, A. R. Marshall, P. Schluz, M. Suri, N. C. Anderson, J. A. Christians, D. Nordlund, D. Sokaras, T. Kroll, S. P. Harvey, J. J. Berry, L. Y. Lin, J. M. Luther, Targeted ligand-exchange chemistry on cesium lead halide perovskite quantum dots for high-efficiency photovoltaics, *J. Am. Chem. Soc.*, 2018, 140(33), 10504–10513.

38. S. Cho, J. Kim, S. M. Jeong, M. J. Ko, J-S. Lee, Y. Kim, High-voltage and green-emitting perovskite quantum dot solar cells via solvent miscibility-induced solid-state ligand exchange, *Chem. Mater.*, 2020, 32(20), 8808–8818.

39. P. Mahajan, B. Padha, S. Verma, V. Gupta, R. Datt, W. C. Tsoi, S. Satapathi, S. Arya, Review of current progress in hole-transporting materials for perovskite solar cells, *J. Energy Chemistry*, 2022, 68, 330–386.

40. J. Xue, J-W. Lee, Z. Dai, R. Wang, S. Nuryyeva, M. E. Liao, S-Y. Chang, L. Meng, D. Meng, P. Sun, O. Lin, M. S. Goorsky, Y. Yang, Surface ligand management for stable FAPbI3 quantum dot solar cells, *Joule*, 2018, 2(9), 1866–1878.

41. S. Ding, M. Hao, C. Fu, T. Lin, A. Baktash, P. Chen, D. He, C. Zhang, W. Chen, A. K. Whittaker, Y. Bai, L. Wang, In Situ bonding regulation of surface ligands for efficient and stable $FAPbI_3$ quantum dot solar cells, *Adv. Sci.*, 2022, 9(35), 2204476.

42. S. Ding, M. Hao, T. Lin, Y. Bai, L. Wang, Ligand engineering of perovskite quantum dots for efficient and stable solar cells, *J. Energy Chem.*, 2022, 69, 626–648.

43. W. Fan, Q. Gao, X. Mei, D. Jia, J. Chen, J. Qiu, Q. Zhou, X. Zhang, Ligand exchange engineering of $FAPbI_3$ perovskite quantum dot solar cells, *Fron. Optoelect.*, 2022, 15, 1–12.

44. Y. Xu, H. Li, S. Ramakrishnan, D. Song, Y. Zhang, M. Cotlet, Q. Yu, Ion-assisted ligand exchange for efficient and stable inverted FAPbI3 quantum dot solar cells, *Appl. Energy Mater.*, 2022, 5(8), 9858–9869.

45. M. Hao, Y. Bai, S. Zeiske, L. Ren, J. Liu, Y. Yuan, N. Zarrabi, N. Chen, M. Ghasemi, P. Chen, M. Lyu, D. He, J-H. Yun, Y. Wang, S. Ding, A. Armin, P. Meredith, G. Liu, H-M. Cheng, L. Wang, Ligand-assisted cation-exchange engineering for high-efficiency colloidal $Cs_{1-x}FA_xPbI_3$ quantum dot solar cells with reduced phase segregation, *Nature Energy*, 2020, 5, 79–88.

46. X. Zhang, H. Huang, L. Jin, C. Wen, Q. Zhao, C. Zhao, J. Guo, C. Cheng, H. Wang, L. Zhang, Y. Li, Y. M. Maung, J. Yuan, W. Ma, Ligand-assisted coupling manipulation for efficient and stable FAPbI$_3$ colloidal quantum dot solar cells, *Angew. Chem.*, 2023, 62(5), e202214241.

47. K. Hills-Kimball, H. Yang, T. Cai, J. Wang, O. Chen, Recent advances in ligand design and engineering in lead halide perovskite nanocrystals, *Adv. Sci.*, 2021, 23, 2100214.

48. H. Lin, Q. Wei, K. W. Ng, J-Y. Dong, J-L. Li, W-W. Liu, S-S. Yan, S. Chen, G-C. Xing, X-S. Tang, Z-K. Tang, S-P. Wang, Stable and efficient blue-emitting CsPbBr$_3$ nanoplatelets with potassium bromide surface passivation, *Small*, 2021, 17(43), 2101359.

49. J. Kim, B. Koo, W. H. Kim, J. Choi, C. Choi, S. J. Lim, J-S. Lee, D-H. Kim, M. J. Ko, Y. Kim, Alkali acetate-assisted enhanced electronic coupling in CsPbI$_3$ perovskite quantum dot solids for improved photovoltaics, *Nano Energy*, 2019, 66, 104130.

50. A. Soosaimanickam, H. P. Adl, V. Chirvony, P. J. Rodriguez-Canto, J. P. Martinez-Pastor, R. Abargues, Effect of alkali metal nitrate treatment on the optical properties of CsPbBr$_3$ nanocrystal films, *Mater. Lett.*, 2021, 305, 130835.

51. S. Lim, S. Han, D. Kim, J. Min, J. Choi, T. Park, Key factors affecting the stability of CsPbI$_3$ perovskite quantum dot solar cells: a comprehensive review, *Adv. Mater.*, 2023, 35(4), 2203430.

52. J. Khan, I. Ullah, J. Yuan, CsPbI$_3$ perovskite quantum dot solar cells: opportunities, progress and challenges, *Mater. Adv*, 2022, 3, 1931–1952.

53. A. Swarnkar, A. R. Marshall, E. M. Sanehira, B. D. Chernomordik, D. T. Moore, J. A. Christians, T. Chakrabarti, J. M. Luther, Quantum dot-induced phase stabilization of α-CsPbI$_3$ perovskite for high-efficiency photovoltaics, *Science*, 2016, 354(6308), 92–95.

54. F. Li, S. Zhou, J. Yuan, C. Qin, Y. Yang, J. Shi, X. Ling, Y. Li, W. Ma, Perovskite quantum dot solar cells with 15.6% efficiency and improved stability enabled by an α-CsPbI$_3$/FAPbI$_3$ bilayer structure, *ACS Energy Lett.*, 2019, 4(11), 2571–2578.

55. S. Lim, J. Kim, J. Y. Park, J. Min, S. Yun, T. Park, Y. Kim, J. Choi, Suppressed degradation and enhanced performance of CsPbI$_3$ perovskite quantum dot solar cells via engineering of electron transport layers, *ACS Appl. Mater. Interfaces*, 2021, 13(5), 6119–6129.

56. K. Chen, Q. Zhong, W. Chen, B. Sang, Y. Wang, T. Yang, Y. Liu, Y. Zhang, H. Zhang, Short-chain ligand-passivated stable α-CsPbI$_3$ quantum dot for all-inorganic perovskite solar cells, *Adv. Fun. Mater.*, 2019, 29(24), 1900991.

57. J. Shi, F. Li, J. Yuan, X. Ling, S. Zhou, Y. Qian, W. Ma, Efficient and stable CsPbI$_3$ perovskite quantum dots enabled by in situ ytterbium doping for photovoltaic applications, *J. Mater. Chem. A*, 2019, 7, 20936–20944.

58. X. Huang, J. Hu, C. Bi, J. Yuan, Y. Lu, M. Sui, J. Tian, B-site doping of CsPbI$_3$ quantum dot to stabilize the cubic structure for high-efficiency solar cells, *Chem. Eng. J*, 2021, 421, 127822.

59. S. Lim, J. Kim, J. Y. Park, J. Min, S. Yin, T. Park, Y. Kim, J. Choi, Suppressed degradation and enhanced performance of CsPbI$_3$ perovskite quantum dot solar cells via engineering of electron transport layers, *ACS Appl. Mater. Interfaces*, 2021, 13(5), 6119–6129.

60. S. Wang, C. Bi, A. Portniagin, J. Yuan, J. Ning, X. Xiao, X. Zhang, Y. Y. Li, S. V. Kershaw, J. Tian, A. L. Rogach, CsPbI$_3$/PbSe heterostructured nanocrystals for high-efficiency solar cells, *ACS Energy Lett.*, 2020, 5(7), 2401–2410.

61. J. Kim, S. Han, G. Lee, J. Choi, M. J. Ko, Y. Kim, Single-step-fabricated perovskite quantum dot photovoltaic absorbers enabled by surface ligand manipulation, *Chem. Eng. J*, 2022, 448, 137672.
62. J. Yuan, C. Bi, S. Wang, R. Guo, T. Shen, L. Zhang, J. Tian, Spray-coated colloidal perovskite quantum dot films for highly efficient solar cells, *Adv. Fun. Mater.*, 2019, 29(49), 1906615.
63. H. Song, J. Yang, W. H. Jeong, J. Lee, T. H. Lee, J. W. Yoon, H. Lee, A. J. Ramadan, R. D. J. Oliver, S. C. Cho, S. G. Lim, J. W. Jang, Z. Yu, J. T. Oh, E. D. Jung, M. H. Song, S. H. Park, J. R. Durrant, H. J. Snaith, S. U. Lee, B. R. Lee, H. Choi, A universal perovskite nanocrystal ink for high-performance optoelectronic devices, *Adv. Mater.*, 2022, 2209486 (1–12).
64. X. Zhang, H. Huang, Y. M. Maung, J. Yuan, W. Ma, Aromatic amine-assisted pseudo-solution-phase ligand exchange in $CsPbI_3$ perovskite quantum dot solar cells, *Chem. Commun.*, 2021, 57, 7906–7909.
65. Y. Wang, J. Yuan, X. Zhang, X. Ling, B. W. Larson, Q. Zhao, Y. Yang, Y. Shi, J. M. Luther, W. Ma, Surface ligand management aided by a secondary amine enables increased synthesis yield of $CsPbI_3$ perovskite quantum dots and high photovoltaic performance, *Adv. Mater.*, 2020, 32(32), 2000449.
66. A. Soosaimanickam, H. P. Adl, V. Chirvony, P. J. Rodriguez-Canto, J. P. Martinez-Pastor, R. Abargues, Effect of alkali metal nitrate treatment on the optical properties of $CsPbBr_3$ nanocrystal films, *Mater. Lett.*, 2021, 305, 130835.
67. J. Kim, B. Koo, W. H. Kim, J. Choi, C. Choi, S. J. Lim, J-S. Lee, D-H. Kim, M. J. Ko, Y. Kim, Alkali acetate-assisted enhanced electronic coupling in $CsPbI_3$ perovskite quantum dot solids for improved photovoltaics, *Nano Energy*, 2019, 66, 104130.
68. X. Ling, S. Zhou, J. Yuan, J. Shi, Y. Qian, B. W. Larson, Q. Zhao, C. Qin, F. Li, G. Shi, C. Stewart, J. Hu, X. Zhang, J. M. Luther, S. Duhm, W. Ma, 14.1% $CsPbI_3$ perovskite quantum dot solar cells via cesium cation passivation, *Adv. En. Mater.*, 2019, 9(28), 1900721.
69. L. M. Wheeler, E. M. Sanehira, A. R. Marshall, P. Schuz, M. Suri, N. C. Anderson, J. A. Christians, D. Nordlund, D. Sokaras, T. Kroll, S. P. Harvey, J. J. Berry, L. Y. Lin, J. M. Luther, Targeted ligand-exchange chemistry on cesium lead halide perovskite quantum dots for high-efficiency photovoltaics, *J. Am. Chem. Soc.*, 2019, 140(33), 10504–10513.
70. S. Lim, G. Lee, S. Han, J. Kim, S. Yun, J. Lim, Y-J. Pu, M. J. Ko, T. Park, J. Choi, Y. Kim, Monodisperse perovskite colloidal quantum dots enable high-efficiency photovoltaics, *ACS Energy Lett.*, 2021, 6(6), 2229–2237.
71. J. Kim, S. Cho, F. Dinic, J. Choi, C. Choi, S. M. Jeong, J-S. Lee, O. Voznyy, M. J. Ko, Y. Kim, Hydrophobic stabilizer-anchored fully inorganic perovskite quantum dots enhance moisture resistance and photovoltaic performance, *Nano Energy*, 2020, 75, 104985.
72. X. Zhang, H. Huang, X. Ling, J. Sun, X. Jiang, Y. Wang, D. Xue, L. Huang, L. Chi, J. Yuan, W. Ma, Homojunction perovskite quantum dot solar cells with over 1 μm-thick photoactive layer, *Adv. Mater.*, 2022, 34(2), 2105977.
73. M. A. Uddin, J. K. Mobley, A. A. Masud, T. Liu, R. L. Calabro, D-Y. Kim, C. I. Richards, K. R. Graham, Mechanistic exploration of dodecanethiol-treated colloidal $CsPbBr_3$ nanocrystals with photoluminescence quantum yields reaching near 100%, *J. Phys. Chem. C*, 2019, 123(29), 18103–18112.
74. L. Ruan, W. Shen, A. Wang, A. Xiang, Z. Deng, Alkyl-thiol ligand-induced shape- and crystalline phase-controlled synthesis of stable perovskite-related $CsPb_2Br_5$ nanocrystals at room temperature, *J. Phys. Chem. Lett.*, 2017, 8(16), 3853–3860.

75. Z. Liu, Y. Bekenstein, X. Ye, S. C. Nguyen, J. Swabeck, D. Zhang, S-T. Lee, P. Yang, W. Ma, A. P. Alivisatos, Ligand mediated transformation of cesium lead bromide perovskite nanocrystals to lead depleted Cs_4PbBr_6 nanocrystals, *J. Am. Chem. Soc.*, 2017, 139(15), 5309–5312.

76. H. Song, J. Yang, S. G. Lim, J. Lee, W. H. Jeong, H. Choi, J. H. Lee, H. Y. Kim, B. R. Lee, H. Choi, On the surface passivating principle of functional thiol towards efficient and stable perovskite nanocrystal solar cells, *Chem. Eng. J.*, 2023, 454, 140224.

77. J. Shi, F. Li, Y. Lin, C. Liu, B. Cohen-Kleinstein, S. Yuan, Y. Li, Z-K. Wang, J. Yuan, W. Ma, In situ ligand bonding management of $CsPbI_3$ perovskite quantum dots enables high-performance photovoltaics and red light-emitting diodes, *Angew. Chem.*, 2020, 132(49), 22414–22421.

78. Q. Wang, Y. Xu, L. Zhang, P. Niu, R. Zhou, M. Lyu, H. Lu, J. Zhu, Aromatic carboxylic acid ligand management for $CsPbBr_3$ quantum dot light-emitting solar cells, *ACS Appl. Nano Mater.*, 2022, 5(8), 10495–10503.

79. J. Khan, X. Zhang, J. Yuan, Y. Wang, G. Shi, R. Patterson, J. Shi, X. Ling, L. Hu, T. Wu, S. Dai, W. Ma, Tuning the surface-passivating ligand anchoring position enables phase robustness in $CsPbI_3$ perovskite quantum dot solar cells, *ACS Energy Lett.*, 2020, 5(10), 3322–3329.

80. J. Chen, D. Jia, J. Qiu, R. Zhuang, Y. Hua, X. Zhang, Multidentate passivation crosslinking perovskite quantum dots for efficient solar cells, *Nano Energy*, 2022, 96, 107140.

81. R. Han, Q. Zhao, A. Hazarika, J. Li, H. Cai, J. Ni, J. Zhang, Ionic liquids modulating $CsPbI_3$ colloidal quantum dots enable improved mobility for high-performance solar cells, *ACS Appl. Mater. Interfaces*, 2022, 14(3), 4061–4070.

82. H. Huang, X. Zhang, R. Gui, C. Zhao, J. Guo, Y. M. Maung, H. Yin, W. Ma, J. Yuan, High-efficiency perovskite quantum dot photovoltaic with homogeneous structure and energy landscape, *Adv. Fun. Mater.*, 2023, 2210728.

83. Z. Ding, S. Li, Y. Jiang, D. Wang, D. Wang, M. Yuan, Open-circuit voltage loss in perovskite quantum dot solar cells, *Nanoscale*, 2023, 15, 3713–3729.

84. J. Yuan, C. Bi, J. Xi, R. Guo, J. Tian, Gradient-band alignment homojunction perovskite quantum dot solar cells, *J. Phys. Chem. Lett.*, 2021, 12(3), 1018–1024.

85. D. Jia, J. Chen, R. Zhuang, Y. Hua, X. Zhang, Inhibiting lattice distortion of $CsPbI_3$ perovskite quantum dots for solar cells with efficiency over 16.6%, *Energy Environ. Sci.*, 2022, 15, 4201–4212.

86. D. Jia, J. Chen, X. Mei, W. Fan, S. Luo, M. Yu, J. Liu, X. Zhang, Surface matrix curing of inorganic $CsPbI_3$ perovskite quantum dots for solar cells with efficiency over 16%, *Energy Environ. Sci.*, 2021, 14, 4599–4609.

87. K. Li, J. Yuan, F. Li, Y. Shi, X. Ling, X. Zhang, Y. Zhang, H. Lu, J. Yuan, W. Ma, High-efficiency perovskite quantum dot solar cells benefiting from a conjugated polymer-quantum dot bulk heterojunction connecting layer, *J. Mater. Chem. A*, 2020, 8, 8104–8112.

88. C. Bi, X. Sun, X. Huang, S. Wang, J. Yuan, J. X. Wang, T. Pullerits, J. Tian, Stable $CsPb_{1-x}Zn_xI_3$ colloidal quantum dots with ultralow density of trap states for high-performance solar cells, *Chem. Mater.*, 2020, 32(14), 6105–6113.

89. L. Zhang, C. Kang, G. Zhang, Z. Pan, Z. Huang, S. Xu, H. Rao, H. Liu, S. Wu, X. Wu, X. Li, Z. Zhu, X. Zhong, A. K. –Y. Jen, All-inorganic $CsPbI_3$ quantum dot solar cells with efficiency over 16% by defect control, *Adv. Fun. Mater.*, 2021, 31(4), 2005930.

90. Y. Liu, Q. Li, W. Zhang, Z. Yang, S. Zhao, W. Chen, $CsPbI_3$ perovskite quantum dot solar cells with both high efficiency and phase stability enabled by Br doping, *ACS Appl. Energy Mater.*, 2021, 4(7), 6688–6696.

91. K. Chen, W. Jin, Y. Zhang, T. Yang, P. Reiss, Q. Zhong, U. Bach, Q. Li, Y. Wang, H. Zhang, Q. Bao, Y. Liu, High efficiency mesoscopic solar cells using $CsPbI_3$ perovskite quantum dots enabled by chemical interface engineering, *J. Am. Chem. Soc.*, 2020, 142(8), 3775–3783.

92. Y. Liu, X. Zhao, Z. Yang, Q. Li, W. Wei, B. Hu, W. Chen, $Cu_{12}Sb_4S_{13}$ quantum dots with ligand exchange as hole transport materials in all-inorganic perovskite $CsPbI_3$ quantum dot solar cells, *ACS Appl. Energy Mater.*, 2020, 3(4), 3521–3529.

93. S. Han, J. Kim, D. E. Kim, M. J. Ko, J. Choi, S-W. Back, Y. Kim, A small-molecule-templated nanostructure back electrode for enhanced light absorption and photocurrent in perovskite quantum dot photovoltaics, *J. Mater. Chem. A*, 2022, 10, 8966–8974.

94. L. Hu, Q. Li, Y. Yao, Q. Zeng, Z. Zhou, C. Cazorla, T. Wan, X. Guan, J-K. Huang, C-H. Lin, M. Li, S. Cheong, R. D. Tilley, D. Chu, J. Yuan, S. Huang, T. Wu, F. Liu, Perovskite quantum dot solar cells fabricated from recycled lead-acid battery waste, *ACS Mater. Lett.*, 2022, 4(1), 120–127.

95. M. M. Tavakoli, M. Nasilowski, J. Zhao, M. G. Bawendi, J. Kong, Efficient semitransparent $CsPbI_3$ quantum dots photovoltaics using graphene electrode, *Small*, 2019, 3(12), 1900449.

96. L. Hu, Q. Zhao, S. Huang, J. Zheng, X. Guan, R. Patterson, J. Kim, L. Shi, C-H. Lin, Q. Lei, D. Chu, W. Tao, S. Cheong, R. D. Tilley, A. W. Y. Ho-Baillie, J. M. Luther, J. Yuan, T. Wu, Flexible and efficient perovskite quantum dot solar cells via hybrid interfacial architecture, *Nat. Commun.*, 2021, 12, 1–9.

97. J. Yuan, N. Rujisamphan, W. Ma, J. Yuan, Y. Li, S-T. Lee, Perspective on the perovskite quantum dots for flexible photovoltaics, *J. Energy Chemistry*, 2021, 62, 505–507.

98. J. Chen, D. Jia, R. Zhuang, Y. Hua, X. Zhang, Highly oriented perovskite quantum dot solids for efficient solar cells, *Adv. Mater.*, 2022, 34(37), 2204259.

99. D. Jia, J. Chen, M. Yu, J. Liu, E. M. J. Johansson, A. Hagfldt, X. Zhang, Dual passivation of $CsPbI_3$ perovskite nanocrystals with amino acid ligands for efficient quantum dot solar cells, *Small*, 2020, 16(24), 2020, 2001772.

100. D. Jia, J. Chen, J. Qiu, H. Ma, M. Yu, J. Liu, X. Zhang, Tailoring solvent-mediated ligand exchange for $CsPbI_3$ perovskite quantum dot solar cells with efficiency exceeding 16.5%, *Joule*, 6(7), 1632–1653.

101. J. Yuan, X. Zhang, J. Sun, R. Patterson, H. Yao, D. Xue, Y. Wang, K. Ji, L. Hu, S. Huang, D. Chu, T. Wu, J. Hou, J. Yuan, Hybrid perovskite quantum dot/non-fullerene molecule solar cells with efficiency over 15%, *Adv. Fun. Mater.*, 2021, 31(27), 2101272.

102. H. Huang, X. Zhang, R. Gui, C. Zhao, J. Guo, Y. M. Maung, H. Yin, W. Ma, J. Yuan, High-efficiency perovskite quantum dot photovoltaic with homogeneous structure and energy landscape, *Adv. Fun. Mater.*, 2023, 2210728.

103. D. Ghosh, Md. Y. Ali, D. K. Chaudhary, S. Bhattacharyya, Dependence of halide composition on the stability of highly efficient all-inorganic cesium lead halide perovskite quantum dot solar cells, *Solar Energy Mat. Solar Cells*, 2018, 185, 28–35.

104. D. Ghosh, Md. Y. Ali, A. Ghosh, A. Mandal, S. Bhattacharyya, Heterovalent substitution in mixed halide perovskite quantum dots for improved and stable photovoltaic performance, *J. Phys. Chem. C*, 2021, 125(10), 5485–5493.

105. X. Zhang, Y. Qian, X. Ling, Y. Wang, Y. Zhang, J. Shi, Y. Shi, J. Yuan, W. Ma, α-$CsPbBr_3$ perovskite quantum dots for application in semitransparent photovoltaics, *ACS Appl. Mater. Interfaces*, 2020, 12(24), 27307–27315.

106. S. Panigrahi, S. Jana, T. Calmeiro, D. Nunes, R. Martins, E. Fortunato, Imaging the anomalous charge distribution inside $CsPbBr_3$ perovskite quantum dots sensitized solar cells, *ACS Nano*, 2017, 11(10), 10214–10221.

107. J. Ye, M. M. Byranvand, C. O. Martinez, R. L. Z. Hoye, M. Saliba, L. Polavarapu, Defect passivation in lead-halide perovskite nanocrystals and thin films: toward efficient LEDs and solar cells, *Angew. Chem.* 2021, 60(40), 21636–21660.

108. H. Xu, H. Yuan, J. Duan, Y. Zhao, Z. Jiao, Q. Tang, Lead-free $CH_3NH_3SnBr_{3-x}I_x$ perovskite quantum dots for mesoscopic solar cell applications, *Electrochem. Acta*, 2018, 282, 807–812.

109. H. Xu, J. Duan, Y. Zhao, Z. Jiao, B. He, Q. Tang, 9.13%-Efficiency and stable inorganic $CsPbBr_3$ solar cells. Lead-free $CsSnBr_{3-x}I_x$ quantum dots promote charge extraction, *J. Power Sources*, 2018, 399, 76–82.

110. F. Liu, C. Ding, Y. Zhang, T. S. Ripolles, T. Kamisaka, T. Toyoda, S. Hayase, T. Minemoto, K. Yoshino, S. Dai, M. Yanagida, H. Noguchi, Q. Shen, Colloidal synthesis of air-stable alloyed $CsSn_{1-x}Pb_xI_3$ perovskite nanocrystals for use in solar cells, *J. Am. Chem. Soc.*, 2017, 139(46), 16708–16719.

111. Y. Wang, J. Tiu, T. Li, C. Tao, X. Deng, Z. Li, Convenient preparation of $CsSnI_3$ quantum dots, excellent stability, and the highest performance of lead-free inorganic perovskite solar cells so far, *J. Mater. Chem. A*, 2019, 7, 7683–7690.

112. L-J. Chen, C-R. Lee, Y-J. Chuang, Z-H. Wu, C. Chen, Synthesis and optical properties of lead-free cesium tin halide perovskite quantum rods with high-performance solar cell application, *J. Phys. Chem. Lett.*, 2016, 7(24), 5028–5035.

113. X. Yang, W. Wang, R. Ran, W. Zhou, Z. Shao, Recent advances in $Cs_2AgBiBr_6$-based halide double perovskites as lead-free and inorganic light absorbers for perovskite solar cells, *Energy Fuels*, 2020, 34(9), 10513–10528.

114. R. Ahmad, G. V. Nutan, D. Singh, G. Gupta, U. Soni, S. Sapra, R. Srivatsava, Colloidal lead-free $Cs_2AgBiBr_6$ double perovskite nanocrystals: synthesis, uniform thin-film fabrication and application in solution-processed solar cells, *Nano Research*, 2021, 14, 1126–1134.

115. A. Kumar, S. K. Swami, S. S. Rawat, V. N. Singh, O. P. Sinha, R. Srivastava, Mixed bismuth-antimony-based double perovskite nanocrystals for solar cell application, *I. J. Energy Res.*, 2021, 45(11), 16769–16780.

116. F. Bai, Y. Hu, Y. Hu, T. Qiu, X. Miao, S. Zhang, Lead-free air-stable ultrathin $Cs_3Bi_2I_9$ perovskite nanosheets for solar cells, *Solar Energy Mat. Solar Cells*, 2018, 184, 15–21.

117. L-J. Chen, Synthesis and optical properties of lead-free cesium germanium halide perovskite quantum rods, *RSC Adv.*, 2018, 8, 18396–18399.

118. S. Aina, B. Villacampa, M. Bernechea, Earth-abundant non-toxic perovskite nanocrystals for solution processed solar cells, *Mater. Adv.*, 2021, 2, 4140–4151.

5 Effect of Nanomaterials on the Production of Hydrogen from the Electrochemical Process of Water Splitting

Chinmay Deheri, Binayak Pattanayak, and Abinash Mahapatro
Institute of Technical Education and Research, India

Bhagiratha Mishra
Altafuel Pvt. Ltd., India

Saroj Kumar Acharya
Institute of Technical Education and Research, India

5.1 INTRODUCTION

The global energy demand is increasing due to fast urbanization, vehicle growth, emergent markets, and economies. This demand is fulfilled by the conventional route such as burning fossil fuels causing emissions and greenhouse effects. Additionally, renewable energy such as solar, biomass, wind, and others have mitigated the demand for power in the last two decades. The storage and transfer of such energy from one point to another are a vital task for researchers, although the emission is negligible by this energy conversion process. The above-mentioned problems from conventional and renewable routes attract researchers for an alternative energy source with insignificant emissions. Hydrogen is widely considered fuel in powering non-polluting vehicles, domestic heating, and aircraft. The abundant water supply and sunlight offer us a reasonable alternative resource to generate hydrogen apart from fossil fuels and biomass. Scalability is the major issue with hydrogen production from biomass through gasification technology. The hydrogen produced through photo-electrochemical water splitting is one of the most effective methods for pure solar hydrogen generation and is used in minute to large-scale hydrogen producers. This technique was reported in the year 1972 by Honda and Fujishima [1] using a single titanium dioxide (TiO_2) crystal and platinum (Pt) as photo-anode and cathode, respectively.

DOI: 10.1201/9781003364825-5

5.2 PHOTO-ELECTROCHEMISTRY OF WATER DECOMPOSITION

The principle of photo-electrochemistry involves the process of production of electricity from light energy in a cell. The cell contains two electrodes that are submerged in an aqueous electrolyte. One of the electrodes in the cell is of semiconductor type and can absorb light. The electricity hence produced is utilized for water electrolysis. In practice, there exist three possible arrangements intended for the photo-electrodes in the photo-electrolysis assembly [2–5].

i. n-type semiconductor used for the photo-anode and metal used as cathode.
ii. n-type semiconductor used for the photo-anode and p-type semiconductor used as photo-cathode.
iii. p-type semiconductor used as photo-cathode and metal used as anode.

Although the arrangements of the electrodes are different, the performance principles of all three arrangements are the same [6].

5.2.1 REACTION MECHANISM

Several processes are involved within the photo-electrode and at the photoelectrode/electrolyte interface for water photo-electrolysis using photo-electrochemical cells (PECs). These processes can be step by step analysed as follows:

- Ionization of the semiconductor material, i.e., the photo-anode using the induced light, which results in the electronic charge carrier formation. These electronic charge carriers mean the quasi-free electrons and electron holes.
- Electron hole causes the oxidation of water at the photo-anode.
- Transportation of H^+ ions in the electrolyte from photo-anode to cathode and electrons from the cathode to photo-anode through an external circuit.
- Hydrogen ion reduction at the cathode by electrons.

When light falls on the n-type semiconducting material used as a photo-anode, the intrinsic ionization results in the generation of electrons in the conduction region and holes in the valence region. It can be represented as:

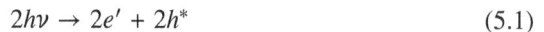

$$2h\nu \rightarrow 2e' + 2h^* \tag{5.1}$$

where h represents the Plank's constant, ν represents the frequency, e' represents the electron, and h^* represents the hole.

The above reaction is initiated when the photon's energy exceeds the bandgap. This also requires an electric field at the electrode/electrolyte interface to avoid the recombination of these charge carriers. The formation of light-induced electron holes results in the separation of water molecules into an oxygen molecules and hydrogen ions.

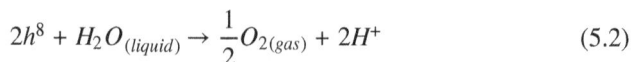

$$2h^8 + H_2O_{(liquid)} \rightarrow \frac{1}{2}O_{2(gas)} + 2H^+ \tag{5.2}$$

The above reaction occurs at the interface of the electrolyte and photo-anode. The gaseous oxygen is liberated at the anode, and the H^+ ion travels to the cathode via electrolyte. The electron produced at the photo-anode, as represented in equation 5.1, moves through the external circuit to the cathode. At the cathode, the electron unites with the hydrogen ion, and gaseous hydrogen is liberated.

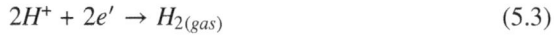

$$2H^+ + 2e' \rightarrow H_{2(gas)} \tag{5.3}$$

If we combine all three equations, the complete outcome of the PEC can be represented as:

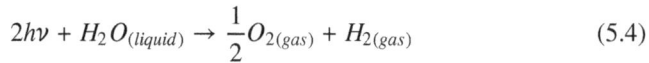

$$2h\nu + H_2O_{(liquid)} \rightarrow \frac{1}{2}O_{2(gas)} + H_{2(gas)} \tag{5.4}$$

The condition of occurrence of photo-electrochemical water splitting is that the energy of the photons absorbed by the photo-anode must be greater than the threshold energy.

$$E_i = \frac{\Delta G^0_{(H_2O)}}{2N_A} \tag{5.5}$$

where $\Delta G^0_{(H_2O)}$ represents the free enthalpy per mole of reaction represented in equation 5.4 = 237.141 kJ/mol and N_A represents the Avogadro's number = 6.022 × 10^{23} mol^{-1}. This leads to the value of threshold energy as $E_i = h\nu = 1.23$ eV.

Hence, this leads to a conclusion that photo-electrochemical splitting of water is feasible once the electromotive force is greater than or equal to 1.23 eV. A PEC for the photo-electrolysis of water is shown in Figure 5.1.

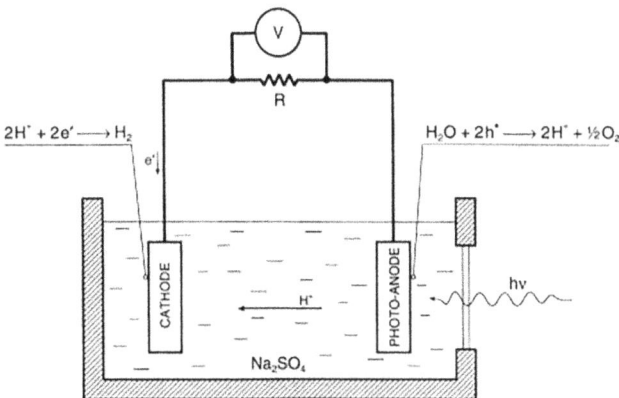

FIGURE 5.1 Representation of photo-electrochemical cell (PEC) for water photo-electrolysis [6].

5.3 EFFECT OF NANOMATERIALS ON HYDROGEN PRODUCTION FROM ELECTROCHEMICAL WATER SPLITTING

The water-splitting reaction that generates hydrogen might be photochemical or photo electrochemical. At the operating and counter electrodes, the water-splitting reaction creates hydrogen and oxygen, respectively. Nanomaterials including oxides, nitrides, and sulfides are required for the splitting reaction. Nonetheless, semiconductor polymers are a novel class of photocatalytic materials that have several advantages, including inexpensive, elevated quantum yield, vast exterior region, and ease of fabrication. One of the key polymers discussed is polyaniline and its derivatives. Other advantages of this type of polymer are good compatibility, stability, porosity, toxicity, inexpensive, redox state, and low bandgap. Khalafalla et al. [7] studied hydrogen generation during water splitting reaction using poly-aniline (PANI) supported accumulation for lead sulphide (PbS) nanocomposite across the polymer-supported ionic adsorption technique. The accumulation procedure of polymer compound was conducted on antimony-doped tin oxide (ATO) glass, in which the ATO/PANI/PbS complex was created. The inexpensive ATO/PANI/PbS nanocomposite was applied as an operational photocathode in a three-conductor cell for H_2 production from wastewater (three-stage management sewage water) deprived of any forgoing mediator under a solar simulator mechanism (xenon light). The response of the electrode to light energy is significant. The current density (J_{ph}) under dark and light is found to be in the range of 10^{-6} to 0.13 mA.cm^{-2}, respectively. The stability and reproducibility of the electrode is found to be in the required limits. In this process, 0.1 mmol/cm^2.h of hydrogen molecules are generated. The produced ΔS^* and ΔH^* were 273.4 J/mol.K, and 7.3 kJ/mol, respectively. ATO/PANI/PbS nanocomposite was produced and employed as a photocathode enabling H_2 generation after sewage water was valorized. The polymer was used to aid PbS deposit in two steps: Pb^{2+} absorption and subsequently PbS deposit employing thiourea and heating. The sewage water was converted to H_2 gas fuel with excellent efficiency using the manufactured electrode. The electrode was sensitive to light, with the current density changing from 10^{-8} to 0.13 mA.cm^{-2} during dark and light conditions, respectively. The schematic diagram for electrochemical H_2 generation is shown in Figure 5.2.

Electrochemical water splitting has been seen as an environmentally friendly, efficient, and clean technique of producing pure hydrogen as a replacement for depleting fossil fuels. Significant efforts have been made to increase reaction rate kinetics in order to reduce over-potential by inventing innovative catalysts with appropriate active material combinations. To maximize the over-potential advantages from the activated polarization of both the hydrogen evolution reaction (HER) and the oxygen evolution reaction (OER), the catalysts should indeed be catalysed to speed multi-electron transfer processes and boost reaction kinetics.

OER requires a rather high over potential since it is a complicated four-electron transfer mechanism with sluggish kinetics. The entire effectiveness of water splitting is heavily dependent on both HER and OER, and effective catalysts only serve as a key to reducing HER and OER over-potential. As a result, finding suitable catalysts is highly desirable in constructing effective water splitting devices for the future.

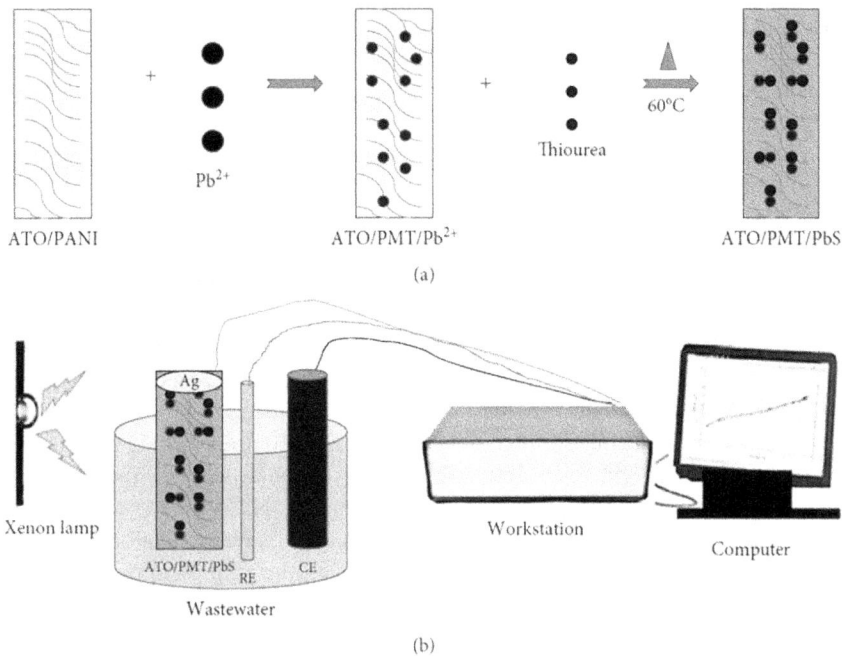

FIGURE 5.2 (a) Schematic deposition of PbS on ATO/PANI and (b) the three-electrode cell for electrochemical H_2 generation from wastewater [7].

Commercially accessible state-of-the-art electro catalytic activity for water splitting comprise of metals including platinum for HER and ruthenium, iridium for OER, but their high cost and rarity limit their substantial practical usage on an industrial scale.

As a result, the development of new active materials based on Earth-abundant, non-precious metals, metal composites, or even metal-free materials with the advancement of existing electrocatalysts has remained a key focus for researchers. To avoid acidic corrosion, industrial-scale water splitting is typically performed in alkaline media. For large-scale applications, a new HER catalyst that functions effectively in alkaline medium is required. Limited investigations on noble free alloys have been documented due to the complex and inconvenient synthesis techniques, particularly a systematic study between metal/metal oxide alloy compositions anchored on a self-supported conductive substrate such as copper foam (CF) for overall water splitting. In comparison to other metals (Co, Ni, and Mo) as conductive substrates, Cu is more appealing due to its abundance on Earth and low cost. Cu-based catalysts for genuine prospective energy applications benefit from the Earth's abundance, decreased toxicity, and profitable redox characteristics. Cu-based compounds have previously been suggested as active catalysts for water oxidation and hydrogen generation. Thus, a simple thermal treatment procedure was used to create a Ni-Co alloy and $NiCoO_2$ nano-hetero-structures implanted on CF.

Qazi et al. [8] reported $NiCo-NiCoO_2$ nano-hetero-structures embedded on the oxidized surface of CF ($NiCo-NiCoO_2@Cu_2O@CF$) as an efficient bi-functional electrocatalyst for overall water splitting in 1 M KOH electrolyte solution. Bimetallic thin-layered nano-hetero-structures of NiCo-NiCoO$_2$@Cu$_2$O@CF exhibit a synergic effect of doubly active metals Ni and Co to achieve remarkable small over potentials of 133 to achieve a current density of 10 mA cm^{-2} for HER. By surface modification of CF, thin sheets of bimetallic NiCo-NiCoO$_2$@Cu$_2$O@CF as a self-supported electrode were successfully obtained and proposed as a bi-functional electrocatalytic material for total water oxidation and hydrogen production. In alkaline media (1 M KOH), the prepared NiCo-NiCoO$_2$@Cu$_2$-O@CF electrode material demonstrated dual perform-ance for HER and OER with a lower over-potential of 133 and 327 mV to attain a current density of 10 mA cm^2, respectively. A simple electrolyzer was successfully constructed using a NiCo-NiCoO$_2$@Cu$_2$O@CF bi-functional electrocatalyst as an anode and cathode electrode. The results demonstrated good electrocatalytic per-formance when compared to other non-precious electro catalysts, which required a cell voltage of 1.69 V to achieve current density more than 10 mA cm^2. Furthermore, the electrolyzer activity remained steady in a strong alkaline solution for more than 12 hours. This low-cost, Earth-abundant bimetallic NiCo-NiCoO$_2$@Cu$_2$O@CF electro-catalyst is suitable for practical demonstrations of future alkaline water oxi-dation and hydrogen production devices due to its outstanding performance and en-durance. A proposed mechanism for overall water oxidation and hydrogen production is shown in Figure 5.3.

The vanadate compound with metallic materials achieves prominence in the realm of photocatalysis for the creation of hydrogen with a lowered potential of 1.23 V. It is explored that metallic vanadate appeared as a broad and promising material for usage as a photo anode with superior features in an electrolyte solution to generate hydrogen energy for water splitting under solar light irradiation.

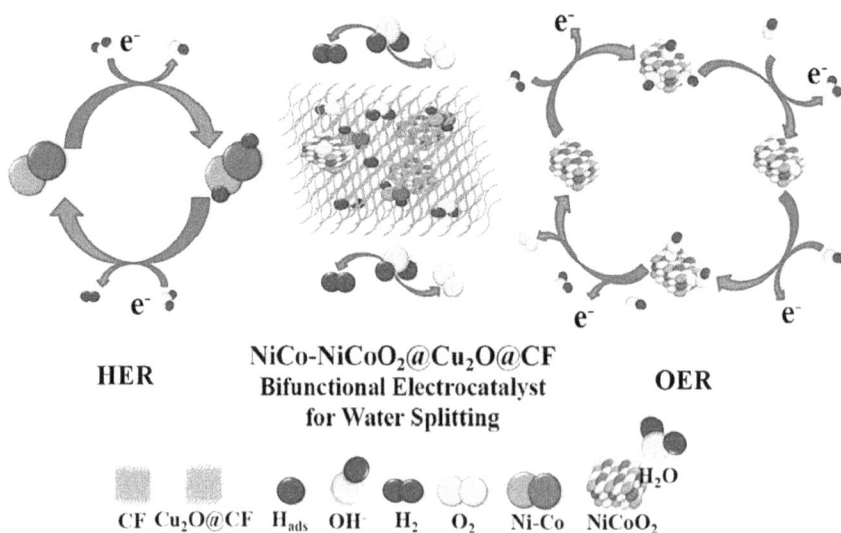

FIGURE 5.3 Proposed mechanism for overall water oxidation and hydrogen production over NiCoeNiCoO$_2$@Cu$_2$O@CF as-prepared bi-functional electrocatalyst [8].

Researchers were drawn to vanadium-based materials with high photocatalytic properties such as $Cu_3V_2O_8$ copper-vanadium oxides (CVO). Vanadium-based materials are an intriguing layered crystalline material that consists of a copper layer in octahedral coordination with oxygen that is joined by vanadium tetrahedral lattice sites to establish monoclinic and triclinic crystal phases with a coordinated layer nanostructured copper vanadate that are extensively used in applications such as electrochemical properties and catalyst. Vanadium-based materials, such as $Cu_3V_2O_8$ copper-vanadium oxides, may have suitable features for water oxidation such as a band hole energy of 2 eV and a prolonged transporter lifetime.

Iqbal et al. [9] studied chromium-incorporated copper vanadate nano-materials for hydrogen evolution by water splitting. Copper vanadate ($Cu_3V_2O_8$) is a promising photocatalyst with 2.0 eV indirect forbidden bandgap, low redox potential, and suitable chemical stability to capture major region (Visible) of solar spectrum from the metal halide lamp 400 W. Chromium metal is incorporated with pure form of copper vanadate material ($Cr:Cu_3V_2O_8$) to get more production of hydrogen energy from the splitting of water molecules.

Chromium atoms were synthesized via hydrothermal technique from the precursors including sodium vanadate (NH_4VO_3) and chromium chloride ($CrCl_3$) at 180°C for 24 h which further incorporated (1.0%, 1.5%, 2.0%, and 2.5%) in copper vanadate-nanostructured photocatalyst. It was found that $Cr:Cu_3V_2O_8$ up to specific value (2.0%) has tremendous optimized photocatalytic performance towards the production of hydrogen energy from the splitting of water molecules through photoelectrochemical and photochemical photocatalysis. The optimal combination of the catalytic activities of water splitting to release hydrogen energy, which decreases with increased chromium atom presence, is 2.5%.

Because of the strong influence of pH on the electrocatalytic activity of the catalyst, some electrocatalysts can only work well under specified acidic or alkaline circumstances. As a result, the pH value restricts the practical applicability of many electrocatalysts and impedes the generation of electrochemical catalysts.

Fu et al. [10] successfully synthesized a series of different proportion bimetallic uploaded nitrogen-doped graphene materials (Fe_2O_3-Co NPs-N-GR) through a simple, green, and cost-effective method for efficient HER. Among the series of catalysts, $Fe_2O_{3(1)}$-$Co_{(1)}$ NPs-N-GR exhibits better HER performance than other catalysts. $Fe_2O_{3(1)}$-$Co_{(1)}$ NPs-N-GR exhibits significant catalytic activity and excellent durability for HER in a wide pH range. The electrocatalytic performance of $Fe_2O_{3(1)}$-$Co_{(1)}$ NPs-N-GR towards HER is better in acidic solution (0.5 M H_2SO_4) than in alkaline solution (1.0 M NaOH), the onset over-potential is 0.36 V, Tafel slope is 66 mV dec^{-1}, and current densities of 10 mA cm^{-2} at over-potential is 0.39 V. The excellent HER catalytic performance of $Fe_2O_{3(1)}$-$Co_{(1)}$ NPs-N-GR stems from its unique composition and structural properties by combining Fe_2O_3-Co nanoparticles and N co-doped graphene. Through a simple, green, and cost-effective technique, a series of composition-adjustable bimetallic N-doped graphene carbon materials were successfully synthesized as a highly efficient and long-lasting HER electrocatalyst in alkaline and acidic conditions.

Li et al. [11] reviewed the recent research works conducted on hydrogen generation from photoelectrochemical water splitting using nanomaterials especially focused on metal oxides. It was observed that the morphology and architecture of the carbon hybrids play a vital role to regulate the HER performance. The usage of single-atom catalysts is a promising strategy to realize a high-performance water-splitting application.

Nickel and nickel alloys are widely recognized for their electrocatalytic activity in the HER and their use in numerous electrochemical processes. The alloying metal (s) used and the electrodeposition circumstances affect the physical and chemical properties of the resulting Ni-based alloy electrodes, which affects their electro-activity for the HER.

Nady and Negm [12] performed a study on nano-crystalline alloy in which Ni, Cu, and Ni-Cu nano-crystalline alloys were electrochemically deposited on a Cu electrode (Cu/Ni-Cu) by the galvanostatic technique and ultrasound waves in view of their possible applications as electrocatalytic materials for HER. It was observed that the electrocatalytic activity of the prepared electrodes depended on the morphology and the microstructure. Ni-Cu surfaces exhibited an enhanced catalysis for HER with respect to Ni and Cu cathode, which is mainly attributed to the high surface area of the developed electrode. Ni-Cu deposits with a Cu content of 49% manifest the highest intrinsic activity for HER as a consequence of the synergetic combination of Ni and Cu.

The traditional solid-state reaction production of Montmorillonite ($LaFeO_3$) perovskite photocatalyst with its large particle size has some disadvantages, including a poor surface area. Tijare et al. [13] investigated on perovskite (ABO_3)-type photocatalyst with $LaFeO_3$ composition for hydrogen generation during water splitting through a sol-gel method. The photocatalytic activity of $LaFeO_3$ was investigated for hydrogen generation through sacrificial donor-assisted photo-catalytic water splitting reaction by varying conditions in feasible parametric changes using visible light source, ethanol as a sacrificial donor, and Pt solution of H_2PtCl_6 as a co-catalyst. The rate of photocatalytic hydrogen evolution was observed to be 3315 μmol g^{-1} h^{-1} under optimized conditions and using 1 mg dose of photocatalyst with reaction time of 4 h and illumination of 400 W. The schematic diagram for photocatalytic reactor for hydrogen generation is shown in Figure 5.4.

Deposition of Ag nanoparticles on TiO_2 has repeatedly been attempted to limit the chance of electron–hole recombination, increase the absorption of visible light, and increase the rate of electron transfer to the oxidant, resulting in an increase in TiO_2 photocatalytic activity. Furthermore, doping with appropriate transitional metals can widen its light absorption to the visible spectrum and improve the separation efficiency of photo-induced electrons and holes. Fan et al. [14] investigated Ag-deposited and Fe-doped TiO_2 nanotube arrays for photocatalytic hydrogen production by water splitting reaction. The photocatalytic activity of Ag–Fe/TiO_2 nanotube was evaluated through the experiment of water splitting. The average maximum H_2 production rate through water splitting was 1.35 μmol/(cm^2 h) with 0.2 mM Ag–0.3 mM Fe/TiO_2 nanotubes as catalyst. The schematic diagram for H_2 production by water splitting reaction is shown in Figure 5.5.

Perovskite-type oxides stand out among the large majority of metal oxide photocatalysts due to their wide range of characteristics. The ideal perovskite-type cubic structure of ABO_3 allows for the content of A and B sites to be varied to

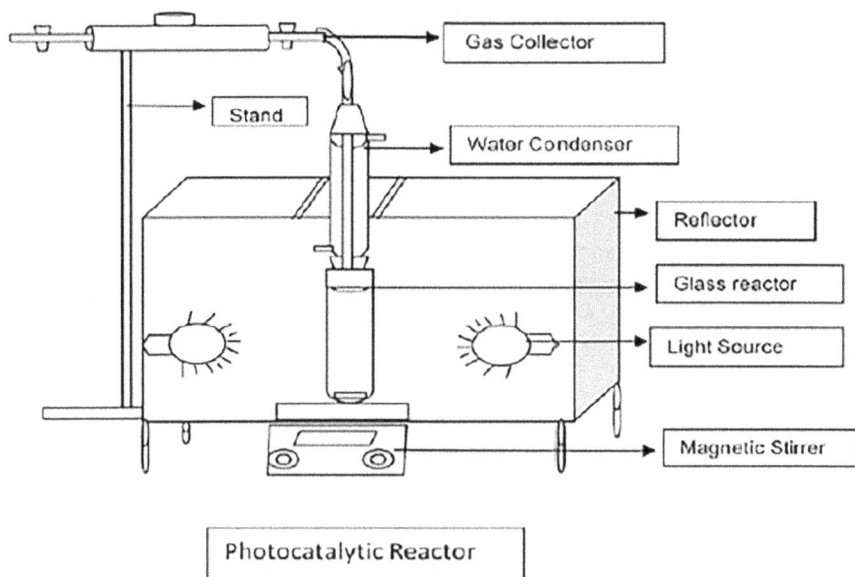

FIGURE 5.4 Photocatalytic reactor for hydrogen generation study [13].

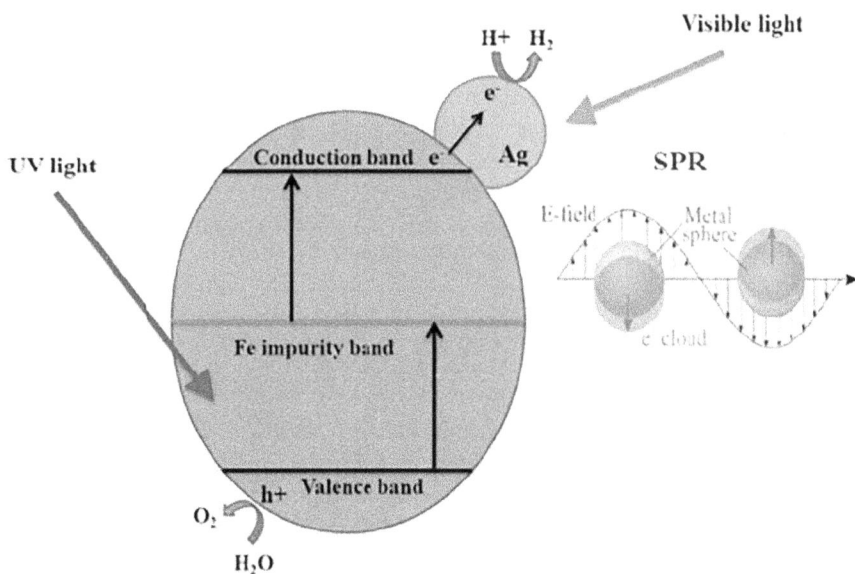

FIGURE 5.5 Mechanism of H_2 production by water splitting over Ag–Fe/TiO$_2$ NTs under both UV and visible light irradiation [14].

generate substituted perovskites. The oxygen anions are 12-fold coordinated with the A cation and 6-fold coordinated with the B cation. The skeleton of the structure is formed by corner-sharing BO_6 octahedra, with the A cation occupying the centre position. Strontium titanate ($SrTiO_3$) is a dielectric substance that is employed as a photocatalyst to decompose organic pollutants or as a water splitter to create hydrogen and oxygen.

Liu et al. [15] developed a novel carbon-TiO_2 nanotube aiming to enhance the solar-driven hydrogen production through water splitting. The newly developed nanotube achieved a cumulative hydrogen production of 43.75 mmol during 50 h of solar irradiation at a rate of 38.66 ± 0.655 mmol/h.g, which is 1.5 folds higher than the maximum rate reported for pure TiO_2-based photocatalyst.

5.4   EFFECT OF TEMPERATURE IN HYDROGEN PRODUCTION FROM WATER BY PHOTO-ELECTROLYSIS

The operation of PEC is based on the temperature as the charge transfer, catalytic activity, electrode kinetics, ionic mobility, electrode stability, diffusion, and conductivity of electrolytes vary considerably with temperature. Hence, photoelectrode stability and PEC cells performance based on the temperature are vital for the improvement of industrial appliances. There are two different effects of temperature on PECs. One is narrowing the thermal bandgap of the photo anode and the second is the reduction of potential essential for water electrolysis. In semiconductor materials, the excitation energy is disturbed by an energy bandgap of width E_g. Generally, n-type semiconductors are used as photoanodes. In such semiconductor materials, conduction band remains empty, whereas the valence bond remains near completely filled. To excite the electron from the valence band to the conduction band, the bandgap energy must be supplied from an external source. An increase in temperature reduces the energy bandgap of the semiconductor. This leads to an increase in the generated photocurrent and absorption of lower energy photons [16,17]. The relationship of bandgap energy (E_g) with temperature (T) is specified by the Varshni model [18] as shown in equation 5.6.

$$E_g(T) = E_g(0) - \frac{\alpha' T^2}{T + \beta'}$$ (5.6)

where $E_g(0)$ represents the semiconductor bandgap at 0 K, α' limit of the gap entropy, and β' is an analogous parameter with Debye temperature. For this temperature function, quadratic behaviour is more applicable at small temperature limits, whereas linear performance is exhibited in the elevated temperature area [18].

The semiconductors that operate at minimal temperatures will obtain limited numbers of holes and electrons for charge transportation. Heating increases the intrinsic charge carriers significantly. The number of charge carriers (n_{int}) as a function of the temperature is given by equation 5.7 [19].

$$n_{int}^2 \approx \exp\left(-\frac{E_g}{k_B T}\right) \tag{5.7}$$

where k_B refers to the Boltzmann constant.

As suggested by Einstein's relation [19], mass transport also increases by an increase in the temperature as shown in equation 5.8.

$$D_i = \frac{k_B T}{q}\mu_i \tag{5.8}$$

where q refers to elementary charge and μ_i refers to the mobility of species i

The temperature increase will further lead to enlarge the current density generated by the PEC cell as suggested by the Butler–Volmer equation [19]:

$$j_{cell} = j_0\left[\frac{n_{H_2O}(b)}{n_{H_2O}^{ref}(b)}\exp\left(\frac{\beta n q \eta_{Pt}}{k_B T}\right) - \frac{n_{H_2O}(b)n_{OH^-}^2(b)}{n_{H_2}^{ref}(b)\left(n_{OH^-}^{ref}\right)^2(b)}\exp\left(\frac{-(1-\beta)nq\eta_{Pt}}{k_B T}\right)\right] \tag{5.9}$$

where j_0 refers to the exchange current density at the platinum counter electrode, n_i is the density of the parameter in the ith reaction, reference particle density is represented by n^{ref}, number of electrons transferred is represented by n, elementary charge by q, and over-potential at the platinum counter electrode by η^{PT} is used for the over-potential at the platinum counter electrode.

The thermodynamics of a photochemical cell is also influenced by temperature. The relation of electrochemical cell potential E with temperature is outlined by the Nernst equation [20]:

$$E = E^0 - \frac{RT}{nF}\ln\left(\frac{\pi a_{products}^{v_i}}{\pi a_{reactants}^{v_i}}\right) \tag{5.10}$$

Here, E^0 is used to refer to standard reversible voltage, R is the universal gas constant, F is used for the Faraday constant, a refers to the activities of reactants and product species, and v_i is the stoichiometric coefficients.

5.5 CONCLUSION

Hydrogen is considered a cleaner fuel, an alternative to other categories of fuels that will provide energy effectively without affecting the environment. There are many different methods for the production of hydrogen. The photoelectrochemical process of water splitting is one such method for hydrogen production. This chapter describes the principle of hydrogen production, the effect of nanomaterials, and operating condition like temperature on the production of hydrogen process. The use of nanomaterials like graphene, and carbon nanotubes leads to a faster electrochemical splitting process and enhances hydrogen production. The increase in temperature was also found to result in an increase in hydrogen production. This chapter describes the relationship of temperature with various parameters such as bandgap energy, and exchange current density which is connected with the hydrogen production process.

REFERENCES

1. Honda K. In: Ohta T, editor. *Solar-Hydrogen Energy Systems*. Oxford: Pergamon Press, pp. 137–169, 1979.
2. Seraphin BO. In: Seraphin BO, editor. *Solar Energy Conversion*. Berlin: Springer, pp. 5–56, 1979.
3. Chandra S. *Photoelectrochemical Solar Cells*. New York: Gordon and Breach, 1985.
4. Morrison SR. *Electrochemistry at Semiconductor and Oxidized Metal Electrodes*. New York: Plenum Press, pp. 1–401, 1980.
5. Nozik AJ. In: Heller A, editor. Semiconductor liquid-junction solar cells. *Proceedings of the Conference on the Electrochemistry and Physics of Semiconductor Liquid Interfaces under Illumination*. Virginia: Airlie, pp. 272–292, 1977.
6. Bak, T., Nowotny, J., Rekas, M., & Sorrell, C. C. Photo-electrochemical hydrogen generation from water using solar energy. Materials-related aspects. *International Journal of Hydrogen Energy*, 27(10), 991–1022, 2002.
7. Khalafalla, M. A., Hadia, N. M. A., Elsayed, A. M., Alruqi, M., El Malti, W., Shaban, M., & Rabia, M. ATO/polyaniline/PbS nanocomposite as highly efficient photoelectrode for hydrogen production from wastewater with theoretical study for the water splitting. *Adsorption Science & Technology*, 2022. https://doi.org/10.1155/2022/5628032
8. Qazi, U. Y., Javaid, R., Zahid, M., Tahir, N., Afzal, A., & Lin, X. M. Bimetallic NiCo–NiCoO$_2$ nano-heterostructures embedded on copper foam as a self-supported bifunctional electrode for water oxidation and hydrogen production in alkaline media. *International Journal of Hydrogen Energy*, 46(36), 18936–18948, 2021.
9. Iqbal, T., Hassan, A., Ijaz, M., Salim, M., Farooq, M., Zafar, M., & Tahir, M. B. Chromium incorporated copper vanadate nano-materials for hydrogen evolution by water splitting. *Applied Nanoscience*, 11(5), 1661–1671, 2021.
10. Fu, M., Liu, Y., Zhang, Q., Ning, G., Fan, X., Wang, H.,... ... & Wang, H. Fe2O3 and Co bimetallic decorated nitrogen doped graphene nanomaterial for effective electrochemical water split hydrogen evolution reaction. *Journal of Electroanalytical Chemistry*, 849, 113345, 2019.
11. Li, W., Wang, C., & Lu, X. Integrated transition metal and compounds with carbon nanomaterials for electrochemical water splitting. *Journal of Materials Chemistry A*, 9(7), 3786–3827, 2021.
12. Nady, H., & Negem, M. Ni–Cu nano-crystalline alloys for efficient electrochemical hydrogen production in acid water. *RSC Advances*, 6(56), 51111–51119, 2016.
13. Tijare, S. N., Joshi, M. V., Padole, P. S., Mangrulkar, P. A., Rayalu, S. S., & Labhsetwar, N. K. Photocatalytic hydrogen generation through water splitting on nano-crystalline LaFeO$_3$ perovskite. *International Journal of Hydrogen Energy*, 37(13), 10451–10456, 2012.
14. Fan, X., Fan, J., Hu, X., Liu, E., Kang, L., Tang, C.,... ... & Li, Y. Preparation and characterization of Ag deposited and Fe doped TiO2 nanotube arrays for photocatalytic hydrogen production by water splitting. *Ceramics International*, 40(10), 15907–15917, 2014.
15. Liu, Y., Xie, L., Li, Y., Yang, R., Qu, J., Li, Y., & Li, X. Synthesis and high photocatalytic hydrogen production of SrTiO3 nanoparticles from water splitting under UV irradiation. *Journal of Power Sources*, 183(2), 701–707, 2008.

16. Nelson, J. Analysis of the p-n junction, in: *The Physics of Solar Cells*. London: Imperial College Press, pp. 145–176, 2003.

17. Dias, P., Lopes, T., Andrade, L., & Mendes, A. Temperature effect on water splitting using a Si-doped hematite photoanode. *Journal of Power Sources*, 272, 567–580, 2014.

18. Gupta, L., Rath, S., Abbi, S.C., & Jain, F.C., Temperature dependence of the fundamental band gap parameters in cadmium-rich ZnxCd1-xSe using photoluminescence spectroscopy, *Pramana*, 61, 729–737, 2003.

19. Andrade, L., Lopes, T., Mendes, A. Dynamic Phenomenological Modeling of Pec Cells for Water Splitting Under Outdoor Conditions, *Energy Procedia*, 22, 23–34, 2012.

20. O'Hayre, R., Cha, S.-W., Colella, W., & Prinz, F.B. *Fuel Cell Fundamentals*. New York: John Wiley & Sons, 2006.

6 Nano-MOFs and MOF-Derived Nanomaterials for Electrocatalytic Water Splitting

Subhradeep Mistry
Hemvati Nandan Bahuguna Garhwal University (A Central University), India

Sourav Laha
National Institute of Technology Durgapur, India

6.1 INTRODUCTION

The consumption of energy will grow considerably in the coming years with the increase in population and development works. To date, the source of our energy demand largely comes from contemporary energy sources which are processed from fossil fuels (coal, oil, natural gas, etc.). However, using these energy sources to meet our energy demand has a terrible environmental cost as it emits greenhouse and other poisonous gases. In addition, the reserve for such fossil fuels is ever decreasing. Thus, we need a renewable clean source of energy for a sustainable future [1]. Research is growing in this area to find a better solution for the conversion and storage of energy. Among them, one promising approach that has drawn much attention recently is the generation of hydrogen from electrocatalytic water splitting.

Electrochemical water splitting (EWS) or electrochemical water electrolysis comprises two critical half-cell reactions, namely hydrogen evolution reaction (HER) at the cathode and oxygen evolution reaction (OER) at the anode. A potential of 1.23 V (vs. reversible hydrogen electrode, RHE) at 25°C and 1 atm is required to overcome the thermodynamic barrier for the EWS and the value corresponds to $\Delta G_o = +237.1$ Kj mol^{-1} [2,3]. However, in reality, due to inherent catalytic activities, the kinetic becomes sluggish and a significantly higher potential is required to surpass the intrinsic activation barriers. This excess potential is generally termed over-potential, η. Thus, we require good electrocatalysts which can help in effective water-splitting process.

There are mainly three ways by which catalytic activities of a catalyst can be improved, namely improving the intrinsic properties, electrical conductivity and specific surface area. The intrinsic properties of a catalyst are predominantly

DOI: 10.1201/9781003364825-6

governed by its constituent elements and hence the composition may be changed to achieve better catalytic activities. On the other hand, enhancement of catalytic activities and electrical conductivities may also be realized by increasing the surface area and morphology of a given composition. Like other catalysis, the electro-catalytic activities largely take place at the active catalytic centres which are located at the interface between the catalysts and the electrolytes. Increasing the surface area brings larger number of electrolyte species in contact with higher numbers of active sites, which in turn leads to improved and more efficient electrocatalysis.

Noble metal-based (Pt, Ru, Ir, etc.) catalysts were largely explored to date due to their appreciable performance for HER and OER. Though these noble metal-based electrocatalysts can exhibit the higher input potential required for water splitting, the high cost of these metals and their scant availability inhibit the prospect of their large-scale applicability. In addition, these catalysts often agglomerate and during electrocatalysis exhibit poor tolerance for poisoning [4]. Thus, the current focus is on finding catalysts that are non-precious metals and highly abundant in nature.

Metal-organic frameworks (MOFs) are a class of highly crystalline porous ma-terials that consist of metals which act as a node and organic linkers. The flexibility offered by the vast possibilities of compositions and structures of MOF-based ma-terials gives us a plethora of possible combinations for metal ion/cluster nodes and coordinating organic linkers to accomplish materials having some of the best intrinsic electrocatalytic activities [5]. It is generally observed that MOF-based materials containing the combination of two first-row transition metals deliver superior electrocatalytic activities than their mono-metallic counterparts. Synergistic effect between the bimetallic catalytic centres is believed to reduce the activation barriers by optimizing the electronic structures and charge transfer rates [6]. Therefore, regu-lating the composition gives us the opportunity to tune the intrinsic electrocatalytic activities of the MOF-based materials.

Another advantage of MOFs and MOF-based materials is their high specific surface area (range from 1000 to 10,000 m^2/g), which has a crucial role in enhancing their electrocatalytic activities [7]. However, the biggest bottleneck of employing pristine MOFs as a catalyst in its bulk crystalline form is the inaccessibility to large number of active sites [8]. Thus, downsizing the MOF crystals is an important step to enhance its catalytic efficiency. For example, Quian et. al. show that trimetallic MOF nanomaterials ($(Ni_2Co_1)_{0.925}Fe_{0.075}$-MOF-NF) with foam-like architecture demon-strate extraordinary OER activity together with high durability in alkaline conditions [9]. The nanostructuring also leads to improvement of charge transport properties, which also has a definite contribution in the OER activity.

During the preparation of MOF-derived materials from MOFs, the derivatives generally retain the porous structures of the pristine MOFs, however with reduced pore sizes and specific surface area. These reductions which are caused by the decomposition of the linker ligands and breaking the coordination bonds between the ligands and the metal centres are not necessarily detrimental to the electro-catalytic activities of the derivatives. MOF-derived products, obtained by thermal decomposition of MOFs, generally have specific structure, composition and mor-phology. For example, Aijaz et. al. reported that pyrolysis of Co-MOF (ZIF-67) under reductive (H_2) atmosphere leads to the formation of core-shell $Co@Co_3O_4$

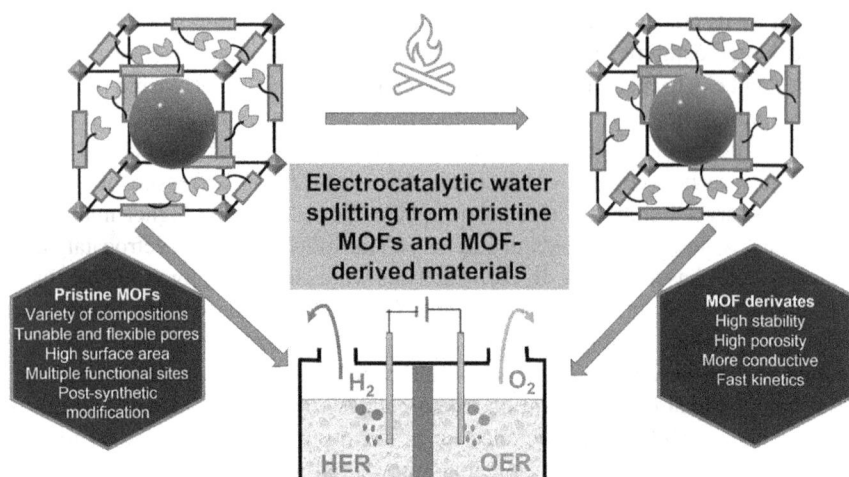

FIGURE 6.1 Schematic diagram of electrocatalytic water splitting from pristine MOFs and MOF-derived materials.

nanoparticles encapsulated in carbon nanotube-grafted nitrogen-doped carbon polyhedral having Brunauer–Emmett–Teller (BET) surface area of only 76 m^2/g. However, it outperforms the state-of-art electrocatalysts Pt/C, IrO_2 and RuO_2 towards electrocatalytic OER activities in alkali medium [10].

Previous research indicates that non-precious metal-based carbides, oxides, sulphides, selenides, nitrides, and phosphides nanomaterials can be excellent electrocatalyst [11]. These materials can act as bifunctional catalyst which can catalyze overall water splitting. Among the non-metal catalysts, various types of carbon materials (porous carbons, N-doped carbons, graphitic carbon nitrides, etc.) are well known. Porosity coupled with well-dispersed doped atoms (including metals) in these carbon materials is an added advantage as it resists the agglomeration of the metal sites to promote the single-site catalytic feature [12]. N/P co-doping proved to be a profitable strategy for the HER as it provides the lowest overpotential, whereas for alkaline media OER N-doped porous carbons can be useful. More importantly, all the above types of materials can be obtained from MOF-based precursors. In maximum cases, the materials obtained after the carbonization of MOFs are in nanometers in size. Therefore, both nano MOFs and MOF-based nanomaterials have drawn significant research interest towards development of improved catalysts for EWS and for the development of clean technologies for green hydrogen-driven planets (Figure 6.1).

6.2 SYNTHESIS AND CHARACTERIZATION

6.2.1 Synthesis of Nano-MOFs

Several synthetic techniques have been employed so far to synthesize MOFs. These diverse techniques, composition of participating metal and ligands, and reaction conditions such as temperature, pH, time and pressure have a huge impact on the final structure and morphology. MOFs are synthesized in solutions applying different temperatures. At low temperature, employing a mild condition slows down the reaction. At this condition, standing the reaction vessel for long time generates bigger MOF crystals. Conversely, reaction in a constant stirring condition produces microcrystals [13]. When the same solution phase reactions are carried out at higher temperatures, it is called solvo-/hydro-thermal method of synthesis. When water is used as a solvent, it is called hydrothermal and for the rest it is solvothermal. The reaction temperature is decided based on the solvent used. Usually, it is kept near or just above the boiling point of the solvent as at that temperature inside a closed reaction vessel it will create an autogenous pressure. This method is the most used technique to synthesize MOFs as it has many advantages. This method gives high yield, produces highly crystalline MOFs at relatively lower time.

However, our focus here is to discuss the design and synthetic routes of MOFs which have more accessible active sites as this increases their electrocatalytic capability. Therefore, controlling the dimensionality as well as the pore sizes becomes two key factors for the stabilization of more active materials. Majority of the metal active sites of MOFs in their bulk form are unavailable due to their small pores and potential interference with the organic ligands. Thus, researchers tried to control its dimensionality and prepared it in thin 2D forms (nanofilm or nanosheet). This has the following advantages: (a) the ultrathin (nanometer range) thickness of the MOF layers will ensure fast mass transfer and electron transfer; (b) will expose the metal active sites more for better interaction as in this form the specific surface area will increase too (c); easy to analyze the structure-property relationship and modulate the performance accordingly [14]. A majority of the reported MOFs which are capable of splitting water are therefore in ultrathin 2D layered form. The preparation of MOFs in thin 2D forms can be done by employing either top-down or bottom-up methods. MOFs in bulk form can be converted to 2D nanosheet form by exfoliation technique which is a top-down method. The bottom-up technique is more advantageous during the synthesis of 2D MOFs as it provides more control on the final topological molecular architecture. For this π-conjugated organic linkers with rigid backbone, metals with square-planar bonding capability are chosen.

6.2.2 MOF-Derived Nanomaterials

MOFs can be an excellent precursor to generate nanomaterials after their pyrolysis. This circumvents the biggest limitation of pristine MOFs which is their stability in presence of harsh chemicals and conditions. Besides their improved stability, these materials also acquire some of the traits of the precursor MOFs even after the pyrolysis such as pore structure, morphology, metals and other atoms present in organic linkers. Diverse synthetic conditions such as heating temperatures, rate and atmosphere (air, Ar, N_2, etc.) and pristine MOF structure can directly influence the size, morphology (nanotubes, nanocubes, nanosheets, nanorods or their hybrid structures), composition (metal/ metal oxides/ hydroxides, nitrides, chalcogenides, phosphides, phosphates, carbides, or their hybrids and their carbon composites) and structure (porous, hollow, hollow, yolk-shell, frame, hierarchical, etc.) of the final materials. In addition, growing some of these materials directly on conductive substances such as carbon paper, Ti foil, Ni foam, Cu foam and carbon clothes to form a hybrid composite can drastically enhance the water-splitting performance of the materials. Mixed metal MOFs (both normal and core-shell form) and MOF-inclusion composites can also be calcined at diverse conditions to obtain MOF-derived nanomaterials.

A common observation is graphitization increases at higher calcination temperature, but it reduces the doping percentage of non-metallic atoms N/S/O/P. Metals like Zn can be removed if we heat above 900°C and for that reason, Zn-based pristine MOFs are often employed for the synthesis of metal-free porous carbon catalysts. In presence of inert gas like Ar at high-temperature calcination, N/S/O/P-doped carbon matrix can be formed, whereas in the presence of air or oxygen metal oxidizes to form metal oxides [15]. Furthermore, researchers have found a solution retaining the original shape of the parent MOFs and forming porous nanoparticles by a process called the two-step calcination method. In this method, MOF samples are generally heated in the presence of inert gases at two different temperatures to obtain a high surface area as well as MOF shape [16].

Among the 0D nanostructures, polyhedra, hollow nanostructures and core-shell nanostructures are important. The most commonly observed morphology of MOF-derived nanomaterials is polyhedra shaped. ZIF (zeolitic imidazolate framework), a class of well-researched MOF, can produce nanomaterials of such morphology after calcination. This type of shapes has many edges and corners, interconnected pores which promote mass transport as well as electrocatalytic activity. Hollow nanostructures can be synthesized by external-templating strategy and later removing the sacrificial template by heating. It has higher loading capacity as it possesses a large surface area. Core-shell nanostructures is another important nano-morphology which can be prepared from core-shell MOF precursors. In this morphology, synergistic interaction between the core and the shell parts is visible which improves the material's performance [17].

In 1D nanostructures, the nanorods and nanotubes morphologies are commonly observed. Self-templating and external-templating strategies are employed to stabilize 1D MOF-derived nanomaterials. As there are only a few 1D MOFs available, an external-templating method is predominant for the production of nanorods. Similarly, nanotubes-like morphology can also be obtained via the external-templating method. Among nanotubes, carbon nanotubes (CNTs) are very well explored due to their large surface area, mechanical strength and flexibility. 2D nanomaterials are also available in different morphologies such as nanofilms, nanosheets, nanoflakes, nanoplates, etc. Such types of nanomaterials are synthesized through template-directed methods and exfoliation.

6.2.3 CHARACTERIZATION OF NANO-MOFs AND MOF-BASED NANOMATERIALS

MOF and MOF-derived nanomaterials are routinely characterized by X-ray diffraction (PXRD), transmission electron microscopy (TEM), scanning electron microscopy (SEM), BET surface area, atomic force microscopy (AFM), etc. By PXRD we can ascertain the bulk phase, atomic and molecular structure of the formed material. In addition, this method indicates material purity as well as the particle size (by Scherrer equation) [18].

Electron microscopic techniques (SEM and TEM) are the important techniques to generate the images of the nanomaterials. In SEM, an electron beam of ~ 5 keV scans the surface of the material to produce its image, which indicates the materials' morphology. For TEM, even a higher energy electron beam (~ 200 keV) is used, and the sample suspensions are generally drop casted over a carbon grid. Apart from generating the sample images TEM provides the diffraction pattern information of the sample. From the later information, we can get the crystal structure of nanomaterials.

BET surface area measurement is an important characterization technique for porous materials like MOFs. By this technique, we can determine the surface area of the sample and it further indicates the nature of the pores (micropores, mesopores or macropores). AFM determines the surface structure of the nanomaterials down to the atomic scale. This technique helps to obtain 3D topographic information of the sample by probing its surface using a sharp tip. Apart from these regular characterization techniques, some of the new emerging techniques are also applied for better characterization of the nanomaterials. One such technique is high-angle annular dark-field imaging (HAADF), which indicates whether nanoparticles are situated inside another material in composite system through tomographic 3D reconstruction.

Inductively coupled plasma mass spectrometry (ICP-MS) technique is employed to determine the metal content in a material. Energy-dispersive X-ray spectroscopy (EDS), eventually called energy dispersive X-ray analysis (EDXA) or also energy dispersive X-ray microanalysis (EDXMA), is the elemental analysis technique for nanomaterials. X-ray photoelectron spectroscopy (XPS) helps us to understand the oxidation state of the metals present in the nano-samples [12].

6.3 APPLICATION IN ELECTROCATALYTIC WATER SPLITTING

The potential equations for the half-cell reactions for an EWS can be written as

$$E_{HER} = E^{\circ}_{HER} + iR + \eta_{HER}$$
$$E_{OER} = E^{\circ}_{OER} + iR + \eta_{OER}$$

where E_{HER} and E_{OER} are the applied potentials for each half cell reactions, $E^{\circ}_{HER} = 0$ V (vs. RHE), $E^{\circ}_{OER} = 1.23$ V (vs. RHE), iR is the Ohmic potential drop of the system and η is the overpotential.

Among the two half-cell reactions, the HER is relatively easier involving a two-electron transfer process, whereas the OER which involves a four-electron transfer process under a harsh oxidizing environment is the bottleneck for the entire process to run smoothly and satisfactorily. Depending upon the reaction environments, the two half-cell reactions can be presented in the following ways:

In alkaline electrolyte:

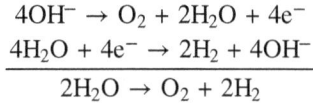

$$\frac{\begin{array}{l} 4OH^- \rightarrow O_2 + 2H_2O + 4e^- \\ 4H_2O + 4e^- \rightarrow 2H_2 + 4OH^- \end{array}}{2H_2O \rightarrow O_2 + 2H_2}$$

In acidic electrolyte:

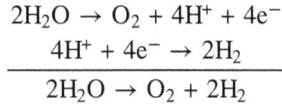

$$\frac{\begin{array}{l} 2H_2O \rightarrow O_2 + 4H^+ + 4e^- \\ 4H^+ + 4e^- \rightarrow 2H_2 \end{array}}{2H_2O \rightarrow O_2 + 2H_2}$$

Both alkaline and acidic environment water electrolysis have their own advantages and limitations. On the one hand, alkali media EWS generally employ non-noble metal based and relatively inexpensive 3d-transition metal-based catalyst. They have longer durability, and the produced hydrogen has higher purity due to lower gas diffusivity. However, the alkaline electrolytes are prone to form carbonate by reacting with carbon dioxide and have relatively lower ionic conductivity. On the other hand, acidic electrolytes have higher ionic conductivities and fewer side reactions but they are expensive as they predominantly employ noble metal-based catalysts and are susceptible to higher dissolutions. Therefore, significant research interests are devoted to improving the performances and eliminating the drawbacks of both alkali and acidic media EWS devices.

6.3.1 HER with Nano-MOFs and MOF-Based Nanomaterials

Zhu *et al.* started with a graphene-like coordination polymer (CP) compound containing benzenehexathiol (BHT) ligand and Cu(II) metal [19]. In their earlier report, the same group had shown that the CP thin film (TF) has very high electrical conductivity ~ 1580 S.cm^{-1} which was till then the highest reported value. Unlike other reported 2D metal-dithiolene complexes, this compound is not porous as here the Cu(II) and the BHT ligand form Kagome layers which are packed densely which ends the prospect of having a porosity in it. To make its active sites more available for catalytic activity, the team prepared Cu-BHT nanocrystals (NC-1) and nanoparticles (NP-1). The nanocrystals are prism-shaped, whereas the nanoparticles are irregular in their shape. The NP form has a diameter of several nanometers and is much smaller in size compared to the NC form (Figure 6.2).

The nanoparticle form displays improved overpotential value during the HER compared to the nanocrystalline form. For NC-1, the overpotential value was observed to be 760 mV to reach a current density value of 10 mA·cm^{-2}. For NP-1, the same was found to be 450 mV, which indicates its superiority as a hydrogen evolution catalyst (Figure 6.3). The NP-1 form retains its electrocatalytic performance even after conducting over 2000 cycles of potential sweeps. The DFT study further indicates that the 'Cu edge' site on the (100) surface is most active and for NP-1 they are more exposed than the NC-1 form.

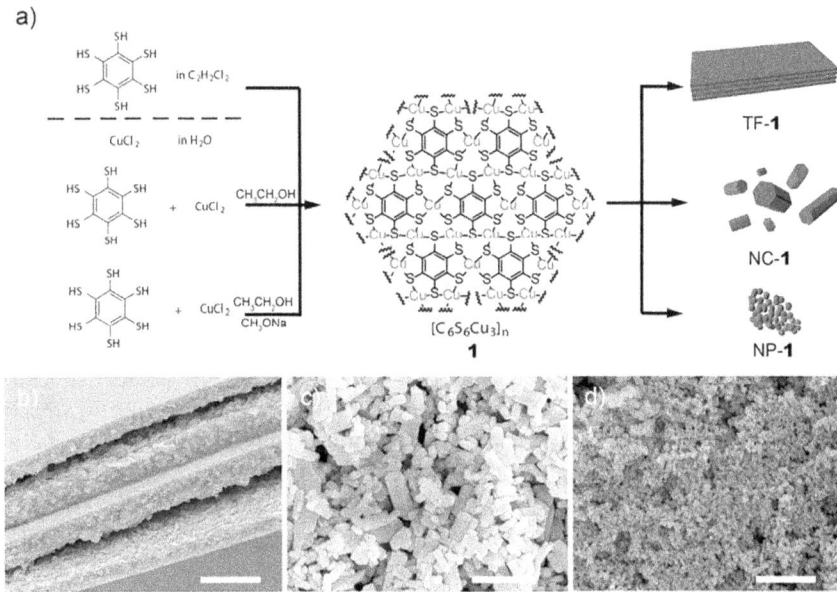

FIGURE 6.2 (a) Scheme from different synthesis methods for synthesizing Cu-BHT is shown. SEM images of (b) thin film (c), nanocrystal, (d) nanoparticle forms are shown. Reprinted (adapted) with permission from Reference 19. Copyright 2017 American Chemical Society.

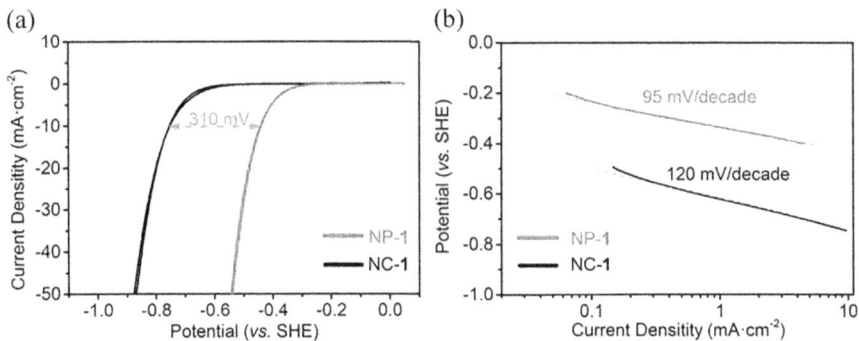

FIGURE 6.3 (a) CV curves of GCE/NC-1 (black line) and GCE/NP-1 (red line) (b) Tafel plot based on the CV curves and green line is the linear fitting line. Reprinted (adapted) with permission from Reference 19. Copyright 2017 American Chemical Society.

The existing studies indicate that the 2D MOFs exhibit promising electro-catalytic behaviour as they possess more open active sites, especially metals. A 2D MOF was stabilized employing a conjugated ligand hexaiminohexaaza-trinaphthalene (HAHATN) with a general formula of $(M2_3(M1_3 \cdot HAHATN)_2)$ (M = Ni) [20]. $Ni_3 \cdot HAHATN$ metalloligand precursor was used to prepare the MOF $Ni_3(Ni_3 \cdot HAHATN)_2$ which is mesoporous. The SEM image indicates that the MOF has petaloid morphology and is made up of thin-layered nanosheets. This MOF is found to be highly conductive in nature as it has a narrow bandgap (0.19 eV) and an in-plane porous structure. The overpotential for $Ni_3(Ni_3 \cdot HAHATN)_2$ was found to be 115 mV at a current density of 10 mA cm^{-2}. The Ni–N2 unit in the MOF plays an important role in the observed HER activity. Tafel slope that indicates the HER kinetics reaches up to 45.6 mV dec^{-1} for Ni-MOF. The MOF sample was found to be durable as it shows marginal deviation even after 1000 cycles of CV tests carried out in a 0.1 M KOH solution.

To find a possible alternative to costly Pt catalysts, Fend et al. adopted an interesting approach of mixing two different building units. A 2D MOF was synthesized by employing the ligands 2,3,6,7,10,11-triphenylenehexathiol (THT) and 2,3,6,7,10,11-triphenylenehexamine (THA) and metals (Co and Ni) [21]. The notable structural feature is that the group has been successful in incorporating the molecular MS_xN_y [x/y are 2/2 (Metal Dithiolene-Diamine), 0/4 {metal bis(diamine)}, and 4/0 {metal bis(dithiolene)}, respectively] complexes into the 2D carbon-rich MOFs. Langmuir-Blodgett (LB) method was used to obtain a single-layer nanosheet of the THTA-Co 2D MOF. The thickness of a free-standing single nanosheet is 0.8±0.1 nm where Co, N, S and C atoms are homogeneously distributed. The electrocatalytic activity of the 2D MOF was tested employing the rotating disk electrode (RDE) technique in a 0.5 M H_2SO_4 solution. The overpotential was found to be 283 mV for THTA-Co at an operating potential of 10 mA cm^{-2} and the Tafel slope was 71 mV dec^{-1}. The catalyst was found to be durable as operating overpotential increased by 12 mV after 300 CV cycles keeping the other condition the same.

Cu-based MOF NENU-5 nano-octahedrons were employed as a starting material to prepare porous MoO_2-Cu-C [22]. Later, the Cu nanoparticles were etched out using an aqueous $FeCl_3$ solution. The Mo metals present in the MOF precursor oxidize MoO_2 at high temperature and the carbon part was derived from the organic linkers. Finally, the MoO_2-C nano-octahedrons were annealed at 650°C for 6 hours under ammonia flow to prepare porous MoN-NC nano-octahedrons (Figure 6.4). The particle size of the final material was found to be near 800 nm from SEM images. It further indicates that it contains a large number of nanopores. The synthesized MoN-NC nano-octahedrons have shown remarkable HER activity and stability under acidic

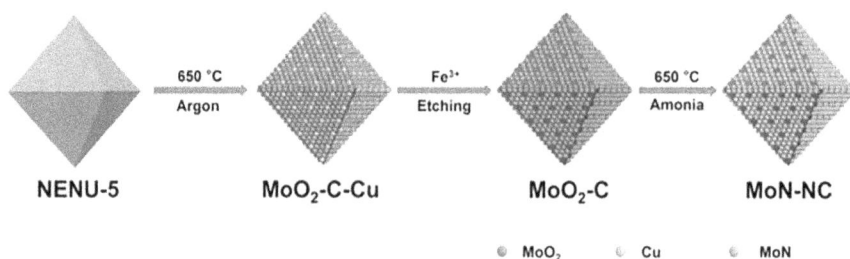

FIGURE 6.4 Synthesis route of MoN-NC nano-octahedrons from MOF (NENU-5) is shown. Reprinted (adapted) with permission from Reference 22. Copyright 2017 American Chemical Society.

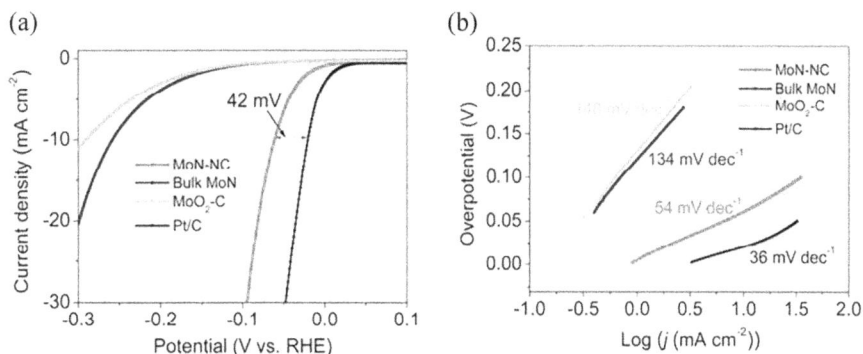

FIGURE 6.5 (a) Polarization curve; (b) Tafel plots of the MoN-NC nano-octahedrons, intermediate MoO$_2$-C, bulk MoN and 20% Pt/C catalysts. Reprinted (adapted) with permission from Reference 22. Copyright 2017 American Chemical Society.

conditions (0.5 M H$_2$SO$_4$). The material exhibits a very low overpotential value of 62 mV at the current density of 10 mA cm^{-2} as well as a low Tafel slope of 54 mV dec^{-1}. The observed values are comparable with the 20% Pt/C commercial catalyst but achieved by employing a non-precious metal-based electrocatalyst (Figure 6.5). The high electrocatalytic activity of MoN-NC nano-octahedrons can be attributed to the uniform distribution of MoN nanoparticles in an N-doped porous carbon matrix and the synergistic relationship between them. A similar CoP nanoparticle encapsulated ultrathin N-doped porous carbon-based electrocatalyst CoP@NC was prepared using a MOF precursor [23]. This material is active in both acidic (0.5 M H$_2$SO$_4$) and alkaline conditions (1 M KOH). The observed overpotential in acidic and alkaline conditions was found to be 78 mV and 129 mV, respectively.

In another study, Chen *et al.* demonstrated that IrCo nanoalloys encapsulated N-doped carbon materials (IrCo@NC-T, T is carbonization temperature) can be outstanding electrocatalysts for HER [24]. This material was prepared from an Ir-doped Co-based MOF precursor which was later carbonized at 500°C, 600°C, 700°C and 800°C under N$_2$ gas. An incredibly low overpotential value of 24 mV was achieved in acidic conditions (0.5 M H$_2$SO$_4$) with the material IrCo@NC-500 at a current density of 10 mA cm^{-2}. Theoretical calculations indicate that the strong and benign interactions between the IrCo nanoalloys and the nitrogen-enriched graphene-based carbon materials facilitate the transfer of electrons from the alloy core to accumulate charge on the nitrogen centres. This, in turn, lowers the reaction-free energy and significantly enhances the HER activities to an extent that almost exceeds the performance of the noble metal-based state-of-the-art commercial Pt/C catalysts.

Layered double hydroxides (LDHs) is an important class of compounds owing to their versatile layered structure, high specific surface area and interesting electronic structure [25]. LDHs containing multiple metal ions of different valencies have shown good electrocatalytic features. Lin *et al.* recently synthesized Pt@CuFe-LDHm a Pt-doped bimetallic (Cu and Fe) LDH material derived from CuFe(dobpdc) MOF [dobpdc = 4,4′-dioxidobiphenyl-3,3′-dicarboxylate]. To obtain Pt@CuFe-LDHm, first CuFe-LDHm was prepared from the precursor MOF CuFe(dobpdc) by immersing the same in KOH solution. Then, the CuFe-LDHm was kept in a solution containing H_2PtCl_6 and $NaBH_4$ to form Pt@CuFe-LDHm (Pt content is ~1.1 wt%). The FESEM and TEM studies indicate that the final material forms a nanosheet of ~800 to 1200 nm. This material was tested as an electrocatalyst for the HER and found to be active in both alkaline and neutral medium. Pt@CuFe-LDHm requires extremely low overpotentials of 33, 47 and 120 mV in 1.0 M KOH, 0.1 M KOH and 1.0 M PBS, respectively, to reach a current density of 10 mA cm^{-2}. The ultrahigh surface area and presence of well-exposed active catalytic sites are the reasons behind this excellent electrocatalytic behaviour.

A nanocomposite (MoS_2/Co-MOF) was prepared with the help of a 2D Co-MOF and MoS_2 has also been explored for the HER [26]. The HER activity of this composite was tested in an acidic medium (0.5 M H_2SO_4) and the overpotential was found to be 262 mV. The lower overpotential and the Tafel slope (51 mV.dec^{-1}) indicate good HER activity. This behaviour is the outcome of a large active surface area which is electrochemically active and the synergistic interaction between the Co-MOF and MoS_2.

In recent work, Wu et al. have adopted an interesting strategy to stabilize a MOF-based nano-composite material for HER [27]. The team took nanorods of nickel zeolite imidazolate framework (Ni-ZIF) and prepared ultrathin nanosheets of Ni-ZIF/Ni-B by employing a simple room-temperature boronization strategy. This step makes the active surface of the material more exposed for better electrocatalytic performance. The morphology, composition and microstructure of the compound can further be tuned by changing the reaction time. The Ni-ZIF/Ni-B nanosheets were supported on nickel foam (NF) to form Ni-ZIF/Ni-B@NF. Ni-ZIF/Ni-B@NF-4 was prepared after 4 hours of boronization which has the thinnest and more exposed layers among other variations. HER was conducted in an alkaline condition with 1.0 m KOH electrolyte. Ni-ZIF/Ni-B@NF-4 has the lowest overpotential of 67 mV at the current density of 10 mA cm^{-2} and the lowest Tafel slope (108 mV dec^{-1}) if we compared it with other forms of the same materials (Figure 6.6). The same material was also employed for OER and overall water splitting.

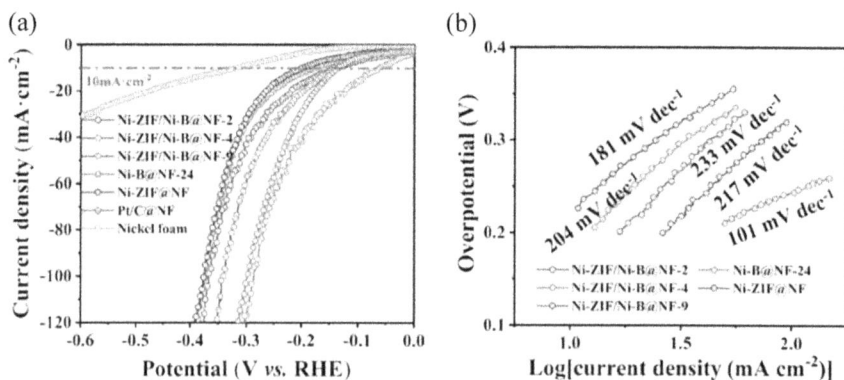

FIGURE 6.6   (a) HER LSV curves recorded at a scan rate of 1 mV s^{-1}, (b) Tafel slopes. Reprinted (adapted) with permission from Reference 27. Copyright 2020 John Wiley & Sons – Books.

In search of a better electrocatalyst for hydrogen production, Chen *et al.* prepared a bi-metallic (Ru/Co) nanocomposite material with nitrogen-doped graphene [28]. Ru is another Pt-group metal which is available in a much cheaper price but it exhibits similar chemical inertness to Pt. The bimetallic materials were developed from a bimetallic MOF precursor. An earlier reported MOF $Co_3[Co(CN)_6]_2$ was taken initially which was doped with Ru metal through a liquid-phase ion-exchange reaction. The exchange of Co^{3+} ions with Ru^{3+} was confirmed by XPS spectra. The final nanocomposite material RuCo@NC was obtained via a one-step annealation of Ru-doped MOF. Electron microscopic studies indicate the mean diameter of the formed particles which are ~30 nm encapsulating in carbon layers. HRTEM further indicates that the carbon layers are an N-doped graphene layer which is ~6–15 layers thick. Electrochemical studies were carried out in 1 M KOH electrolyte using a three-electrode electrochemical cell. Different composite materials with varying amounts of Ru doping were prepared for electrochemical studies. The lowest overpotential of 28 mV was obtained with the S4 form and a trend of increased overpotential value with Ru amount was observed. However, after a certain limit, the trend reverses. The Tafel slope for S4 was found to be 31 mV dec^{-1}.

6.3.2   OER with Nano-MOFs and MOF-Based Nanomaterials

Zhang *et al.* synthesized a NiCoFe-based MOF nanofoam ($(Ni_2Co_1)_{1-x}Fe_x$-MOF-NF) and employed the same as a superior catalyst for OER [29]. The MOF was prepared at room temperature by adding Ni, Co and Fe-acetate salts together along with stoichiometric 1,4-bezene-dicarboxylate acid (1,4-BDC) in the DMF solvent. The metal ratio in MOF can also be controlled by controlling the metal precursor amount. The TEM study indicates the formation of hierarchical networks by interconnected nanofibres that form an overall foam-like nanostructure. The pristine MOF exhibits notable OER activity in an alkaline medium (1.0 KOH) at room temperature. The lowest overpotential and Tafel slope of 257 mV and 41.3 mV dec^{-1} was attained from the $(Ni_2Co_1)_{0.925}Fe_{0.075}$-MOF-NF. The role of Fe may be vital for the MOF's OER activity as Fe is known for exhibiting a partial-charge transfer activation effect on Ni and Co that enhances conductivity. To ascertain this, the same MOF was prepared with only Fe (Fe-MOF) and its OER catalytic activity was studied. It performs poorly in comparison to $(Ni_2Co_1)_{1-x}Fe_x$-MOF-NF, which indicates that a possible synergistic effect among the three metals gives birth to the observed OER catalytic activity. This MOF shows a minor increment of overpotential value after 3000 cycles at the current density of 10 mA cm^{-2}, which indicates the stability of the material.

The design strategy of electrocatalytically active MOFs was broadly to make the active sites more available. Thus, an ultrathin layered structured MOF was designed by Tang *et al.* for OER application [30]. Bimetallic MOF nanosheets (NiCo-UMOFNs) were prepared from hydrothermal synthesis at 140°C. AFM images indicate that the NiCo-UMOFNs layers have thickness of ~3.1 nm and surface area of 209.1 m^2 g^{-1}. The OER measurement was carried out in alkaline condition (1 M KOH) with an electrode where NiCo-UMOFNs was homogeneously deposited over a glassy-carbon supporting electrode. For NiCo-UMOFNs, a dramatic increase of anodic current response can be observed at 1.42 V (onset potential) which is better compared to Ni-UMOFNs, Co-UMOFNs and bulk NiCo-MOFs. The material also possesses remarkably low overpotential of 250 mV at 10 mA cm^{-2}. Interestingly, the performance of NiCo-UMOFNs further improved when deposited on conductive copper foam. The E_{onset} becomes 1.39 V, overpotential 189 mV at 10 mA cm^{-2}. The reason behind this outstanding OER performance for this material can be opined due to its ultrathin layered structure, coordinatively unsaturated metal centres and coupling effect between the metal centres (Co and Ni).

Another bimetallic MOF series (CTGU-10a1–d) containing Co and Ni metals has reported for electrocatalysing OER [31]. The material takes shape of hierarchical nanospheres formed by nanosheets of ~ 1.11 nm. Among them, CTGU-10c2 (CoNi$_2$-MOF) was found to be the best performing for its OER catalysis. The reported overpotential value is 240 mV at a current density of 10 mA cm^{-2} and a Tafel slope of 58 mV dec^{-1} (Figure 6.7). The hierarchical nanobelt structure, coordination unsaturation of metals and coupling effect present between the metals are identified as potential reasons behind this electrocatalytic behaviour. The investigating team further found based on the DFT studies that (a) the performance of bimetallic MOFs are better in comparison to the single metal MOFs, (b) performance of Co is superior to Ni and (c) cluster containing CoNi$_2$ unit can have best overpotential value.

Song *et al.* demonstrated the importance of modulating structural and electronic structure of a conductive MOF to bring out better OER catalytic efficiency [32]. They have been able to replace the Ni-O$_4$ sites with the Fe-O$_4$ sites in a bimetallic MOF series denoted as NiPc–NiFe$_x$. This bimetallic MOF forms nanosized cubic crystallites of 100–500 nm. NiPc–NiFe$_{0.09}$ exhibits the least overpotential value of 300 mV at 10 mA cm^{-2} and a Tafel slope of 55 mV dec^{-1}. The DFT studies indicate that the electronic interaction among the Ni–O$_4$ and Fe–O$_4$ units remarkably increases the intrinsic activity. This study emphasizes the importance of understanding structure-property relationship in a material and tune its property accordingly.

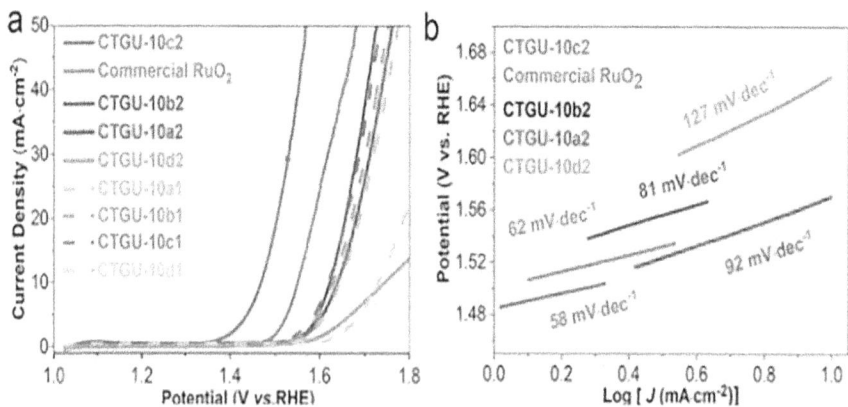

FIGURE 6.7 (a) LSV curves and (b) Tafel plots of RuO$_2$ and the CTGU electrocatalysts in the OER in 0.1 M KOH. Reprinted (adapted) with permission from Reference 31. Copyright 2019 John Wiley & Sons – Books.

Xu *et al.* explored OER activity with bimetallic phosphide hollow nanocubes $(Ni_{0.62}Fe_{0.38})_2P$ which was derived from a Ni-Fe-based bimetallic MOF [33]. The target of the team was to find a new electrocatalyst with non-noble bimetallic phosphides as such materials have rarely been studied. OER activity of this material was explored in an alkaline medium (1 M KOH). The current density of 10 mA cm^{-2} was obtained at 1.52 V (*vs.* RHE, overpotential η_{10} = 290 mV). The Tafel plot was also found to be lower which was 44 mV dec^{-1} and this performance is better in comparison to the benchmark IrO_2 catalyst.

A similar mixed non-noble metal-based porous oxyphosphide material (Co_3FeP_xO) was stabilized from MOFs [34]. The porous structure of the material helps the diffusion of oxygen gas bubbles and induces electron coupling that facilitates electron transfer during electrocatalysis. The overpotential value of 291 mV at 10 mA cm^{-2} was found when Co3FePxO was employed as an electrocatalyst for OER in an alkaline medium.

Yang *et al.* synthesized a MOF-derived material, Ni nanoparticles embedded in N-doped carbon nanotubes (Ni NPs@N-CNTs), to explore its electrocatalytic performance for OER (Figure 6.8) [35]. The design idea behind this material is that carbon-based materials are well-known electrical conductors and porous nature is advantageous for mass and electron transport. One Ni-based MOF was employed as a precursor and heated at 600°C for 3 hours to prepare Ni NPs@N-CNTs. During the OER electrocatalytic experiment, the onset potential was found to be 1.49 V (*vs.* RHE) in an alkaline medium. The overpotential at 10 mA cm^{-2} was found to be 0.46 V and the Tafel slope value is 106 mV dec^{-1} (Figure 6.9).

Yu *et al.* explored the catalytic activity for OER with cheaper non-precious metals like Pd [36]. Pd@PdO–Co_3O_4 nanocubes were prepared by using ZIF-67 (a Co-based zeolite-type MOF) as a sacrificial agent. Pd metal was incorporated into the porous MOF by allowing slow diffusion of $Pd(CH_3COO)_2$ and the guest included MOF after pyrolysis forms Pd@PdO–Co_3O_4 nanocubes. The nanocubes during OER catalytic study produce a current density of 10 mA cm^{-2} at a potential of 1.54 V (vs. reversible RHE). The Tafel slope of Pd@PdO–Co_3O_4 nanocubes was found to be 70 mV dec^{-1} which is lower than Co_3O_4 and Pt/C (86 and 177 mV dec^{-1}) indicating its superiority as a catalyst. This also indicates that the incorporation of Pd increases the availability of active sites which may improve the electrocatalytic activity of this material.

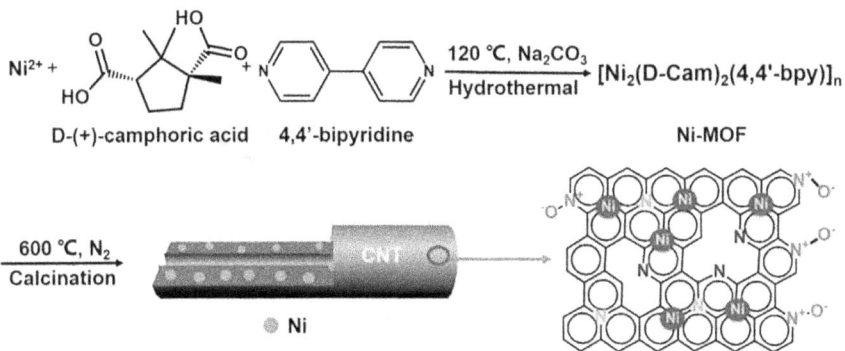

FIGURE 6.8   Scheme for synthesizing Ni NPs@N-CNTs. Reprinted (adapted) with permission from Reference 35. Copyright 2017 Elsevier Science & Technology Journals.

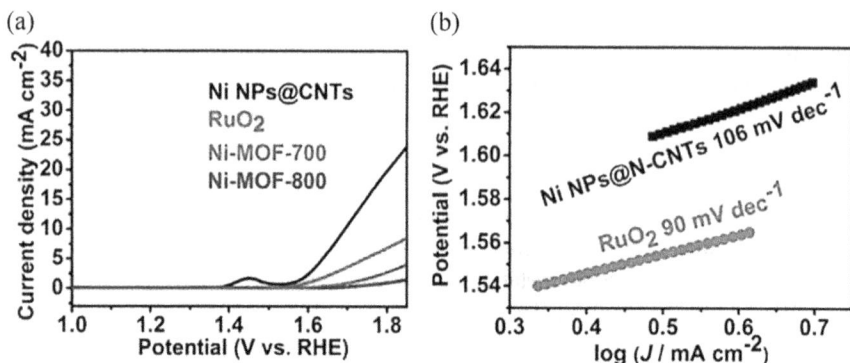

FIGURE 6.9   (a) LSV curves for the OER of Ni NPs@N-CNTs, Ni-MOF-700, Ni-MOF-800 and RuO_2 catalysts in N_2-saturated 0.1 M KOH (900 rpm and 5 mV s^{-1}), (b) Tafel plots of the Ni NPs@N-CNTs and RuO_2 catalysts. Reprinted (adapted) with permission from Reference 35. Copyright 2017 Elsevier Science & Technology Journals.

A new OER electrocatalyst material (Co_9S_8/NSCNFs), which is Co_9S_8 nanoparticles-embedded N/S-codoped carbon nanofibres, has been synthesized MOF-wrapped CdS nanowires [37]. Presynthesized CdS nanowires were submerged in zinc nitrate/cobalt nitrate, 2-methylimidazole and PVP in a methanol solution to prepare the precursor CdS@ZIF. This CdS@ZIF was pyrolysed in N_2 atmosphere at 850 °C to obtain Co_9S_8/NSCNFs. The FESEM and TEM studies indicate that Co_9S_8/NSCNFs have 1D nanofibre-like morphology and the Co_9S_8 nanoparticles are embedded in the carbon nanofibre matrix. The nitrogen adsorption–desorption study suggests that Co_9S_8/NSCNFs-850 has a BET surface area of 558 m^2 g^{-1}. This porous nano-composite structure encouraged the team to explore its electrocatalytic performance for water splitting. At a current density of 10 mA cm^{-2}, Co_9S_8/NSCNFs-850 exhibits the overpotential of 302 mV in alkaline conditions (1.0 M KOH) (Figure 6.10).

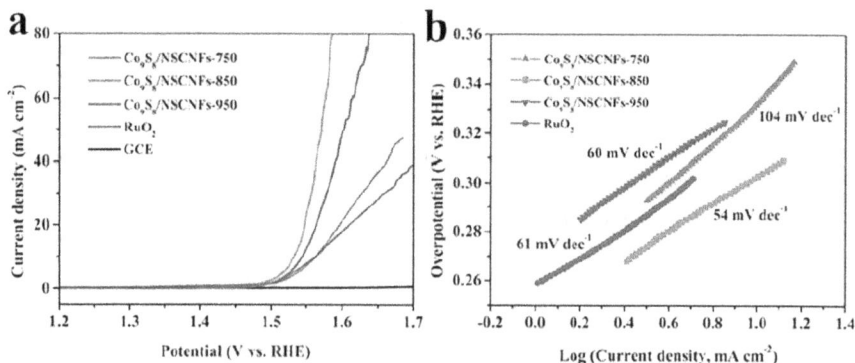

FIGURE 6.10   (a) LSV curves of Co_9S_8/NSCNFs-750, Co_9S_8/NSCNFs-850, Co_9S_8/NSCNFs-950, RuO_2 and bare GCE in 1.0 m KOH solution. (b) Tafel plots of Co_9S_8/NSCNFs-750, Co_9S_8/NSCNFs-850, Co_9S_8/NSCNFs-950 and RuO_2. Reprinted (adapted) with permission from Reference 37. Copyright 2018 John Wiley and Sons.

6.3.3 OVERALL WATER SPLITTING WITH NANO-MOFS AND MOF-BASED NANOMATERIALS

Zhao *et al.* utilized one more 2D ultrathin MOF nanosheet for the electrocatalytic water splitting. Earlier we discussed the advantages of employing thin-layered MOFs for such applications [38]. This 2D MOF can be grown over different substrates via convenient one-step dissolution–crystallization technique. The metals (M = Ni, Fe and Cu) in the MOF form a MO_6 unit that are connected by the organic ligand 2,6-naphthalenedicarboxylic to form the 2D layered structure. Large-size NiFe-MOF nanosheets can be grown in 20 h from the above method (Figure 6.11). The NiFe-MOF can be grown on nickel foam to form NiFe-MOF/NF and the resulting material has macroporosity (pore size in between 200 and 400 μm). The NiFe-MOF nanosheets have a lateral size of several hundred nanometers and pores are in the range of 2.5–18 nm. NiFe-MOF is capable of splitting water at the applied cell voltage of 1.6 V. The experiment was done in a two-electrode cell when both electrodes are prepared with NiFe-MOF. The identity of the H_2 and O_2 gases evolving from the cathode and anode, respectively, was confirmed via gas chromatography. This electrocatalytic cell can produce a current density of 10 mA cm^{-2} at a voltage of 1.55 V and the Tafel slope of NiFe-MOF was found to be 256 mV dec^{-1}. These above-mentioned values indicate the superiority of 2D MOF as an effective electrocatalyst as they surpass the performance of commercial Pt/C cathode and IrO_2 anode system.

A bifunctional electrocatalyst was stabilized from a monolithic zeolitic imidazolate framework@layered double hydroxides (ZIF@LDH) precursor grown over Ni-foam which is simultaneously active in HER, OER and the overall water

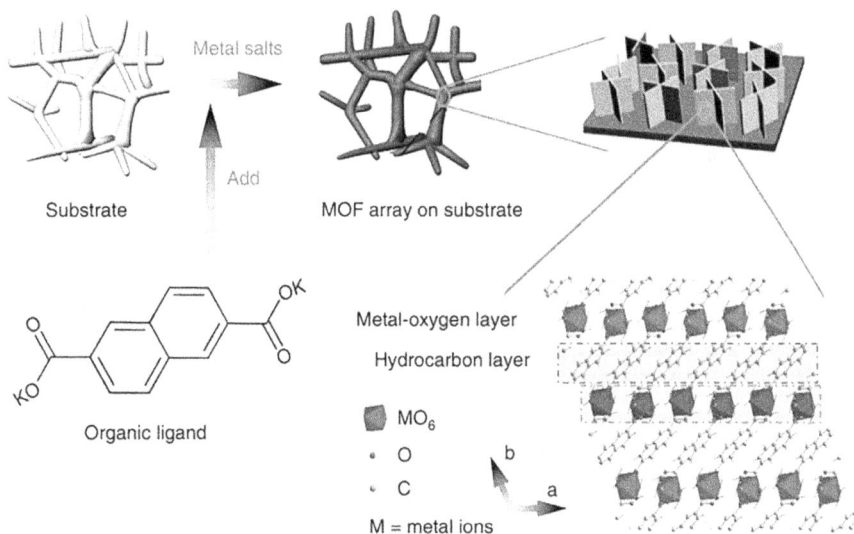

FIGURE 6.11 Synthetic routes of NiFe-MOF nanosheets. Reprinted (adapted) with permission from Reference 38. Copyright 2017 Springer Nature.

splitting (WS) [39]. Upon calcination at various temperatures, the ZIF@LDH@Ni forms Co_3O_4 nanoparticles (3–5 nm) well dispersed on the N-doped carbon matrix. During OER, the overpotential value is found to be of 318 mV at a current density of 10 mA cm^{-2} whereas for HER it is −106 mV at −10 mA cm^{-2}. Encouraged by the above finding, overall water splitting was attempted employing the same catalyst (ZIF@LDH@Ni foam-600) in both anode and cathode using a two-electrode cell. 1.59 V of voltage was applied during the overall water splitting in alkaline medium (1 M KOH) to attain a current density of 10 mA cm^{-2}.

Liu *et al.* reported another bifunctional electrocatalyst ($CoSe_2$–NC) for efficient overall water splitting obtained from ZIF-67 MOF precursor [40]. At first Co-NC hybrid was obtained by pyrolysing followed by acid leaching of ZIF-67 MOF. Later Co-NC-800 hybrid powder was taken along with Se powder and heated together to form $CoSe_2$-NC hybrid. $CoSe_2$–NC hybrids contain mesopores as in BET surface area measurement it exhibits a typical type-IV isotherm. The $CoSe_2$ nanoparticles are well dispersed and confined by a few layers of carbon shell which helps to avert the self-agglomerations. The best-performing $CoSe_2$–NC catalyst displays an overpotential of 234 mV at 10 mA cm^{-2} during HER and a small Tafel slope value of 95 mV dec^{-1}. During the OER, the same catalyst potential of 1.59 V is required to generate the current density of 10 mA cm^{-2}. The above two results indicate that the material has good potential to be used as a bifunctional catalyst for overall water splitting. The $CoSe_2$–NC catalyst with an applied voltage of only 1.73 V attains a current density of 50 mA cm^{-2} in the alkaline electrolyte (1 M KOH).

A material containing lots of heteroatoms which is a S-doped CoWP nano-particle embedded in S- and N-doped carbon matrix (S-CoWP@(S,N)-C) was derived from a MOF and its electrocatalytic property was explored [41]. It is a well-known fact that carbon-based core-shell material containing heteroatoms exhibit better electrocatalytic activity [42]. It has been observed that heteroatoms doping with carbon materials modify its intrinsic electronic structure. A Hofmann-type MOF precursor ($\{[W(CN)_8](SCN)_3Co_3(C_{13}H_{14}N_2)_6\}_n$) in its nanowire form was employed to synthesize S-CoWP@(S,N)-C. S-CoWP@(S,N)-C retains the nano-wire morphology of the precursor MOF and have diameters ranging from 100 to 300 nm (Figure 6.12). HER activity of S-CoWP@(S,N)-C was determined in acidic medium, which indicates a current density of −10 mA cm^{-2} at a potential of −19 mV (*vs.* RHE). OER activity of the same material was measured in alkaline medium (1 M KOH) and a current density of 10 mA cm^{-2} can be obtained at 1.66 V. When the overall water splitting activity of S-CoWP@(S,N)-C was measured, the onset potential was found to be 1.50 V and the current density of the device reaches 10 mA cm^{-2} at a potential of 1.65 V. This material also found to be durable as it can maintain stable current density of 10 mA cm^{-2} for over 12 hours (Figure 6.13).

Mu *et al.* developed a bifunctional catalyst Co-NC@Mo_2C from MOF which is efficient for overall water splitting [43]. The material was prepared by annealing ZIF-67 MOF and ammonium molybdate tetrahydrate mixture together at 700°C for 3 h under Ar gas atmosphere. The SEM and TEM study indicates that Co-NC@Mo_2C adopts a rhombic dodecahedron structure. Due to the synergistic effect present between the Mo_2C and Co-NC units, the material is capable of electrocatalyze

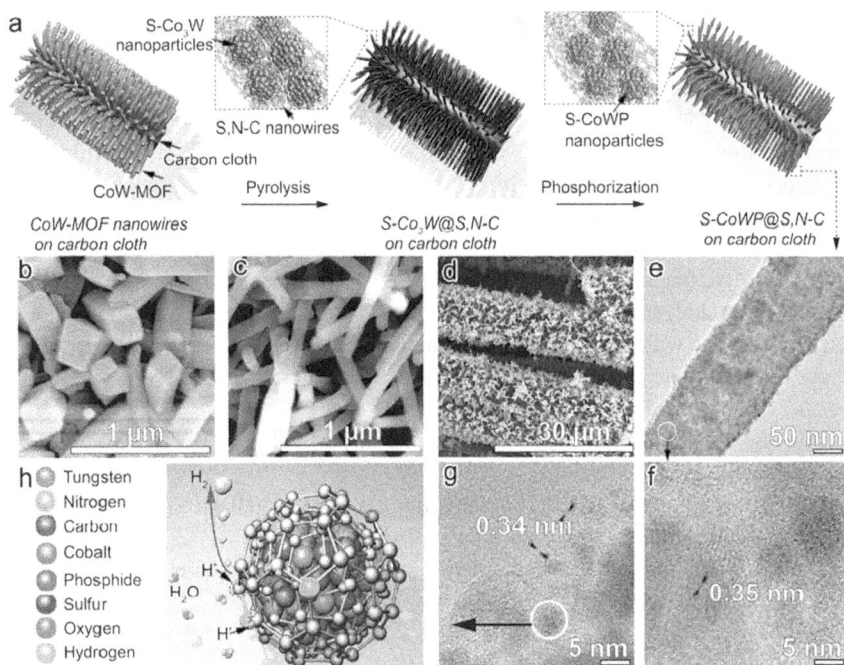

FIGURE 6.12 Synthetic scheme (a) and morphologies observed in SEM, TEM and HRTEM for CoW-MOF and S-CoWP@S,N-C nanowires (b–f). Reprinted (adapted) with permission from Reference 41. Copyright 2018 American Chemical Society.

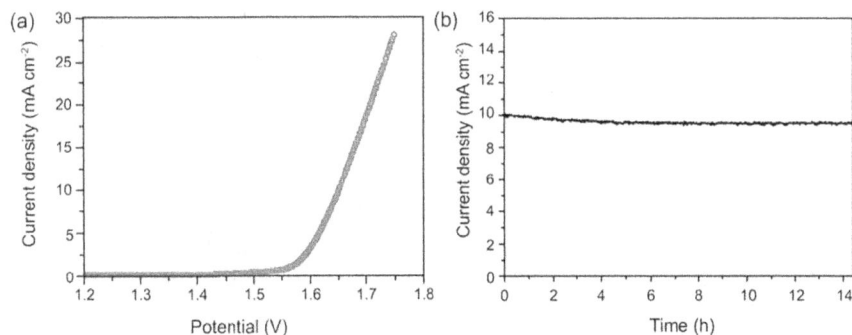

FIGURE 6.13 (a) J–V curve of a two-electrode system composed of S-CoW@S,N-C and S-CoWP@S,N-C nanowire electrocatalysts. (b) Durability test for the two-electrode system at 1.65 V in a 1.0 M KOH solution. Reprinted (adapted) with permission from Reference 41. Copyright 2018 American Chemical Society.

splitting of water. It achieves a current density of 10 mA cm^{-2} in alkaline medium at overpotentials 99 mV and 347 mV for HER and OER, respectively. For overall water splitting, the catalyst Co-NC@Mo$_2$C was applied on both cathode and anode and the current density of 10 mA cm^{-2} can be attained with a low cell voltage of 1.685 V.

In another report, a new efficient MOF-derived electrocatalyst material (Co@Ir/NC-x) is stabilized where Co@Ir core-shell nanoparticles are embedded in nitrogen-doped porous carbon [44]. The core-shell structure of Co@Ir is confirmed from HR-TEM, HAADF-STEM and EDX studies. The best catalytic performance was exhibited by the catalyst with composition Co@Ir/NC-10% which has dodecahedron-like morphology with a diameter of ~600–800 nm. The overpotential of −121 mV at a current density of 10 mA cm^{-2} can be observed for alkaline medium (1 M KOH) whereas during OER the overpotential is 280 mV for the same current density. Inspired from the above results, Co@Ir/NC-10% was explored for overall water catalysis in 1 M KOH. The gases from the cathode and anode can be seen coming out with an applied potential 1.7 V vs RHE. There could be several factors that facilitate the overall efficiency of the material that are

a. Ir can offer multiple active sites which is particularly effective for OER performance.
b. the synergistic interaction between the Co and Ir shell paves the electron transfer kinetics.
c. the possible metal-carbon interaction can heavily excel the electrocatalytic performance of the material.

6.4   CONCLUSION

Renewable energy-driven EWS is one of the best ways to realize sustainable green hydrogen infrastructure. Robust, cost-effective and efficient electrocatalysts have a crucial role to play in making the entire process economically viable. Generally, earth-abundant non-noble metal-based electrocatalysts have lower activities. Hence, significant research works have been devoted to imitating or even exceeding the performance of scare and precious noble metals. Expanding surface area and therefore increasing accessibility by means of nano-structuring is one of the most effective ways to significantly improve catalytic activities. This book chapter summarizes the EWS activities of nano-MOFs and MOF-derived nanomaterials. It is evident from the above discussion that employing nano-MOFs and MOF-derived materials as an electrocatalyst has several advantages over others. MOFs bring different functional moieties such as metals and organic moieties into the same material in addition to their unique porous structure. Likewise, MOF-derived structures give birth to some of the most distinctive and complex materials that are otherwise difficult to synthesize in conventional methods. Multiple metals mixed with non-metallic elements such as C, N, S and P. in the same material give rise to the possibility of synergistic interaction between them. However, more understanding is required of the EWS mechanism of MOF-derived nanomaterials.

Therefore, both nano-MOFs and MOF-derived nanomaterials are being considered as a potential alternative to the traditional noble-metal-based electrocatalysts for EWS. As considerable research is going on in this particular area, we believe MOFs or MOF-based materials will have a significant impact towards the development of clean technologies in the near future.

REFERENCES

1. B. You, Y. Sun, Innovative Strategies for Electrocatalytic Water Splitting. *Acc. Chem. Res.* 2018, *51*, 7, 1571–1580.
2. B. Zhu, R. Zou, Q. Xu, Metal–Organic Framework Based Catalysts for Hydrogen Evolution. *Adv. Energy Mater.* 2018, 1801193.
3. W. Zhang, L. Cui, J. Liu, Recent advances in cobalt-based electrocatalysts for hydrogen and oxygen evolution reactions. *J. Alloys Compd.* 2020, *821*, 153542.
4. Y. Li, Y. Sun, Y. Qin, W. Zhang, L. Wang, M. Luo, H. Yang, S. Guo, Recent Advances on Water-Splitting Electrocatalysis Mediated by Noble-Metal-Based Nanostructured Materials. *Adv. Energy Mater.* 2020, 1903120.
5. B. Zhu, R. Zou, Q. Xu, Metal–Organic Framework Based Catalysts for Hydrogen Evolution. *Adv. Energy Mater.* 2018, *8*, 1801193.
6. S. Sanati, A. Morsali, H. García, First-row transition metal-based materials derived from bimetallic metal–organic frameworks as highly efficient electrocatalysts for electrochemical water splitting. *Energy Environ. Sci.* 2022, *15*, 3119–3151.
7. H. Furukawa, K. E. Cordova, M. O'Keeffe, O. M. Yaghi, The Chemistry and Applications of Metal-Organic Frameworks. *Science* 2013, *341*, 1230444.
8. Y.-L. Liu, X.-Y. Liu, L. Feng, L.-X. Shao, S.-J. Li, J. Tang, H. Cheng, Z. Chen, R. Huang, H.-C. Xu, J.-L. Zhuang, Two-Dimensional Metal–Organic Framework Nanosheets: Synthesis and Applications in Electrocatalysis and Photocatalysis. *ChemSusChem* 2022, *15*, e202102603.
9. Q. Qian, Y. Li, Y. Liu, L. Yu, G. Zhang, Ambient Fast Synthesis and Active Sites Deciphering of Hierarchical Foam-Like Trimetal–Organic Framework Nanostructures as a Platform for Highly Efficient Oxygen Evolution Electrocatalysis. *Adv. Mater.* 2019, *31*, 1901139.
10. A. Aijaz, J. Masa, C. Rösler, W. Xia, P. Weide, A. J. R. Botz, R. A. Fischer, W. Schuhmann, M. Muhler, Co@Co3O4 Encapsulated in Carbon Nanotube-Grafted Nitrogen-Doped Carbon Polyhedra as an Advanced Bifunctional Oxygen Electrode. *Angew. Chem. Int. Ed.* 2016, *55*, 4087.
11. Y. Pan, K. Sun, S. Liu, X. Cao, K. Wu, W.-C. Cheong, Z. Chen, Y. Wang, Y. Li, Y. Liu, D. Wang, Q. Peng, C. Chen, Y. Li, Core–Shell ZIF-8@ZIF-67-Derived CoP Nanoparticle-Embedded N-Doped Carbon Nanotube Hollow Polyhedron for Efficient Overall Water Splitting. *J. Am. Chem. Soc.* 2018, *140*, 7, 2610–2618.
12. Q. Wand, D. Astruc, *Chem. Rev.* 2020, State of the Art and Prospects in Metal–Organic Framework (MOF)-Based and MOF-Derived Nanocatalysis. *120*, 2, 1438–1511.
13. R. P. Paitandi, Y. Wan, W. Aftab, R. Zhong, R. Zou, Pristine Metal–Organic Frameworks and their Composites for Renewable Hydrogen Energy Applications. *Adv. Funct. Mater.* 2022, 2203224.
14. L.Fan, Z. Kang, M. Li, D. Sun, Recent progress in pristine MOF-based catalysts for electrochemical hydrogen evolution, oxygen evolution and oxygen reduction. *Dalton Trans.* 2021, *50*, 5732–5753.
15. L. Yang, X. Zeng, W. Wang, D. Cao, Recent Progress in MOF-Derived, Heteroatom-Doped Porous Carbons as Highly Efficient Electrocatalysts for Oxygen Reduction Reaction in Fuel Cells. *Adv. Funct. Mater.* 2018, 1704537.
16. Y. Wang, B. Wang, F. Xiao, Z. Huang, Y. Wang, C. Richardson, Z. Chen, L. Jiao, H. J. Yuan, Facile synthesis of nanocage Co3O4 for advanced lithium-ion batteries. *Power Sources* 2015, *298*, 203–208.
17. S. Dang, Q.-L. Zhu, Q. Xu, Nanomaterials derived from metal–organic frameworks. *Nat. Rev. Mater.* 2018, *3*, 17075.

18. P. Hirschle, T. Preiß, F. Auras, A. Pick, J. Völkner, D. Valdepérez, G. Witte, W. J. Parak, J. O. Rädler, S. Wuttke, Exploration of MOF nanoparticle sizes using various physical characterization methods – is what you measure what you get? *CrystEngComm* 2016, *18*, 4359–4368.

19. X. Huang, H. Yao, Y. Cui, W. Hao, J. Zhu, W. Xu, D. Zhu, Conductive Copper Benzenehexathiol Coordination Polymer as a Hydrogen Evolution Catalyst. *ACS Appl. Mater. Interfaces* 2017, *9*, 46, 40752–40759.

20. H. Huang, Y. Zhao, Y. Bai, F. Li, Y. Zhang, Y. Chen, Conductive Metal–Organic Frameworks with Extra Metallic Sites as an Efficient Electrocatalyst for the Hydrogen Evolution Reaction. *Adv. Sci.* 2020, *7*, 2000012.

21. R. Dong, Z. Zheng, D. C. Tranca, J. Zhang, N. Chandrasekhar, S. Liu, X. Zhuang, G. Seifert, X. Feng, Immobilizing Molecular Metal Dithiolene-Diamine Complexes on 2D Metal-Organic Frameworks for Electrocatalytic H2 Production. *Chem. Eur. J.* 2017, *23*, 10, 2255–2260.

22. Y. Zhu, G. Chen, X. Xu, G. Yang, M. Liu, Z. Shao, Enhancing Electrocatalytic Activity for Hydrogen Evolution by Strongly Coupled Molybdenum Nitride@Nitrogen-Doped Carbon Porous Nano-Octahedrons. *ACS Catal.* 2017, *7*, 5, 3540–3547.

23. F. Yang, Y. Chen, G. Cheng, S. Chen, W. Luo, Ultrathin Nitrogen-Doped Carbon Coated with CoP for Efficient Hydrogen Evolution. *ACS Catal.* 2017, *7*, 6, 3824–3831.

24. P. Jiang, J.Chen, C. Wang, K. Yang, S. Gong, S. Liu, Z. Lin, M. Li, G. Xia, Y. Yang, J. Su, Q. Chen, Tuning the activity of carbon for electrocatalytic hydrogen evolution via an iridium-cobalt alloy core encapsulated in nitrogen-doped carbon cages. *Adv. Mater.* 2018, 1705324.

25. J. Song, J.-L. Chen, Z. Xu, R. Y.-Y. Lin, Metal–organic framework-derived 2D layered double hydroxide ultrathin nanosheets for efficient electrocatalytic hydrogen evolution reaction. *Chem. Commun.* 2022, *58*, 10655–10658.

26. M. Zhu, Q. Ma, S.-Y. Ding, Y.-Z. Zhao, W.-Q. Song, H.-P. Ren, X.-Z. Song, Z.-C. Miao, A molybdenum disulfide and 2D metal-organic framework nanocomposite for improved electrocatalytic hydrogen evolution reaction. *Mater. Lett.* 2019, *239*, 155–158.

27. H. Xu, B. Fei, G. Cai, Y. Ha, J. Liu, H. Jia, J. Zhang, M. Liu, R. Wu, Boronization-induced ultrathin 2D nanosheets with abundant crystalline–amorphous phase boundary supported on nickel foam toward efficient water splitting. *Adv. Energy Mater.* 2019, 1902714.

28. J. Su, Y. Yang, G. Xia, J. Chen, P. Jiang, Q. Chen, Ruthenium-cobalt nanoalloys encapsulated in nitrogen-doped graphene as active electrocatalysts for producing hydrogen in alkaline media. *Nat. Commun.* 2017, *8*, 14969.

29. Q. Qian, Y. Li, Y. Liu, L. Yu, G. Zhang, Ambient fast synthesis and active sites deciphering of hierarchical foam-like trimetal–organic framework nanostructures as a platform for highly efficient oxygen evolution electrocatalysis. *Adv. Mater.* 2019, 1901139.

30. S. Zhao, Y. Wang, J. Dong, C.-T. He, H. Yin, P. An, K. Zhao, X. Zhang, C. Gao, L. Zhang, J. Lv, J. Wang, J. Zhang, A. M. Khattak, N A Khan, Z. Wei, J. Zhang, S. Liu, H. Zhao, Z. Tang, Ultrathin metal–organic framework nanosheets for electrocatalytic oxygen evolution. *Nat. Energy* 2016, 16184.

31. W. Zhou, D.-D. Huang, Y.-P. Wu, J. Zhao, T. Wu, J. Zhang, D.-S. Li, C. Sun, P. Feng, X. Bu, Stable Hierarchical Bimetal–Organic Nanostructures as HighPerformance Electrocatalysts for the Oxygen Evolution Reaction. *Angew. Chem. Int. Ed.* 2019, *58*, 1–6.

32. J. Li, P. Liu, J. Mao, J. Yan, W. Song, Structural and electronic modulation of conductive MOFs for efficient oxygen evolution reaction electrocatalysis. *J. Mater. Chem. A* 2021, *9*, 11248–11254.

33. H.-H. Zou, C.-Z. Yuan, H.-Y. Zou, T.-Y. Cheang, S.-J. Zhao, U. Y. Qazi, S.-L. Zhong, L. Wang, A.-W. Xu, Bimetallic phosphide hollow nanocubes derived from a prussian-blue-analog used as high-performance catalysts for the oxygen evolution reaction. *Catal. Sci. Technol.* 2017, *7*, 1549–1555.

34. D. D. Babu, Y. Huang, G. Anandhababu, M. A. Ghausi, Y. Wang, Mixed-metal–organic framework self-template synthesis of porous hybrid oxyphosphides for efficient oxygen evolution reaction. *ACS Appl. Mater. Interfaces* 2017, *9*, 44, 38621–38628.

35. H. Han, S. Chao, X. Yang, X. Wang, K. Wang, Z. Bai, L. Yang, Ni nanoparticles embedded in N doped carbon nanotubes derived from a metal organic framework with improved performance for oxygen evolution reaction. *Int. J. Hydrog. Energy* 2017, *42*, 16149–16156.

36. H.-C. Li, Y.-J. Zhang, X. Hu, W.-J. Liu, J.-J. Chen, H.-Q. Yu, Metal–organic framework templated Pd@PdO–Co₃O₄ nanocubes as an efficient bifunctional oxygen electrocatalyst. *Adv. Energy Mater* 2017, 1702734.

37. L.-L. Wu, Q.-S. Wang, J. Li, Y. Long, Y. Liu, S.-Y. Song, H.-J. Zhang, Co₉S₈ nanoparticles-embedded N/S-codoped carbon nanofibers derived from metal–organic framework-wrapped CdS nanowires for efficient oxygen evolution reaction. *Small* 2018, 1704035.

38. J. Duan, S. Chen, C. Zhao, Ultrathin metal-organic framework array for efficient electrocatalytic water splitting. *Nat. Commun.* 2017, *8*, 15341.

39. Y. Tang, X. Fang, X. Zhang, G. Fernandes, Y. Yan, D. Yan, X. Xiang, J. He, Space-confined earth-abundant bifunctional electrocatalyst for high-efficiency water splitting. *ACS Appl. Mater. Interfaces* 2017, *9*, 42, 36762–36771.

40. H. Lu, Y. Zhang, Y. Huang, C. Zhang, T. Liu, Reaction packaging CoSe2 nanoparticles in N-doped carbon polyhedra with bifunctionality for overall water splitting. *ACS Appl. Mater. Interfaces* 2019, *11*, 3, 3372–3381.

41. B. Weng, C. R. Grice, W. Meng, L. Guan, F. Xu, Y. Yu, C. Wang, D. Zhao, Y. Yan, Metal–organic framework-derived CoWP@C composite nanowire electrocatalyst for efficient water splitting. *ACS Energy Lett.* 2018, *3*, 6, 1434–1442.

42. K. Qu, Y. Zheng, X. Zhang, K. Davey, S. Dai, S. Z. Qiao, Promotion of electrocatalytic hydrogen evolution reaction on nitrogen-doped carbon nanosheets with secondary heteroatoms. *ACS Nano* 2017, *11*, 7, 7293–7300.

43. Q. Liang, H. Jin, Z. Wang, Y. Xiong, S. Yuan, X. Zeng, D. He, S. Mu, Metal-organic frameworks derived reverse-encapsulation Co-NC@Mo2C complex for efficient overall water splitting. *Nano Energy* 2019, *57*, 746–752.

44. D. Li, Z. Zong, Z. Tang, Z. Liu, S. Chen, Y. Tian, X. Wang, Total water splitting catalyzed by Co@Ir core–shell nanoparticles encapsulated in nitrogen-doped porous carbon derived from metal–organic frameworks. *ACS Sustainable Chem. Eng.* 2018, *6*, 4, 5105–5114.

7 Novel Strategies for Improved Catalytic Hydrogen Evolution Using Highly Efficient Nanostructures

V. Harshitha, P.C. Nethravathi, and D. Suresh
Tumkur University, India

7.1 INTRODUCTION

Worldwide populace growth and rising livelihoods have greatly increased energy demand on a worldwide scale [1]. A significant portion of energy sources are derived from fossil fuels used for power generation and the transport segment, which causes a large emission of CO_2 and other greenhouse gases [2,3]. Generating a liable depletion of carbon-based materials may have been used to produce valuable chemicals. Establishing a sustainable alternative resource is crucial to reduce the grave effects of fossil fuel usage such as climate changes, the depletion of energy resources, and financial distress [4–6]. Several alternative energy resources, such as wind, hydroelectric, thermal, and solar electricity, are also common and are more reliable and long-lasting than conventional fuels. Each of these, however, has distinct shortcomings. Wind-generated energy cannot be conserved. Due to the high cost of hydroelectric dams and potential environmental harm, hydroelectric power production is also challenging. The lifespan of geothermal resources is likewise constrained, making it costly to use [7].

Sunlight may generate electric power or heat without requiring a lot of upkeep or the installation of expensive pieces since sunlight is inexhaustible, inexpensive, and limitless. Only a couple moments of the Earth being exposed to solar irradiance might provide all the power used in a year [8–10]. However, because it depends upon topographic location, daylight, period, as well as even seasons, solar light is a discrete power resource which limits the amount of the incoming sunlight. Additionally, the intensity of sunlight is lower per square of ground atmosphere [11]. Therefore, it is crucial to develop an energy resource that is stored, clean, consistent, and durable in order to meet the global energy needs. Hydrogen is a particularly favorable fuel since it can be produced in sufficient quantities through

DOI: 10.1201/9781003364825-7

various sustainable resources (organic matter or water), produces a large amount of energy, is environmentally benign, and offers a significant storing capacity. As a result, many people view hydrogen as an ideal as well as safe alternative to conventional fuels [12–14].

In order to solve the energy issue by splitting water, photocatalysis recently evolved as a greener method that employs semiconducting nanoparticles as catalysts to transform sunlight into chemical energy in the form of a cleaner and sustainable energy transporter like H_2 that also removes impurities from effluent [15–17]. Therefore, the creation of an extremely effective as well as inexpensive catalyst to speed up photocatalytic processes is crucial for the generation of cleaner energy as well as environmental rehabilitation [18]. Numerous research works have been conducted to comprehend the basic methodology for water splitting, sunlight-fuel production, and reduction of carbon dioxide, as well as organic contaminant rehabilitation since the exploration of the photocatalytic water splitting into H_2 as well as O_2 in the vicinity of a TiO_2-modified GCE [15–21].

The development of green and inexhaustible technology is of utmost importance in addressing both the growing energy problem and climate issues simultaneously. According to this viewpoint, sunlight has drawn the attention of everybody since it is clean, inexhaustible, and viable, in addition to the fact that even a small amount that reaches the biosphere is sufficient to utilize fruitfully [22,23]. Nevertheless, due to its sporadic character, sunlight may be transformed into chemical energy enabling storage [24]. A viable method for producing cleaner, highly efficient fuel like H_2 from water is sunlight-driven water splitting, which has the ability to resolve the current energy dilemma [25,26]. Such strategy might be practical if the nanomaterials were potent, stable, as well as affordable and comprised metals that are plentiful on the Earth [27]. H_2 and O_2 are generated as a result of such photocatalyic water splitting, which is often initiated and promoted by absorbing sunlight [28].

$$2H_2O \rightarrow 2H_2 + O_2 \; \Delta G = +237.13 \; KJ/mol$$

$$2H^+ + 2e \rightarrow H_2$$

$$2H_2O \rightarrow O_2 + 4H^+ + 4e^-$$

Numerous nanocatalysts such as sulfides, composites, and other materials have been investigated in the past perhaps to effectively cleave water to evolve hydrogen and oxygen or to release H_2 on its own at the cost of sacrificing reagents [22,25,29–34]. The potentials are too low to be realized in functional gadgets, though. Inadequate absorption, recombination of photoinduced charges, or a paucity of active sites are common causes of poor efficiency. It is essential to enhance visible-light capturing heterogeneous catalysts as almost 43% of solar radiation falls in this area [24,31]. In order to efficiently exploit sunlight, the semiconductors' bandgaps should be small. In contrast, in order to improve reaction rates and deal with significantly high current densities, broad energy gap semiconductors are needed.

Therefore, it seems implausible that one catalyst could accomplish both of these. Nanostructures of two distinct semiconducting materials with Z-scheme electron transfers logical design have a lot of promise to solve problems. It benefits from increased sunlight uptake and excellent oxidation–reduction capabilities of the charge transfer, having a potential optimal performance of 40% [22]. Z-scheme hydrogen production has made considerable strides because of the use of hydrogen evolution photocatalysts (HEP) [22,35].

Photocatalytic water splitting process is a process wherein water redox reaction takes place on a nanoparticle. Three phases constitute the photocatalytic water splitting in this case [23,24]. The first one is the uptake of light with energy larger than that of the photocatalyst's energy gap ($h\nu > E_g$), which excites charges to the conduction band (CB) and leaves voids in the valence band (VB). The second phase is after overcoming the relative bonding potential; both the ion pairs reach the photocatalyst's interface. However, imperfections as well as grain structure unavoidably facilitate the reintegration of the ions. The third phase is electrons in CB reduce and excited holes oxidize water to generate hydrogen and oxygen, respectively [23,24].

The conduction band and valence band margins of the nanocatalyst, respectively, influence the redox potential of ion pairs. Therefore, the conduction edge must be greater than the water redox potential for the reduction of water by the electrons in conduction band (0 V vs. NHE). The valence band margin should be greater than that of the water redox potential (1.23 V vs. NHE) in order to facilitate water oxidation by voids [23]. According to this, a nanocatalyst's energy gap must be at least 1.23 eV, but due to other activating hurdles, it is undoubtedly higher than this value. It is important to keep in mind that the precise band placements matter greater than the size of the energy gap. Unfortunately, because of the wide energy gap, they are essentially ineffective. Their energy gaps could be greatly reduced via doping them with the appropriate metals [36,37]. The possible mechanism of photocatalytic hydrogen evolution (PHE) is shown schematically in Figure 7.1.

7.2   TITANIUM DIOXIDE

Due to its great stability, low toxicity, affordable price, and environmental friendly nature, TiO_2 is regarded as one of the most intriguing photocatalysts for prospective uses in photocatalysis [38–40]. The two forms of TiO_2 that are utilized in photocatalytic activity most typically include rutile and anatase. Further, anatase was proved to be highly effective over rutile for catalyzing reactions in sunlight due to its strong optical absorptivity and wider energy gap (Eg = 3.2 eV) (compared to rutile whose Eg = 3.0 eV) [41]. The increased concentration of hydroxyl ions on the interface and large surface area contribute towards the increased catalytic performance. In photocatalysis, hydroxyl groups are essential to prevent the recombination of photoinduced charge pairs and large surface area of the photocatalyst favorably influences the binding of the reactant on to its interface. Similarly, the photocatalytic performance of TiO_2 is typically controlled by a number of factors such as its specific surface area, crystalline size, and shape. Kho and coworkers investigated the anatase as well as rutile phase constituents of TiO_2

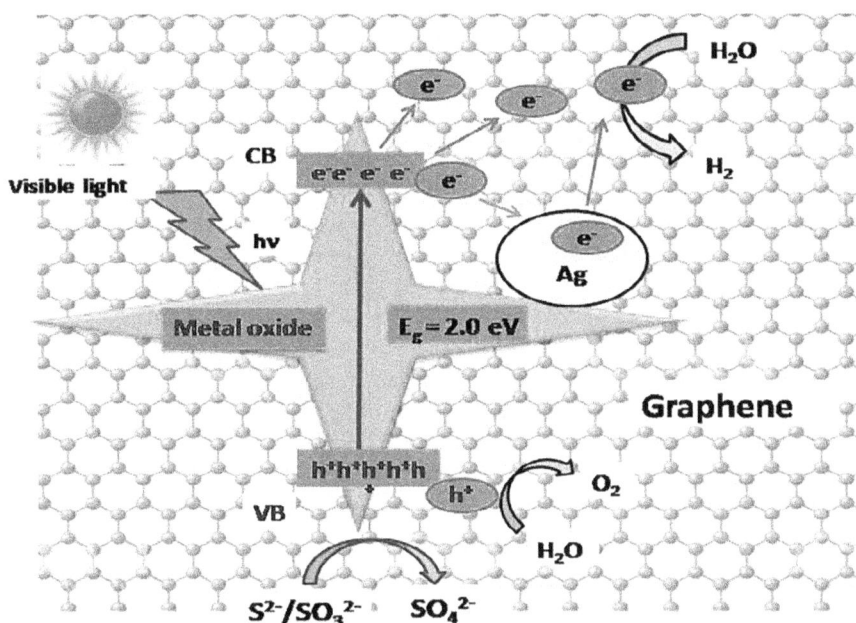

FIGURE 7.1 Schematic representation of the mechanism of charge separation and photocatalytic hydrogen evolution in the presence of Na_2S, Na_2SO_3 sacrificial agents under visible light irradiation.

and its photocatalytic hydrogen generation activity in aqueous TiO_2 mixture containing methanol as a hole trapper [42].

The flame-synthesized, finely crystallized TiO_2 NPs (22–36 m^2g^{-1}) were formed to have 4–95 mol% anatase and the remaining portion of rutile phase. Despite the fact that the quantity of photocurrent produced by imposed potential biases rises with escalating anatase concentration, a distinct pattern was seen for photocatalytic hydrogen production in liquid medium. In comparison, the photocatalytic performance was improved greater than two-fold at 39 mol% of anatase concentration. The optimum electron-hole separation between phase junctions was assumed to be the basis of the collective impact in these composite anatase–rutile phases. Because of limited physical interaction, there were no cumulative impacts for the physically coupled anatase and rutile nanoparticles.

Asjad et al. reported an approach for the modification of anatase TiO_2 to rutile via anatase–rutile phase heterojunction [43]. The results revealed that photocatalytic hydrogen production of mixed-phase was better compared to its pure form (anatase as well as rutile phase). This could be ascribed to the construction of type-II heterojunction, which enables the electron transport over the heterojunction in the mixed phase. Further, Sutiono et al. synthesized the anatase–rutile TiO_2 heterojunction over fluorine-doped tin oxide [44]. Among the synthesized phototcatalysts, low energy (101) anatase–rutile heterojunction was found to be more competent photocatalyst for the evolution of H_2 compared to high energy (001) mixed phase heterojunction.

The enhanced photocatalytic efficiency of the synthesized materials was attributed to the synergistic effects which could be attributed to its change in its Fermi level, band edge location, and energy gap which lead to efficient charge separation. Additionally, the well-faceted crystal facilitates the charge transport among the facets owing to the interface energy variances. Further, Chandra and others demonstrated simple one-step solvo thermal approach for synthesizing anatase–rutile mixed phase TiO_2 nanoparticles for effective photocatalytic hydrogen production [45]. By varying the proportion of water and HCl, they have adjusted the concentration of phase components. The findings demonstrate that the water to HCl proportion is crucial for the development of mixed-phase TiO_2 as well as the transformation of nanoparticles to nanorods in terms of morphology. Optimum water to HCl proportion of 41:59 between anatase and rutile (AR-2) having an urchin-like morphology results in improved hydrogen generation of 5753 $molg^{-1}$.

Although it acquires a reduced surface area compared to anatase nanoparticles, the increased performance of AR-2 is endorsed to the efficient separation of photo-generated charge carriers across the phase junction contact across anatase and rutile. Ysng et al. prepared Au-doped TiO_2 photocatalyst for hydrogen evolution reaction (HER) through electrospinning approach and compared hydrogen evolution activity among P25, TiO_2 nanofibers (NFs), and Au-TiO_2 NFs [46]. The Au-TiO_2 NFs with 9 wt% gold showed significant photocatalytic efficiency by evolving 12440 $\mu molg^{-1}h^{-1}$ which is 21 fold greater than commercial P25 and 10 times greater than pure TiO_2 NFs. The existence of Au in Au-TiO_2 is acting as electron sink to photoinduced electrons by the way it is increasing the lifetime of charge generated. Also, it is generating and transferring the hot electrons to TiO_2 through which the prospective light-harvesting capacities of photocatalysts were considerably improved.

Beasley and others [47] reported the hydrogen evolution activity of pure TiO_2 in comparison with metal-containing TiO_2 ($MTiO_2$) photocatalysts under UV light irradiation. Several metals with varied bandgaps were incorporated to TiO_2, including Pt (5.93 eV), Pd (5.60 eV), Cu (5.10 eV), Ru (4.71 eV), and Ag (4.26 eV). It was observed that the generation of H_2 improved steadily as metal bandgap increased. As the difference in bandgap among the metal and TiO_2 decreases, the probability of electron–hole recombination was found to be increasing. Because of considerably higher work functions and uptrend bands, which confine photoelectrons in particular metal domains which are later utilized during water splitting, Pt and Pd proved extremely effective co-catalysts to TiO_2 promoting H_2 production from water. Owing to their lesser energy gaps, co-catalysts like Ag, Ru, and Cu were less efficient for splitting water because they allowed photoinduced electrons to migrate backwards to TiO_2 through the metal that accelerated charge coupling.

Recently, Erfn et al. [48] reported the synthesis of $CdTiO_3$ nanoparticles and nano fibrous materials using varied cadmium concentrations. The H_2 generation efficiency of Cd-doped TiO_2 nanoparticles was reported to be highly influenced by nano morphology, cadmium concentration and reaction temperature. Because of electron captivity in 0D nanoparticles, nanofibers topology accelerates the amount of H_2 generation about 10 times higher compared to the rate of nanoparticle. For composite of 0.5 wt% Cd, the H_2 generation values for nanoparticle and nanofibers were found to be 0.7 and 16.5 ml/g min, respectively.

In case of pure and Cd-deposited (2 wt%) TiO_2 nanofibers, correspondingly, cadmium loading that increased H_2 generation was 9.6 and 19.7 $mlg^{-1}min^{-1}$, respectively. The principal cause for the demonstrated increase in photocatalysis was the significant suppression of charge carrier coupling due to the establishment of type I heterostructures among the TiO_2 lattice as well as $CdTiO_3$ nanoparticles. In light of this, it is possible to use the suggested heterogeneous catalyst to generate H_2 from a solution devoid of scavengers. The Arrhenius equation is thought to be incompatible with hydrogen evolution over the suggested catalyst, according to changing temperature range. Especially, it was discovered that temperature range had a detrimental effect on photocatalysis. This research demonstrates the potential of employing nano-fibrous $CdTiO_3$ as a co-catalyst in photocatalytic hydrogen production.

Imparato et al. [49] synthesized the TiO_2/C bulk heterostructure which exhibits considerable capturing of visible radiation resulting in efficient hydrogen evolution by water splitting. Two key unique properties characterizing TiO_2-based semiconductors are band edge reduction and Ti^{3+} self-loading that permit the photocatalytic generation of H_2. The capacity to yield maximum hydrogen generation beneath visible light without metal co-catalysts is strongly influenced by the graphene form, carbon radical, as well as oxygen defects containing entrapped electrons (F^+ -centers). Wu et al. prepared the $MoSe_2/TiO_2$ nanocomposite through a hydrothermal method and subjected for H_2 evolution studies [50]. The results revealed that the synthesized material exhibited H_2 production rate of 5.1340 $\mu molh^{-1}$ which is two-fold greater than pristine TiO_2. The observed photocatalytic activity was mainly due to the increased visible light harvesting, creation of heterojunction at the interface of TiO_2 and $MoSe_2$ which facilitate better photoinduced charge separation.

Lalitha and others prepared In_2O_3/nano TiO_2 as well as In_2O_3/nano TiO_2 (P-25) materials and analyzed for phototcatalytic hydrogen evolution activity under irradiation of sunlight [51]. The results revealed that there was significant improvisation in the H_2 production rate due to the incorporation of In_2O_3 when the water-splitting reaction was conducted in methanol-water medium. 5 wt% In_2O_3 doped TiO_2 nanomaterial exhibited the H_2 evolution rate of 4080 $\mu molh^{-1}$ which is two times greater than 2 wt% In_2O_3/nano TiO_2 (P-25) phototcatalyst. Ismael et al. synthesized the pure TiO_2 and TiO_2 doped with varied quantities of Fe^{3+} from 0.05 to 1 mol% [52]. The outcome revealed that 1 mol% Fe incorporated TiO_2 produced hydrogen to a greater extent (2429 $\mu molh^{-1}$) compared to pure TiO_2. The increase in the photocatalytic efficiency is mainly due to reduction in energy gap, increased surface area to volume ratio, and inhibition of quick electron–hole recombination.

The synthesis of β-Ga_2O_3-TiO_2 (TG) nanocomposite and their evaluation for hydrogen generation activity were reported by Navarrete et al. [53]. Among the synthesized composites with different wt% of β-Ga_2O_3, 5%Ga_2O_3/TiO_2 photocatalyst exhibited superior HER (1217 $\mu molg^{-1}$ in 5 h) compared to pristine Ga_2O_3 and Pt/TiO_2 test material. The enhanced photocatalysis is ascribed to its efficient separation of electron–hole pairs, which in turn increases the charge migration at the heterojunction. Reddy et al. decorated CuO quantum dots over TiO_2 and analyzed their PHE activity [54]. The effect of Cu doping in nanocomposite was investigated and results showed that the composite material exhibited nine times greater hydrogen evolution compared to bare TiO_2 nanoparticles under sunlight illumination. It suggested that CuO co-catalyst played a vital role in improving the H_2 generation by enhancing the charge separation.

A new era in science began with the discovery of graphene. The first 2D material featured exceptional qualities including a large specific surface area as well as efficient electron conductivity; graphene's properties reduce the recombination of photoinduced ions, thereby increasing the material's photoelectrochemical efficiency [55]. TiO_2 has a high energy gap compared to graphene's nonexistent energy gap. As a result, the coupling of graphene with TiO_2 is thought to be ideal for photocatalysis. Zhang et al. initially reported this strategy for the photocatalyic hydrogen generation [55].

Corredor and coworkers [56] synthesized the rGO/TiO_2 composites with different graphene oxide proportions (1% to 10%). The activity of composite with 2% of rGO showed better PHE activity of 40 $\mu molg^{-1}$ in 5 hours compared to pure TiO_2 NPs. This enhanced performance is endorsed by the presence of rGO which facilitates better charge separation by accepting the photogenerated electrons from conduction band of TiO_2. Further increase in the rGO proportion beyond 2% reduced the photocatalytic efficiency of nanocomposite due to reduction in the number of active sites available for reaction.

Besides, the transition metal dichalcogenides constitute a different category of 2D compounds that may be used alongside graphene to enhance the efficacy of TiO_2 NPs for photocatalysis. For instance, Zhu's team developed facile mechanochemistry-based MoS_2/TiO_2 photocatalysts containing varying MoS_2 concentrations [57]. The photocatalytic efficiency of these nanocatalysts beneath UV illumination was assessed. In comparison to pristine TiO_2, which merely produced H_2 at a yield of 3.1 $\mu molh^{-1}$, the findings demonstrated that 4%-MoS_2/TiO_2 performed best, achieving reaction kinetics of 150.7 $\mu molh^{-1}$. According to the researchers, MoS_2 functions as a charge carrier collector, preventing the combining of lone pair. Additionally, MoS_2's high conductance makes photogenerated electron–hole separation easier, which enhances photocatalytic activity.

Rhatigan et al. functionalized the TiO_2 surface with metal chalcogenide nanocomposite enabling H_2 evolution employing density functional theory [58]. The nanoclusters are deposited on the interface of rutile (110) phased TiO_2 and possess constituents of M_4X_4 (M = Sn, Zn; X = S, Se). The sulfide-functionalized TiO_2 would more strongly induce HER compared to selenide-loaded TiO_2. The high electronegativity of sulfur (S) compared to selenium (Se) is the reason why hydrogen adsorption on sulfur blogs is generally highly preferential compared to selenium blogs. It is additionally supported by P-band measurement which revealed that sulfide-functionalized TiO_2 possesses smaller p-band compared to selenium-functionalized TiO_2 nanocomposite.

Chalgin and coworkers enhanced the HER performance by preparing the Pd NPs with (100)-facet loaded on anatase TiO_2 nanomaterial [59]. The results revealed that HER activity was significant when (001) faceted TiO_2 was employed as supporting material rather than mixed 101/001 TiO_2 or (101) faceted TiO_2. The findings show that the heterojunction between Pd and various TiO_2 is crucial for reducing the overpotential for hydrogen evolution reaction. This suggests that larger electron flow takes place during the HER when TiO_2 is present, particularly when the 001-facet is subjected. This enhances the hydrogen adhesion as well as accelerates the HER.

Parmar et al. reported that high-quality black anatase TiO_2 films with substitutional N and spatial H loading have been found to significantly influence HER activity in alkaline media. These films undergo electrochemical reduction in the course time of cathodic chronoamperometry producing Ti(III) and Ti(II) that facilitate the diffusion of adsorbed hydrogen resulting in the efficient HER activity [60]. An aqueous TiO_2 over anatase and rutile phase composition with methanol as hole scavenging activity was employed for HER activity. In this regard, as the anatase content was increased, the efficiency of photocatalyst was also increased. In 39 mol% of anatase content, the photocatalytic efficiency was improved by two-fold compared to H_2 production over aqueous TiO_2 solution. The concomitant oxidation of methanol and the generation of strongly reducing hydroxymethyl free radicals effectively infuse more electrons across the TiO_2 conduction band, enhancing potential for heterogeneous hydrogen generation as reported by Rhatigam et al. [61].

Using liquid UV photocatalytic reduction and a sacrificial methanol oxidizer, Montoya and Gillan surface-functionalized Degussa P25-TiO_2 nanocrystals (50 nm, 80:20 anatase–rutile) by 3D metals like Ni, Co, and Cu. These nanoparticles Degussa P25-TiO_2 (1–2 wt % metal to TiO_2) produce visibly colored titanias, which results the UV photocatalytic H_2 production in the rate of 8500 $\mu molh^{-1}g^{-1}$ (7% QY) for Cu(1%):TiO_2 versus 500 $\mu molh^{-1}g^{-1}$ (0.4% QY) for P25-TiO_2 [62]. The MIL-125-NH_2 nanocrystals function as a framework for the fabrication of innovative MOF-based TiO_2 reported by Stavroula Kampour and others. This templated TiO_2 generates H_2 at a rate of around 1394 $molh^{-1}g^{-1}$. This distinctive shape of TiO_2 and its distribution among anatase as well as rutile with regard to the prepared TiO_2 framework were key factors in the enhanced photocatalytic performance of MIL-125-NH_2-derived TiO_2 [63].

Also, Chandra demonstrated crystal phase and morphology of TiO_2 with 3D urchin-like structure and mixed TiO_2 with anatase portion of 41% and rutile portion of 59%, which shows very good photocatalytic H_2 generation of 5753 $\mu mol\ g^{-1}$ and this high performance is due to the charge carrier separation along the phase junction [64]. To overcome certain problems like bandgap narrowing, lower recombination rate of charge carriers, improved charge separation, doping of heterostructures of metal oxide was introduced.

By doping, the photocatalytic efficiency can be improved, which leads to change in its bandgap and optical absorption range. Noble metals like Pt, Pd, Ru, Rh, Au, and Ag are the most efficient metals in photocatalytic HER [65]. According to Gao, Pt has the greatest HER potential for 0.1 Pt/TiO_2 having a 0.1% molar ratio of Pt to Ti, achieving 108.5 fold the standard material produced using commercially available flame-made TiO_2. Remarkably, strong activity is due to the fact that Pt serves as an effective active site for the production of H_2 [66]. Transition metals also serve as dopants for photocatalytic H_2 evolution studies. Huang et al. reported that the photocatalytic H_2 production efficiency of CuO_x-CT400 photocatalyst

(433.3 molh^{-1}) is 56 times higher than those of CT400 (7.7 molh^{-1}) due to TiO$_2$ nanoparticles co-modified with CuO$_x$ (0<x<2). It has been demonstrated that CuO$_x$ particles act as electron donors, enabling sites for proton reduction as well as electron transport. The carbon material serves as an inhibitor to facilitate the flow of electrons between CuO$_x$ entities into TiO$_2$ component [67].

Tahira and coworkers investigated a novel MoS$_2$-decorated TiO$_2$ nanorod composite system as an efficient electrocatalyst for HER activity. They investigated it simultaneously in acidic as well as basic conditions at lower overpotential and a modest Tafel slope. The result was found to be 48 mV/dec and 60 mV/dec in acidic and alkaline environments, respectively [68]. Additionally, Li et al. reported low-cost transition metals such as cobalt-doped TiO$_2$ nanorod clusters (Co-TiO$_2$@Ti(H$_2$)) to exhibit awesome stability even at an extremely high current potential of 480 mAcm^{-2} in 1.0 m Potassium hydroxide solution and a low overpotential of only 78 mV at 10 mAcm^{-2}, a small Tafel plot of 67.8 mV/dec. This is probably due to the fact that oxygen vacancies created in TiO$_2$ interact with Co efficiently, which lowers the energy barrier for water absorption [69].

Ruthenium, a transition metal which shows a good photo absorption in visible light region, was studied by Khorashadizade et al. They detailed the preparation of black Ru-doped TiO$_{2-x}$, which showed a long-term cycling stability and high absorption capacitance. Owing to decreased electron–hole recombination, this photoanode produces hydrogen at a rate which is significantly higher (1.91 molh^{-1}cm^{-2}) than the TiO$_2$ material in its purest form (0.044 molh^{-1}cm^{-2}) [70]. By depositing palladium/strontium nanoparticles, synthesizing Pd/Sr-NPs@P25, Hussain and colleagues reported the photocatalytic efficiency of titania (P25) for hydrogen generation by the water splitting process. Herein, strontium oxide acts as electron donor to the surface of palladium by increasing Fermi energy. Herein, they have demonstrated the factors that influence the photocatalytic HER activity such as method of synthesis, doping quantity, grain size, and chemical state of the metal [71].

Petri and colleagues showed that the ternary composites comprised Ni as well as Ag nanoparticles on the interface of XTiO$_2$ (X: P25, rutile (R)) as an effective visible light-induced nanocatalyst in order to tackle aforementioned limitations. Conversely, RTiO$_2$-based Ni-Ag-RTiO$_2$ exhibits the maximum efficiency, exhibiting a hydrogen production yield of 86 μmolg^{-1}, which is caused by electron–hole recombination in the case of RTiO$_2$ photocatalyst [72]. The P/Ag/Ag$_2$O/Ag$_3$PO$_4$/TiO$_2$ nanocomposite featuring exceptional hydrogen production capacity beneath sun energy was synthesized using a simple one-pot sol-gel/hydrothermal two-step (SH II) technique reported by Zhu and coworkers. Herein, they discussed how the preparation method will affect the photocatalytic performance. Reduced crystallite size, greater total surface area, quicker photogenerated electrons transfer, as well as reduced charge recombination in SH II are some of the factors that contribute to H$_2$ production [73]. The most efficient titanium dioxide dots (TDs) were coupled with a cyano-substituted soluble conjugated polymer to form a unique nanoparticle, as reported by Haofan Yang and colleagues. Due to electron transmission across the photocatalyst materials, the resulting nanomaterial's photocatalytic H$_2$ generation rate was 4.25 fold greater than that of pure polymeric nanomaterials synthesized under identical circumstances [74].

The most recent developments in the hydrogen generation over photocatalysts lacking noble metals have been described by Zhao et al. [75]. There were also additional non-precious metallic photocatalysts such as those based on metal oxide, metal sulfide, metal phosphide, as well as different non-noble metals. It was opined that various strategies like metal or metalloid doping, ion co-doping, creating heterojunctions, structuring core shell and constructing ternary systems may effectively increase the HER activity. Romero et al. have discussed how ruthenium will serve as an alternative to Pt-based photocatalysts. Ternary structured hybrid nonmaterial 4-phenylpyridine-capped Ru nanoparticles (RuPP), TiO_2 nanocrystals, and $[Ru(bpy)_2(4,4'-(PO_3H_2)_2(bpy))]Cl_2$ (RuP) using triethanolamine as a sacrificial electron-donor were used. With turnover numbers exceeding 480 molH$_2$ molRu^{-1} and a turnover frequency of 21.5 molH$_2$ hmolRu^{-1}, solar light hydrogen evolution was accomplished with an overall quantum efficiency of 1.3% [76].

Employing widely scattered nanomaterials, Emran synthesized the significantly improved Sr-TNT/Pd functionalized catalyst on a gold electrode. When relative to undoped TNT/Pd in 0.1 molL^{-1} H$_2$SO$_4$ solution, it exhibits better HER performance with a reduced chemical potential (5.56 kJmol^{-1}) and optimal current density (1.393 mAcm^{-2}). Herein, Sr and Pd act as electron transfer agents that use Volmer–Tafel mechanism. This effective hydrogen evolution efficiency was due to the presence of more active sites of Pd used for the hydrogen binding energies in acidic media [77]. Using a doctor blade approach, Regonin et al. developed uniform TiO_2 films over fluorine-doped SnO_2 (FTO) glass materials. A photocurrent of 0.5 mAcm^{-2} (at 0.23 V against Ag/AgCl) was recorded for composites comprising 5 wt% TiO_2 fibers, compared to just 0.2 mA/cm^2 for nanocomposites comprising 5 wt % P25 NPs. The mesoporosity of nanofibers and the arrangement of the fibers which caused photogenerated electron–hole separation and electron transport across the fiber matrix were the causes of their enhanced activity [78].

7.3 COPPER OXIDE

Copper oxide is a p-type semiconductor that demonstrates numerous applications. CuO shows superior HER activity when it is combined with heterostructures. Kushwaha et al. reported that at a voltage of 0 V v/s reversible hydrogen electrode (RHE), copper oxide (CuO) nanoparticles formed over fluorine-loaded tin oxide (FTO) yield 1.5 mAcm^{-2} of photocurrent and 1.1 mAcm^{-2} for nanosheet electrodes [79]. Moakhar et al. reported carbon-doped CuO dandelions/g-C$_3$N$_4$ with remarkable photocurrent efficiency of 2.85 mAcm^{-2} vs RHE at zero volts [80]. Yang and coworkers reported that Cu$_2$O/CuO bilayered composites were proved to have HER with photocurrent density of 3.15 mAcm^{-2} at a voltage of 0.40 V vs. RHE [81]. Amare et al. successfully synthesized a Cu$_2$O/CuO/CuS photocathode with a remarkable increase in photocurrent density of 5.4 mAcm^{-2} at 0 V compared to RHE [82].

Tahiraand et al. reported proficient H$_2$ production reaction for cobalt oxide-based HER electro-catalysts and it achieves photocurrent density of 10 mAcm^{-2} at a voltage of 0.288 V vs. RHE [83]. Zhang and coworkers with photocatalyst constructed a novel CuO/CuBi$_2$O$_4$ bilayered structured photocatalyst for HER. In neutral condition, the maximum photocurrent efficiency was 2.17 mAcm^{-2} at 0.20 V vs. reversible hydrogen electrode [84]. Muthukumar et al. reported hybrid Cu-Cu$_2$O@C materials whose hydrogen evolution reaction showed enhanced current response (-172 mAcm^{-2}) for hybrid NPs with higher Cu$_2$O ratio [85]. Morales-Guio et al. studied photocurrents of up to 5.7 mAcm^{-2} at 0 V against the RHE that were produced by the fabrication of polymorphic molybdenum sulfide film as a H$_2$ generating photocatalyst over shielded copper(I) oxide films (pH 1.0) [86].

7.4   ZINC OXIDE

Zinc oxide (ZnO) is a versatile semiconducting material used for HER due to its suitable bandgap of around 3.37 eV. Sofianos et al. detailed that spherical morphology of nanoparticles shows highest electrocatalytic activity at lowest potential voltage [87]. Abdullah et al. studied a high efficient hydrogen evolution rate of 213 mmolg^{-1}h^{-1} watt [88]. Patil et al. reported that 2 mol % Ce incorporated ZnO calcined at 600°C evolves 43 μmolh^{-1}g^{-1} of H$_2$, which is 35 folds better compared to pure ZnO [89]. Bermejo et al. reported ternary Co-Zn-B for superior HER electrocatalytic performance with a lower overpotential than the homologous Co-B [90].

Ling et al. made a detailed study on the dual-doped Ni, Zn, and CoO nanorods and achieved current densities of 10 and 20 mAcm^{-2} at overpotentials of 53 and 79 mV [91]. Wang et al. reported a HER catalyst, the Pd/graphene/ZnO/Ni foam (Pd/G/ZnO/NF) nanocomposite that shows an overpotential of -31 mV and Tafel slope of 46.5 mVdec^{-1} in 1 M KOH [92]. Wang et al. successfully synthesized ZnO/CdS heterostructures that showed a good photocatalytic efficiency (1040 μmolg^{-1}h^{-1}) [93]. Pataniya et al. reported the WS$_2$/ZnO nano-heterostructures, which shows visible light-sensitive catalytic activity with overpotential of -182 mV against RHE (reversible hydrogen electrode) at 10 mA/cm^2 [94]. Bare ZnO, Al-ZnO, Cu-ZnO, AlCu@ZnO 0.01 M, AlCu@ZnO 0.02 M, and AlCu@ZnO 0.05 M thin films were synthesized and explored for hydrogen generation by Huner et al. on ITO glass substrates. The outcomes showed that AlCu@ZnO 0.01 M had the minimum electrical resistance [95].

The Volmer–Heyrovsky approach is used by Tseng et al. on the NP (oxide) active site to control the HER activity on NP/ZTO (Pt/ZnO and Au/ZnO) nanostructures [96]. Sumesh reported the heterostructured MoS$_2$.ZnO for HER activity. These composite materials displayed a 239 mV overpotential, a cathodic current density of 10, and an interchange current density of 3.2 mAcm^{-2} [97]. Alhokbany et al. reported that reduced graphene oxide (rGO) supported ZnWO$_4$ NPs for HER and the results showed that the Tafel slope of the nonmaterial was found to be at 149 mV/dec [98]. Zinc oxysulfide nanoparticles were synthesized using zinc acetate dihydrate as zinc source and various amounts of thioacetamide as sulfur source. Herein, for hydrogen evolution to occur, the quantum well containing 3D multi-energy gap photocatalyst was developed and the hydrogen evolution rate was found to be 213 μmol/gh watt in the solar light [99].

Zinc oxide with cerium-doped photocatalyst was prepared for HER performance by Patil et al. Here, photogenerated electron–hole recombination was the key role for H$_2$ generation. Further, holes generated are oxidant and are answerable for splitting of water to H_2. Also, the electrons reduce Ce^{4+} ion to Ce^{3+} on the interface of ZnO and in turn inhibits the electron hole recombination to facilitate HER activity [100]. ZnO/CdS heterostructures were prepared by depositing Cd^{2+} and S^{2-} on the interface of ZnO for effective hydrogen evolution (1040 $\mu molg^{-1}h^{-1}$) under solar light irradiation by Wang and coworkers. With the increase in loading of CdS, the water-splitting activity was found to be enhanced. In the presence of solar light irradiation, the semiconducting CdS excites the photogenerated electron from conduction band to the ZnO's conduction band, which impedes the electron–hole recombination. Consequently, this ZnO/CdS heterostructure reveals superior photocatalytic competence than pure CdS [101].

Ahmad et al. reported silver-doped zinc oxide for HER activity. In this article, it was observed that the water splitting rate enhanced by increasing catalyst content, which was due to increase in active centers on the surface area of Ag-ZnO photocatalyst, in turn increased H_2 evolution. On the other hand, if the NPs concentration was increased further, H_2 production was reduced which was ascribed to the decrease in the surface area owing to aggregation of photocatalysts. Here, in this case, 8% SZO (silver-doped zinc oxide) exhibit reduced water splitting rate paralleled to 6% SZO. This diminished hydrogen evolution rate was due to light scattering and delay of light dispersion over ZnO [102]. Similar observation was reported with Ca-doped ZnO (CZO) [103].

A novel 2D nanocomposite MoS_2-ZnO heterostructure was synthesized by Yuan et al. The MoS_2 nanosheets were deposited on the interface of ZnO, which approves charge transfer and, in turn, overwhelms charge separation among MoS_2 and ZnO. Under sunlight illumination, the electrons from conduction band of ZnO electrons excite to valence band generating holes. The transfer of electrons is from the conduction band of ZnO to the conduction band of MoS_2, which acts as electron sink and also affords effective proton reduction centers. The heterojunction created between the MoS_2 and ZnO results in the upsurge of active sites and suppress electron–hole recombination, leading to higher activity of water splitting [104]. The HER activity can also be shown by a simple metal oxide like ZnO nanoparticles as reported by Archana et al. Herein, the team reported that smaller-sized nanoparticles exhibited improved H_2 evolution rate. By varying catalytic load (10 mg, 20 mg, 30 mg), the (ZnO-10) one with the smaller size showed better HER activity rate of up to 360 $\mu molhg^{-1}$. This advancement of catalytic activity of smaller-sized NPs was due to the oxygen vacancies created, thereby increasing photocatlytic efficiency [105].

Wu et al. reported Pt-metal free PHE. Here, they prepared a porous cluster of ZnO dotted ZnS microsphere, wherein ZnO dotting not only promotes electron–hole separation but also replaces Pt with PHE. After dotting of ZnO, the observed effective separation of charges is ascribed to the development of active catalytic sites (S-Zn-O) in the as-prepared catalysts and also because of small crystallite size on subsequent dotting of ZnO. Thus, this surface sites formation acts as sites for effective water splitting [106]. The composite of ZnO-In_2S_3 nanofiber was fabricated by Chang and team. In this context, they demonstrated the photocatalytic activity of as-prepared nanofibers by using blue light irradiation that has shown the enhanced activity of water splitting. The mechanism involved here is the transfer of photoinduced electrons from valence band of In_2S_3 to the conduction band of In_2S_3. Also, the location of band edge potential of CB in ZnO is higher compared to In_2S_3 thereby transferring the electrons from CB of In_2S_3 into CB of ZnO. These circumstances magnify the charge separation effectively [107].

Ahmad et al. reported a novel Lu-modified ZnO/CNTs composite for efficient water splitting using water–glycerol mixture under solar light illumination. It was observed that the maximum hydrogen evolution rate obtained was about 380 $\mu molh^{-1}$ attributed to improved charge separation, large surface area, and red shift in optical absorption due to fractionation of glycerol by lutetium, owing to the presence of Lu^{3+} in the lattice of ZnO, the bandgap. This Lu-modified ZnO/CNTs composite shows stable and effective potency towards water splitting [108]. Kumar et al. reported a novel method of heterostructure wherein nitrogen-doped ZnO is layered on MoS_2 nanosheets forming an heterostructure. The as-prepared heterostructures with 15 wt% of imperfection-rich MoS_2 nanosheets coated on nitrogen-doped ZnO show about 17.3 $\mu molh^{-1}g^{-1}$. This enhancement in the water splitting rate is due to increase in the number of interfaces between MoS_2 and ZnO, heterojunction designed among ZnO and MoS_2 smoothens charge transfer and existence of rich sulfur edge atoms in defect MoS2 nanosheets which has robust attraction towards H+ ions, thereby increasing HER activity [109].

Girish and coworkers designed a unique carbon nitride-based nanohybrid photocatalyst ZnO/g-C_3N_4. Due to the upsurge in surface area, there is sufficient charge separation. Also, the creation of heterojunction between ZnO and C_3N_4 in turn increases the water-splitting ability [110]. Zhang et al. studied the hydrogen evolution activity of nanorods and mesospheres using a unique structure black ZnO nanorods (NRs)/TiO_2-X mesosphere nanohybrid. This composite serves as a photoanode for HER performance. Under solar light illumination, electrons gathered in the CB of TiO_2-X, whereas holes are shifted from TiO_2 to ZnO. Furthermore, oxygen vacancies created in black TiO_2-X improves light absorption capability, which facilitates the charge separation. Also, the appropriate band alignment can result in good charge separation that enhances photocatalytic water splitting even more [111]. Our group has synthesized ZnO, Ag-ZnO, and Ag-ZnO-rGO and assessed for their hydrogen evolution activity. The results revealed that Ag-ZnO-rGO nanocomposite exhibited superior H_2 evolution activity by exhibiting 3.2 and 1.7 times better photocatalytic activity related to ZnO and Ag-ZnO NPs, respectively [112].

7.5   BI-BASED MATERIALS

Bi_2O_3 is an efficient photocatalyst for HER due to less ideal hydrogen adsorption Gibbs free energy (ΔGH^*). Also, Bi_2O_3 oxygen vacancies play a very close relation towards HER. Bi_2O_3 photocatalyst oxygen void percentage was reported by Wu et al. The results showed improved HER performances with an overpotential of 174.2 mV to achieve 10 $mAcm^{-2}$, a Tafel slope of 80 $mVdec^{-1}$, and an interchange current density of 316 $mAcm^{-2}$ under alkaline medium [113]. Wu et al. also studied the activation of Bi_2O_3, phase engineering strategy in alkaline media. It gives an overpotential of 127 mV (at j = 10 $mAcm^{-2}$) and a Tafel slope of 92 $mVdec^{-1}$ in

1 M KOH [114]. For electrochemical hydrogen generation, Syah et al. developed a novel composite material consisting of reduced Bi_2O_3 to metallic Bi and embedded on a 3D matrix of Ni-foam [115]. It permits high HER of 52 $\mu molg^{-1}$ beneath sunlight [116]. Ru- and Pt-loaded Bi_2O_3 was synthesized by Hsieh et al. Pt/Bi_2O_3-RuO_2 nanocatalyst exhibited the enhanced HER, which is around 11.6 and 14.5 $\mu molg^{-1}h^{-1}$, respectively, by employing 0.03 M Na_2SO_3 and 0.03 M $H_2C_2O_4$ as sacrificial reagents [117].

Kalyan et al. synthesized n-type semiconducting Bi_2O_3 NPs and analyzed for hydrogen evolution activity under sunlight. The results revealed that photocurrent density steadily decreased from 0.95 mA/cm^2 with a bias potential of 1.5 VRHE. [118]. Our group has doped Ag NPs to Bi_2O_3 matrix and prepared the composite with rGO for improving the PHE activity. The results revealed that Ag-Bi_2O_3-rGO nanocomposite exhibited 5.5 and 4.2 times higher H_2 evolution performance compared to pristine and Ag-doped NPs, respectively [119]. The possible mechanism of H_2 evolution is shown in Figure 7.2.

Li et al. constructed a core-shell heterostructured $BiVO_4@Fe_2O_3$ photocatalysts for photocatalytic H_2 generation. Herein, $BiVO_4@Fe_2O_3$ nanocomposite consisting 10 wt% Fe_2O_3 was found to have the highest catalytic activity [120]. Xie et al. reported novel $BiVO_4$ quantum dot/screw-like SnO_2 nanostructures for water

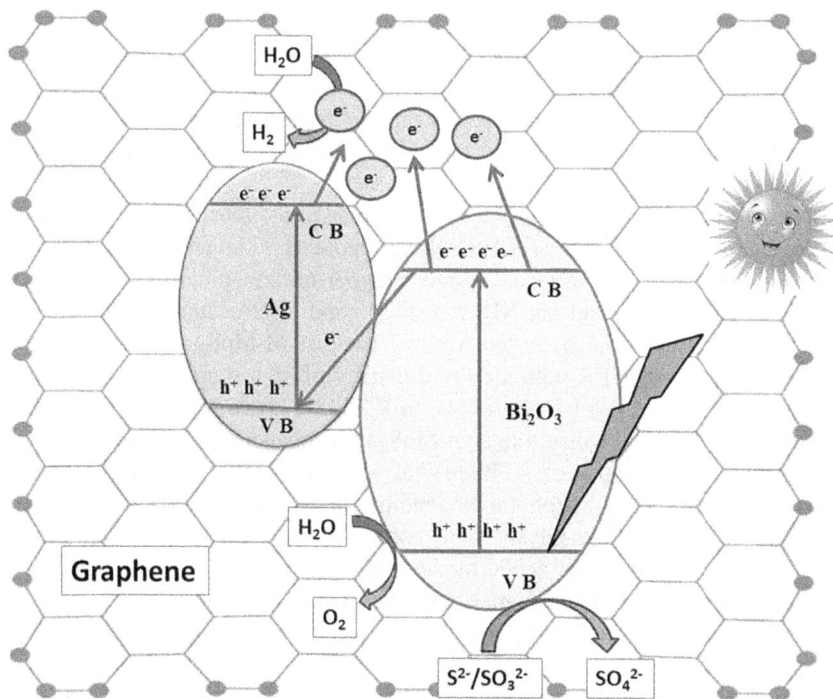

FIGURE 7.2 Schematic representation of the mechanism of charge separation and photocatalytic hydrogen evolution in the presence of Na_2S, Na_2SO_3 sacrificial agents under Visible light irradiation.

splitting with generation rate of up to 1.16 $\mu molh^{-1}cm^{-2}$ [121]. Monfort et al. reported that nanocrystalline bismuth vanadate was deposited on FTO electrodes, which shows maximal HER of 0.15 $\mu mol/h$ beneath electric bias of 1.4 V vs Ag/AgCl plus 0.37 V chemical bias [122]. Tahir et al. reported that reduced graphene oxide-bismuth vanadate nanocomposite to have hydrogen evolution activity. Studies revealed that increase in the content of rGO affects phase transformation of bismuth vanadate from monoclinic to hexagonal and then to orthorhombic. This has resulted in increasing the hydrogen evolution activity up to 97% [123].

7.6   MO-BASED MATERIALS

Cao et al. reported molybdenum disulfide as a promising alternative for HER differing in various aspects like composition and morphology [124]. Yang has demonstrated molybdenum disulfide for an enhanced HER performance in a 2D substoichiometric composition [125]. Campos-Roldán et al. studied the phase transition of MoS_2 from 2 H(hexagonal) to 1 T(trigonal), where 1 T is active for HER activity [126]. Wang et al. reported that ultra-small molybdenum sulfide nanoparticles for HER which exhibit a Tafel slope of 69 mV show extremely high catalytic efficiency [127]. Tuan et al. reported a three-electrode system in a standard acidic medium where MoS_2 NH showed better activity: an overpotential of -230 mV at -10 mAcm^{-2} and a Tafel slope of 64 mVdec^{-1} [128]. Chen et al. studied $MoSe_2$ nanosheets with 2 H/3 R hetero-phases, which revealed superior HER action and also showed reduced overpotential of 164 mV at the current density of 10 mAcm^{-2} with a Tafel slope of 44 mV dec^{-1} [129]. Sayed et al. detailed metal and non-metal dual-doped MoP catalysts that showed overpotentials at -10 mAcm^{-2} of only 65 and 68 mV in 0.5 M H_2SO_4, and 50 and 51 mV in 1.0 M KOH, and increased turnover rate compared to undoped MoP [130].

N-doped, pore-engineered carbon as a two-phase electrocatalyst for H_2 generation was reported by Singh et al (HER). With a reduced initial potential (117 mV), a narrower Tafel slope (94 mV dec^{-1}), and a higher exchange current density (jo = 1.5 102 mAcm^{-2}), the resulting NDCs display good electrocatalytic properties for HER [131]. Xiumei et al. reported a novel catalyst M-MoS$_2$ which works with better action towards HER with a current density of 10 mAcm^{-2} at a low voltage of -175 mV and a Tafel slope of 41 mV [132]. Gyawali et al. successfully synthesized molybdenum disulfide (syn-MoS$_2$) for enhanced bifunctional HER that exhibits low overpotential of -170 mV at -10 mA/cm^2 [133]. Morozan et al. demonstrated a bimetallic iron–molybdenum sulfide bio-inspired electrocatalysts for HER. Herein, electrocatalysts were combined with CNTs and their activity towards HER was studied in acidic media [134].

Sun synthesized MoS_2 and analyzed for HER performance, wherein sulfur vacancies created on the basal plane of MoS_2 boosted HER performance [135]. Ling et al. reported an original electronic structure modifying CoO by Ni and Zn dual doping on molybdenum surface. The dual Ni, Zn-doped CoO nanorods produced current densities of 10 and 20 mAcm^{-2} at overpotentials of, respectively, 53 and 79 mV [136]. Zhu et al. reported novel photocatalyst MoS_2 doped with single (MS@MoS$_2$) and monolayer(ML) metal atoms (MM@MoS$_2$) with current densities of MoS_2 doped by Pd ML and Au ML of 6.208 and 1.109 mA/cm^{-2}, respectively, analogous to Pt(111) [137]. Wang et al. reported that ultra-small molybdenum sulfide nanoparticles for HER which exhibits a Tafel slope of 69 mV showed extremely high catalytic efficiency [138]. Nguyen et al. reported a three-electrode system in a standard acidic medium where MoS_2 NH showed enhanced activity: an overpotential of −230 mV at −10 mAcm^{-2} and a Tafel slope of 64 mVdec^{-1} [139].

Morozan et al. demonstrated a bimetallic iron–molybdenum sulfide bio-inspired electrocatalysts for HER. Herein, electrocatalysts were combined with CNTs and their activity towards HER was studied in acidic media [140]. Marie-Luise et al. reported that molybdenum sulfides (MoSx) are thiomolybdate clusters like $[Mo_3S_{13}]^{2-}$ for HER based on earth-abundant elements [141]. Zhang et al. reported novel amorphous molybdenum tungsten sulfide/nitrogen-doped reduced graphene oxide nanocomposites (a-MoWSx/N-RGO) for use as high-performance HER catalysts [142].

7.7 NICKEL OXIDE

In recent years, nickel oxide was explored as an efficient photocatalyst for HER activity. Herein, we present the emergence of nickel oxide as an effective catalyst for HER. Lu et al. successfully synthesized NiO/C composite by encapsulation of eggshell of NiO through the green synthesis method. In this context NiO/C nanocomposite showed smallest Tafel slope of 77.8 mV dec^{-1} compared to NiO nanoparticles (112.6 mV dec^{-1}) and this was attributed to the synergistic effect between eggshell membrane and NiO, with better robustness even after 500CV cycles [143]. Alaa et al. reported Ni/NiO nanosheet heterostructures maintained with β-Ni(OH)$_2$ having superior HER activity [144]. Gong et al. demonstrated nickel oxide/nickel heterostructures on carbon nanotube sidewalls which are very effective for hydrogen evolution with activity comparable to that of platinum [145]. Emma et al. designed NiO/Ni heterostructures supported on carbon for HER [146]. Faisal et al. reported nickel/nickel oxide heterostructure incorporated on nitrogen-doped graphene nanosheets which showed excellent activity towards HER in acidic and alkaline media [147].

Huo et al. reported incorporated the electrocatalytic water splitting method for H_2 production in low cost with the use of earth-abundant electrocatalysts [148]. Ibupoto et al. adopted a novel $MoS_2@NiO$ composite for HER which results in low overpotential (226 mV) to produce 10 mAcm^{-2} current density [149]. Wang et al. reported N-NiO nanosheet arrays on a nickel foam substrate electrocatalytic HER performance with low overpotential (154 mV at a current density of 10 mAcm^{-2}) and a low Tafel slope of 90 mVdec^{-1} [150]. Alaa et al. reported higher activity of $NiFe_2O_4$ and $NiCo_2O_4$ composition over that of NiO for HER in alkaline media [151]. Another study reported Ni_2P nanowires displayed superior electrocatalytic activity of hydrogen evolution reaction [152]. El-Maghrabi et al. successfully synthesized $Ni/Gd_2O_3/NiO$ coaxial heterostructures that exhibit overpotential of 89 mV which is very close to that of platinum [153].

Zhou et al. reported a novel platinum-based nickel heterostructure PtSA-NiO/Ni as hydrogen evolution performance activity of 20.6 Amg^{-1} for overpotential of

100 mV [154]. NiOx@bamboo-like carbon nanotube hybrids (NiOx@BCNTs) were designed that revealed outstanding catalytic efficiency and considerable durability in alkaline solution. A standard HER current density of 10 $mAcm^{-2}$ was obtained at an overpotential of ~79 mV. It was reported that the N dopants activated the adjacent C atoms to further enhance the catalytic activity. The inherent high NiO level is an important aspect for the exceptional catalytic performance [155]. Srivastava reported N-doped reduced graphene oxide hybrid have the characteristics of high electrical conductivity and larger defects which enhances the HER activity. Herein they reported vanadate of nickel which is fabricated on N-doped reduced graphene oxide in alkaline media, in which NRGO that was efficiently distributed over $Ni_3V_2O_8$ acts as a key role for water reduction. This synthesized heterostructures have a very low overpotential of ~43 mV and high exchange current density (~1.24 $mAcm^{-2}$) of $Ni_3V_2O_8$/NRGO (5.6 wt%) which shows effective HER activity [156].

7.8   CERIUM OXIDE

Cerium oxide is extensively used in various electrocatalysis, organic compounds transformation reaction, and CO oxidation. The reversible conversion between Ce^{3+} and Ce^{4+} promotes electron transfer reaction. Here are some attempts that were made for HER activity by combining cerium oxide with various materials. At present, studies on CeO_2-based HER catalysts are burgeoning and not abundant. Endrodi et al. reported that the cathodic sensitivity for H_2 evolution is increased when cerium (III) salts are added to a hypochlorite mixture [157]. Swathi doped Gd over CeO_2 and compared to pure CeO_2. Gd-doped CeO_2 showed significant HER performance with low overpotential (99 mV for HER) and small Tafel slope (211 mV/dec for HER and 183 mV/dec for OER) [158]. Zhang et al. reported the synthesis of Ni_2P-CeO_2 nanosheet array on Ti mesh functions as a stable catalyst for electrocatalytic H_2 production and can produce 20 $mAcm^{-2}$ at an overpotential of 84 mV in 1.0 M KOH, overperforming all recorded Ni phosphide HER catalysts [159].

   Anantha et al. reported the synthesis of CeO_2 NPs hydrothermally in a single step and encrusted with layers on the interface of graphene oxide (GO) sheets. According to CV measurements [160], CeO_2/GO nanomaterials have greater specific capacitance than individual values at low scan rates of 2 mVs^{-1}. The Ni-rGO/CeO_2 nanocatalyst was designed by Xu et al., utilizing a simple modification technique that involved employing rGO as a conductive substrate for CeO_2 nanocrystals and doped with Ni [161]. Producing very effective bifunctional photocatalytic activity, Ji et al. effectively synthesized Ce-doped $LaCoO_3$ perovskite oxide as electrocatalysts for HER usage. With an overpotential of 305 mV at 10 $mAcm^{-2}$, the resulting LCC4 demonstrated strong electrocatalytic performance against HER [162]. By combining Ce6 secondary building units (SBUs) with 1,3,5-benzenetribenzoate (BTB) linkers, Song developed the first Ce-based metal–organic layer (MOL), known as Ce6-BTB. After being synthesized, Ce6-BTB was post-modified using photosensitizing [(MBA) Ir(ppy)$_2$]Cl or [(MBA)Ru(bpy)2]Cl$_2$ (MBA = 2-(5′-methyl-[2,2′- bipyridin]-5-yl) acetate, ppy = 2-phenylpyridine, bpy = 2,2′-bipyridine) to produce Ce6-BTB. With turnover numbers of 1357 and 484 for Ce6-BTB-Ir and Ce6-BTB-Ru, respectively, and closeness of photosensitizing ligands and Ce6 SBUs in the MOLs, photocatalytic HER is driven beneath sunlight [163].

7.9   METAL-ORGANIC FRAMEWORKS

Using metal-organic frameworks (MOFs) as starters is one of the viable methods for fabricating metal oxide/TiO_2 composites. The development of MOFs marked the beginning of a brand-new era for nanomaterial research and development. Within the class of cage-like nanocomposites made of a mix of metallic clusters as well as organic chemicals, MOFs were recognized as the major supporter. Kudos to its exceptional qualities like wide surface area, pore volume, and versatile structural properties. MOFs have historically been employed in gas sensing, hydrogen generation, photocatalyst, biosensors, and drug carrier uses. Additionally, metal oxide/TiO_2 composites for photocatalysis were made using MOFs as sacrificed substrates. Bala and colleagues created Co_3O_4/TiO_2 photocatalyst hydrogen evolution using Co-based MOF as the starting material [164]. In UV light, 2 wt% Co_3O_4 showed a reactivity of 7 $\mu molg^{-1}h^{-1}$. The fabrication of a photocatalyst and the inclusion of Co_3O_4 as a co-catalyst amplify the charge transport as well as charge carrier segregation, which results in the enhanced photocatalytic efficacy. Analog to this, Mondal and Pal developed a nanocomposite using Cu-based MOF as a sacrifice framework [165]. Because of the development of a narrow heterojunction as well as Cu insertion to the TiO_2 lattice, the optimized $Cu/CuO/TiO_2$ mixed composite achieved a yield of 286 $\mu molg^{-1}h^{-1}$ beneath sun light that was substantially higher compared to a standard CuO/TiO_2 mixed framework. Dekrafft et al. synthesized a $Fe_2O_3@TiO_2$ morphology using a Fe-based metal-organic frame prior to loading Pt upon this interface to carry out a PHE reaction [166].

The composite material demonstrated a significantly higher efficiency of hydrogen production than Fe_2O_3, TiO_2, and related blends. Zhang and coworkers fabricated FePc@Ni–MOF, which is employed as working electrode achieves efficient HER activity owing to strong interaction between FePc and 2D Ni–MOF [167]. Do and coworkers fabricated MoS_x/Ni-MOF-74 which shows the low overpotential of −114 mV in acidic media and 53.1 $mVdec^{-1}$ of a desirable Tafel slope. Here, the NiMoS phase will lower the electrocatalyst's energy required for H_2 evolution [168]. Additionally, Ni-MOFs have the ability to generate a carbon-based structure during high-temperature degradation. Furthermore, high-temperature reduction of Ni-MOFs can result in the formation of a carbon-based structure. Metals can facilitate the graphitization of the carbon framework. As a result, it is possible to increase conductance and stop the metal ion from aggregating and corroding. In line with this, Ni-MOF was proficiently annealed at 300°C by Shao et al. to synthesize NiO/graphite carbon composite materials [169].

Owing to their poor conductance and fragility under basic and acidic conditions, MOFs have often not been used for HER catalysis. Several operations have been done in this respect to enhance the photocatalytic efficiency of unprocessed MOFs for HER without the need for post-treatment. It involves combining MOFs with remarkably conducting materials like graphene oxide, graphene, and others, as well as MOF coupling with nanomaterials. The Cu-MOF nanocatalyst for HER was described by Jahan et al. The catalyst's capacity for HER was measured in an acidic media and was found to be 0.209 V. HER at a current density of 30 mAcm^{-2} and 84 mVdec^{-1} for the Tafel slope. The standard catalyst, 20% Pt/C with a voltage of 0.058 V vs. RHE and a Tafel slope of 30 mVdec^{-1}, exhibits similar properties to these ones. The results revealed that the current density of 30 mAcm^{-2} is lower than the standard value of 10 mAcm^{-2} [170]. It was observed that the type of metal used in the pure MOF can affect the catalytic performance for HER. Additionally, the organic binder that is chosen has a significant impact on the whole MOF structure and that in turn has a significant effect on the MOF-based nanocatalysts for HER [171].

Ruthenium-based HER electrocatalyst loaded on hierarchical carbon materials was described by Qiu et al., with CuRu-MOF serving as a framework throughout the production process. When the Cu nanoparticles were eliminated, the finer Ru nanoparticles might be exposed and could contribute significantly to the HER performance. Interestingly, Ru-HPC achieved a current density of 25 mAcm^{-2} and a turnover rate of 1.79 H$_2$ S^{-1} at 25 mV, which is almost twice as high as that of the commercial Pt/C. Additionally, Ru-HPC showed a maximum reactive site concentration of 0.390×10×10^{-3} mol gmetal^{-1} and an ECSA of 385.57 m^{2}gmetal^{-1}, which supported the idea that the Ru reactive sites were exposed to a great deal of electrocatalysis [172]. MOFs are one of the most efficient multifunctional nanoparticles because of their changing morphology, high surface area, modifications in composition and porosity.

7.10 CONCLUSION

The greener generation of hydrogen as an alternate renewable fuel as well as energy carrier entails the preparation of high-performance and affordable catalysts. Persistent advances in the development of metal oxide-based photocatalysts for effective photocatalytic and photochemical hydrogen evolution had already been accomplished over the last several decades, revitalizing curiosity as well as contradicting the conventional perspective of hydrogen evolution inactive metal oxide nanoparticles by transforming them to prospective HER photocatalysts. Our extensive and relevant survey of metal oxide-based photocatalysts, transition metal oxides, metal chalcogenides, MOFs, and specially structured oxides and doped metal oxides for HER is intended to pique fascination in additional exploration in the field.

Owing to the combinatorial, electrical, and supporting effects, hybrids consisting of metal oxides and other photoactive substances have also been consistently reported as a series of efficient HER photocatalysts. Many of the investigated photocatalysts exhibit performance that is comparable to the standard Pt-based catalysts. In particular, methods to improve the electronic conductivity, optimize the electrical properties, as well as increase the number of highly reactive active sites in metal oxide nanoparticles are addressed in order to improve HER photocatalytic activity. Doping, phase-structure construction, deformation design, morphological construction, crystalline phase construction, valence control, compositional improvement, and hybridization are some of the proposed successful approaches discussed in this chapter. Despite the significant advancements achieved in this area, there remain a number of difficulties. Further work on metal oxide-based photocatalysts should, in our perspective, concentrate on the following.

- It is expected to develop an appropriate method to increase the optimum photocatalytic performance to develop specific surfaces in hybrid photo-catalytic mechanisms with various individual elements to generate optimal interactions including p-n junctions, heterojunctions, and Z-scheme nanomaterials.
- Despite the fact that photo electro-catalytic studies have received a considerable attention, there is still some barrier among laboratory and commercial use. Laboratory tests could only demonstrate the viability and functionality of the photocatalytic process; they cannot contribute for costs, catalyst recycling, energy demand, ecological sustainability, or other difficulties. Nevertheless, there are a number of uncontrolled aspects in the real manufacturing technique in the context of boosting commercial applications; thus, the catalyst synthesis conditions won't be as steady and under control as in the laboratory. To accomplish the commercial use of photocatalysts, consequently, sizable manufacturing procedures that are affordable, practical, and reliable must be developed.
- The prevalent perception of photocatalysts is one of being pristine, greener, using little energy, and being environmentally friendly. Yet, it cannot be denied that we frequently introduce electrons or hole sacrificial reagents to the photocatalytic reaction, including sodium sulfate, sodium sulfide, lactic acid, triethanolamine, and different alcohols. The utilization of such scavengers can significantly increase the photocatalytic performance; however, it also leads to unfavorably damaged products and ecosystem. Efforts need to be devoted to develop environmentally benign strategies.

REFERENCES

1. H. T. Pao, C. M. Tsai, H. T. Pao, C. M. Tsai, *Energy Policy*, 38 (2010) 7850. 10.1016/j.enpol.2010.08.045.
2. S. J. Davis, K. Caldeira, *Proc. Natl. Acad. Sci.*, 107 (2010) 5687. 10.1073/pnas.0906974107
3. D. Dodman, *Environ. Urban.*, 21 (2009) 185. 10.1177/095624780910301
4. J. Byrne, K. Hughes, W. Rickerson, L. Kurdgelashvili, *Energy Policy*, 35 (2007) 4555. 10.1016/j.enpol.2007.02.028
5. S. Solomon, G.-K. Plattner, R. Knutti, P. Friedlingstein, *Proc. Natl. Acad. Sci.*, 106 (2009) 1704. 10.1073/pnas.0812721106
6. P. R. None, H. B, G. V. MA, *Nature*, 406 (2000) 173.
7. E. Barbier, *Renew. Sustain. Energy Rev.*, 6 (2002) 3. 10.1016/S1364-0321(02)00002-3
8. I. Dinçer, C. Zamfirescu, *Sustain. Energy Syst. Appl.*, 1 (2011), 1–810.
9. B. Parida, S. Iniyan, R. Goic, B. Parida, S. Iniyan, R. Goic, *Renew. Sustain. Energy Rev.*, 15 (2011) 1625. 10.1016/j.rser.2010.11.032

10. W. T. Xie, Y. J. Dai, R. Z. Wang, K. Sumathy, W. T. Xie, Y. J. Dai, R. Z. Wang, K. Sumathy, *Renew. Sustain. Energy Rev.*, 15 (2011) 2588. 10.1016/j.rser.2011.03.031

11. C. Zamfirescu, I. Dincer, G. F. Naterer, R. Banica, *Chem. Eng. Sci.*, 97 (2013) 235. 10.1051/e3sconf/202130901032

12. J. A. Turner, *Science*, 305 (2004) 972. 10.1126/science.1103197

13. E. P. Melian, O. G. Diaz, A. O. Mendez, C. R. Lopez, M. N. Suarez, J. M. D. Rodriguez, J. A. Navio, D. F. Hevia, J. P. Pena, *Int. J. Hydrog. Energy*, 38 (2013) 2144. 10.5757/ASCT.2018.27.4.61

14. J. Zhu, M. Zäch, *Curr. Opin. Colloid Interface Sci.*, 14 (2009) 260.

15. Y. Wang, H. Suzuki, J. Xie, O. Tomita, D. J. Martin, M. Higashi, D. Kong, R. Abe, J. Tang, *Chem. Rev.*, 118 (2018) 5201–5241. 10.1021/acs.chemrev.7b00286

16. S. R. Lingampalli, C. N. R. Rao, in: S. Ghosh (Ed.), *Visible Light Active Photocatalysis*, (2018) pp. 365–391.

17. C. N. R. Rao, S. R. Lingampalli, *Small*, 12 (2016) 16–23. 10.1002/smll.201500420

18. K. H. Fujishima, *Nature*, 238 (1972) 37–38. 10.1038/238037a0

19. U. Maitra, S. R. Lingampalli, C. N. R. Rao, *Curr. Sci.*, 106 (2014) 518–527. https://www.jstor.org/stable/24100059.

20. Y. Wang, A. Vogel, M. Sachs, R. S. Sprick, L. Wilbraham, S. J. A. Moniz, R. Godin, M. A. Zwijnenburg, J. R. Durrant, A. I. Cooper, J. Tang, *Nat. Energy*, 4 (2019) 746–760. 10.1039/D1TA03098A

21. R. P. Fujishima, awar, C. S. Lee, *Heterogeneous Nanocomposite-Photocatalysis for Water Purification*, first ed., William Andrew, Elsevier, 2015 eBook ISBN: 9780323393133.

22. S. Suib, New and future developments in catalysis, in: *Solar Photocatalysis*, first ed., Elsevier, 2013 eBook ISBN: 9780444538734.

23. K. Maeda, K. Teramura, D. Lu, T. Takata, N. Saito, Y. Inoue, K. Domen, *Nature*, 440 (2006) 295. 10.1038/440295a

24. L. Palmisano, G. Marci, Heterogeneous photocatalysis, in: *Relationships with Heterogeneous Catalysis and Perspectives*, first ed., Elsevier, 2019. ISBN: 9780444640154.

25. K. Maeda, *ACS Catal.*, 3 (2013) 1486–1503.

26. N. Fajrina, M. Tahir, *Int. J. Hydrog. Energy*, 44 (2019) 540–577. 10.1016/j.ijhydene.2018.10.200

27. X. Li, J. Yu, J. Low, Y. Fang, J. Xiao, X. Chen, *J. Mater. Chem.*, 3 (2015) 2485–2534. 10.1039/C4TA04461D.

28. S. J. A. Moniz, S. A. Shevlin, D. J. Martin, Z.-X. Guo, J. Tang, *Energy Environ. Sci.*, 8 (2015) 731–759. 10.1039/C4EE03271C

29. T. Banerjee, K. Gottschling, G. Savasci, C. Ochsenfeld, B. V. Lotsch, *ACS Energy Lett.*, 3 (2018) 400–409. 10.1021/acsenergylett.7b01123

30. M. A. Fox, M. T. Dulay, *Chem. Rev.*, 93 (1993) 341–357. 10.1021/cr00017a016

31. A. Kudo, Y. Miseki, *Chem. Soc. Rev.*, 38 (2009) 253–278. 10.1039/B800489G

32. Q. Xu, L. Zhang, J. Yu, S. Wageh, A. A. Al-Ghamdi, M. Jaroniec, *Mater. Today Off.*, 21 (2018) 1042–1063. 10.1016/j.mattod.2018.04.008

33. Z. Wang, C. Li, K. Domen, *Chem. Soc. Rev.*, 48 (2019) 2109–2125. 10.1039/C8CS00542G

34. J. K. Stolarczyk, S. Bhattacharyya, L. Polavarapu, J. Feldmann, *ACS Catal.*, 8 (2018) 3602–3635. 10.1021/acscatal.8b00791

35. W. Y. Gang, Z. X. Gang, *Electrochim. Acta*, 49 (2004) 1957–1962.

36. W. Zhou, Z. Yin, Z. Du, X. Huang, Z. Zeng, Z. Fan, H. Liu, J. Wang, H. Zhang, *Small*, 9 (2013) 140–147. 10.1002/smll.201201161 (2013) 140-147.

37. B. Yar, T. Haspulat, V. Üstün, A. Eskizeybek, H. Avcı, S. Kamı Achour *RSC Adv.*, 7 (2017) 29806–29814. 10.1039/C7RA03699J

38. H. Chen, M. Tang, Z. Rui, X. Wang, H. Ji, *Catal. Today*, 264 (2016) 23–30. 10.1016/j.cattod.2015.08.024
39. I. Rossetti, A. Villa, M. Compagnoni, L. Prati, G. Ramis, C. Pirola, C. Bianchi, W. Wang, D. Wang, *Catal. Sci. Technol.*, 5 (2015) 4481–4487. 10.1039/C5CY00756A
40. S. P. Hong, J. S. M. Park, S. Bhat, T. H. Lee, S. A. Lee, K. Hong, M. J. Choi, M. Shokouhimehr, H. W. Jang, *Cryst. Growth Des.*, 18 (2018) 6504–6512. 10.1021/acs.cgd.8b00609
41. A. F. Alkaim, T. A. Kandiel, F. H. Hussein, R. Dillert, D. W. Bahnemann, *Appl. Catal. A* 466 (2013) 32–37. 10.1016/j.apcata.2013.06.033
42. Y. K. Kho, A. Iwase, W. Y. Teoh, L. Madler, A. Kudo, R. Amal, *J. Phys. Chem. C*, 114 (2010) 2821–2829. 10.1021/jp910810r
43. M. Asjad, M. Arshad, N. A. Zafar, M. A. Khan, A. Iqbal, A. Saleem, A. Aldawsari, *Mater. Chem. Phys.*, 265 (2021) 124416. 10.1016/j.matchemphys.2021.124416.
44. H. Sutiono, A. M. Tripathi, H. M. Chen, C. H. Chen, W. N. Su, L. Y. Chen, H. Dai, B. J. Hwang, *ACS Sustain. Chem. Eng.*, 11 (2016) 5963–5971. 10.1021/acssuschemeng.6b01066
45. M. Chandra, D. Pradhan, *ChemSusChem*, 13 (2020) 3005–3016. 10.1002/cssc.202000308
46. X. Yang, X. Wu, J. Li, Y. Liu, *RSC Adv.*, 9 (2019) 29097. 10.1039/c9ra05113a
47. C. Beasley, M. K. Gnanamani, E. S. Jimenez, M. Martinelli, W. D. Shafer, S. D. Hopps, N. Wanninayake, D. Y. Kim, *ChemistrySelect*, 5 (2020) 1013–1019. 10.1002/slct.201904151
48. N. A. Erfan, M. S. Mahmoud, H. Y. Kim, N. A. M. Barakat, *PLoS ONE*, 17 (2022) 0276097. 10.1371/journal.pone.0276097
49. C. Imparato, G. Levolino, M. Fantauzzi, C. Koral, W. Macyk, M. Kobielusz, G. D. Errico, I. Rea, R. D. Girolamo, L. D. Stefano, A. Andreone, V. Vaino, A. Rossi, A. Aronne, *RSC Adv.*, 10 (2020) 12519. 10.1039/d0ra01322f
50. L. Wu, S. Shi, Q. Li, X. Zhang, X. Cui, *Int. J. Hydrog. Energy*, 44 (2019) 720–728. 10.1016/j.ijhydene.2018.10.214
51. K. Lalitha, V. D. Kumari, M. Sunramanyam, *Indian J. Chem.*, 53 (2014) 472–477.
52. M. Ismael, *J. Environ. Chem. Eng.*, 8 (2020) 103676. 10.1016/j.jece.2020.103676
53. M. Navarrete, S. C. Diaz, R. Gomez, *J. Chem. Technol. Biotechnol.*, 94 (2019) 3457–3465. 10.1002/jctb.5967
54. N. L. Reddy, S. Emin, V. D. Kumari, S. M..Venjatakrishnan, *Ind. Eng. Chem. Res.*, 57 (2018) 568–577. DOI: 10.1021/acs.iecr.7b03785
55. X. Y. Zhang, H. P. Li, X. L. Cui, Y. Lin, *J. Mater. Chem.*, 20 (2010) 2801–2806.
56. J. Corredor, M. J. Rivero, I. Ortiz, *Int. J. Hydrog. Energy*, 46 (2021) 17500–17506. 10.1016/j.ijhydene.2020.01.181
57. S. Bala, I. Mondal, A. Goswami, U. Pal, R. Mondal, *J. Mater. Chem. A*, 3, 20288–20296.
58. R. Mondal, U. Pal, *Phys. Chem. Chem. Phys.*, 18 (2016) 4780–4788. 10.1039/C5CP06292F.
59. K. E. Dekrafft, C. Wang, W. Lin, *Adv. Mater.* 24 (2012) 2014–2018. 10.1002/adma.201200330
60. Y. Zhu, Q. Ling, Y. Liu, H. Wang, Y. Zhu, *Phys. Chem. Chem. Phys.*, 17 (2015) 933–940. 10.1039/C4CP04628E.
61. S. Rhatigan, L. Niemitz, M. Nolan, *J. Phys. Energy*, 3 (2021) 025001. 10.1088/2515-7655/abe424
62. A. Chalgin, W. Chen, Q. Xiang, F. Yi Wu, F. Li, C. Shi, P. Song, W. Tao, Shang, *J. ACS Appl. Mater. Interfaces*, 12 (2020) 27037–27044. 10.1021/acsami.0c03742.

63. S. Parmar, T. Das, B. Ray, B. Debnath, S. Gosavi, G. S. Shanker, S. Datar, S. Chakraborty, S. Ogale, *Adv. Energy Sustain. Res.*, 3 (2022) 2100137. 10.1002/aesr. 202100137.

64. Y. K. Kho, A. Iwase, W. Y. Teoh, L. Madler, A. Kudo, R. Amal, *J. Phys. Chem. C*, 114 (2010) 2821–2829. 10.1021/jp910810r.

65. A. T. Montoya, E. G. Gillan, *ACS Omega*, 3 (2018) 2947–2955. 10.1021/acsomega.7b02021

66. S. Kampouri, C. P. Ireland, B. Valizadeh, E. Oveisi, P. A. Schouwink, M. Mensi, K. C. Stylianou, *ACS Appl. Energy Mater.*, 11 (2018) 6541–6548. 10.1021/acsaem.8b01445.

67. M. Chandra, D. D. Pradhan, *ChemSusChem*, 13 (2020) 3005–3016. 10.1002/cssc. 202000308.

68. M. H. Suhag, I. Tateishi, M. Furukawa, H. Katsumata, *J. Compos. Sci.*, 6 (2022) 327. 10.3390/jcs6110327.

69. F. Gao, Z. Xu, H. Zhao, Proc. Combust. Instit., 38 (2021) 6503–6511. 10.1016/j.proci.2020.06.330.

70. X. Huang, M. Zhang, R. Sun, G. Long, Y. Liu, W. Zhao, *Natl. Center Biotechnol. Inf.* ,14 (2019) 0215339. 10.1371/journal.pone.0215339.

71. Z. Tahira, R. Ibupoto, S. Mazzaro, V. You, M. M. Morandi, M. Natile, A. Vagin, Vomiero, *ACS Appl. Energy Mater.*, 2 (2019) 2053–2062. 10.1021/acsaem.8b02119.

72. R. Li, B. Hu, T. Yu, Z. Shao, Y. Wang, S. Song, Electrochemical Transformation of Renewable Compounds5 (2021) 2100246. 10.1002/smtd.202100246.

73. E. Khorashadizade, S. Mohajernia, S. Hejazi, H. Mehdipour, N. Naseri, O. Moradlou, A. Z. Moshfegh, P. Schmuki, *J. Phys. Chem. C*, 125 (2021) 6116–6127. 10.1021/acs.jpcc.1c00459.

74. E. Hussain, I. Majeed, M. A. Nadeem, A. Badshah, Y. Chen, M. A. Nadeem, R. Jin, *J. Phys. Chem. C*, 120 (2016) 17205–17213. 10.1021/acs.jpcc.6b04695.

75. P. M. Leukkunen, E. Rani, A. A. Sasikala Devi, H. Singh, G. King, M. Alatalo, W. Cao, M. Huttulaac, *RSC Adv.*, 60, 10, (2020) 36930–36940. 10.1039/D0RA07078E.

76. Q. Zhu, N. Liu, Q. Ma, A. Sharma, D. Nagai, X. Sun, C. Zhang, Y. Yang, *Mater. Today: Proc.*, 20, (2021) 100648. 10.1016/j.mtener.2021.100648.

77. H. Yang, H. Amari, L. Liu, C. Zhao, H. Gao, A. He, N. D. Browning, M. A. Little, R. S. Sprick, A. I. Cooper, *Nanoscale*, 12 (2020) 24488–24494. 10.1039/D0NR05 801G.

78. W. Zhao, Z. Chen, X. Yang, X. Qian, C. Liu, D. Zhou, T. Suna, M. Zhang, G. Wei, P. D. Dissanayake, Y. S. Ok, *Renew. Sust. Energ. Rev.*, 132 (2020) 110040. 10.1016/j.rser.2020.110040.

79. N. Romero, R. L. Gil, S. Drouet, I. Sanchez, O. Illa, K. Philippot, M. Natali, J. G. Anton, X. Sal, *Sustain. Energy Fuels*, 8 (2020) 4170–4178. 10.1039/D0SE00446D.

80. K. M. Emran, *Int. J. Electrochem. Sci.*, 15 (2020) 4218–4231. 10.20964/2020.05.02.

81. D. Regonini, A. C. Teloeken, A. K. Alves, F. A. Berutti, K. Gajda-Schrantz, C. P. Bergmann, T. Graule, F. Clemens, *ACS Appl. Mater. Interfaces*, 5 (2013) 11747–11755. 10.1021/am403437q.

82. A. Kushwaha, R. S. Moakhar, G. K. L. Goh, G. K. Dalapati, *J. Photochem. Photobiol. A: Chem.*, 337 (2017)54–61. 10.1016/j.jphotochem.2017.01.014.

83. R. Moakhar, S. Morteza, H. Hosseinabad, S. Panah, A. Seza, M. Jalali, H. Fallah-Arani, F. Dabir, S. Gholipour, Y. Abdi, M. B. Hariri, N. R. Noori, Y. Lim, A. Hagfeldt, M. Saliba, *Adv. Mater.*, 33 (2021) 2007285. 10.1002/adma.202007285.

84. Y. Yang, D. Xu, Q. Wu, P. Diao, *Sci. Rep.*, 6 (2016) 35158. 10.1038/srep35158.

85. A. A. Dubale, A. G. Tamirat, H. M. Chen, T. A. Berhe, C. J. Pan, W. N. Su, B. J. Hwang, *J. Mater. Chem. A*, 4 (2016) 2205–2216. DOI 10.1039/C5TA09464J.

86. A. Tahira, Z. H. Ibupoto, M. Willander, O. Nur, *Int. J. Hydrog. Energy*, 44 (2019) 26148–26157. 10.1016/j.ijhydene.2019.08.120.
87. Q. Zhang, B. Zhai, Z. Lin, X. Zhao, P. Diao, *Int. J. Hydrog. Energy*, 46 (2021) 11607–11620. 10.1016/j.ijhydene.2021.01.050.
88. P. Muthukumar, P. S. Kumar, S. P. Anthony, *Mater. Res. Express*, 6 (2019) 025518. 10.1088/2053-1591/aaf204.
89. C. G. M. Guio, S. D. Tilley, H. Vrubel, M. Grätzel, X. Hu, *Nat. Commun.*, 5 (2014) 3059. 10.1038/ncomms4059.
90. V. M. Sofianosa, J. L. Debbie, S. SilvestercPralok, K. SamantabMark, P. Niall, J. EnglishbCraig E. Buckleya, *J. Energy Chem.*, 56, (2021) 162–170. 10.1016/j.jechem.2020.07.051.
91. H. Abdullah, D. H. Kuo, X. Chen, *Int. J. Hydrog. Energy*, 42 (2017) 5638–5648. 10.1016/j.ijhydene.2016.11.137.
92. A. B. Patil, B. D. Jadhav, P. V. Bhoir, *Mater. Renew. Sustain. Energy*, 14 (2021) 954. 10.1007/s40243-021-00199-5.
93. J. Q. Bermejo, S. G. Dali, R. Karthik, R. Canevesi, M. T. Izquierdo, M. Emo, A. Celzard, V. Fierro, *Front. Energy Res.*, 10 (2022) 901395. 10.3389/fenrg.2022.901395.
94. T. Ling, T. Zhang, B. Ge, L. Han, L. Zheng, F. Lin, Z. Xu, W. B. Hu, X. W. Du, K. Davey, S. Z. Qiao, *Adv. Mater. Lett.* 16 (2019) 1807771. 10.1002/adma.201807771.
95. N. Wang, B. Tao, F. Miao, Y. Zang, *RSC Adv.*, 9 (2019) 33814–33822. 10.1039/C9RA05335B.
96. Y. Wang, H. Ping, T. Tan, W. Wang, P. Mab, H. Xie, *RSC Adv.*, 9 (2019) 28165–28170. 10.1039/C9RA04975D.
97. P. M. Pataniya, D. Late, C. K. Sumesh, *ACS Appl. Energy Mater.*, 4 (2021) 755–762. 10.1021/acsaem.0c02608.
98. B. Huner, M. Farsak, E. Telli, *Int. J. Hydrog. Energy*, 15 (2018) 7381–7387. 10.1016/j.ijhydene.2018.02.186.
99. W. C. Tseng, C. W. Chang, C. C. Kaun, Y. H. Su, *Int. J. Hydrog. Energy*, 47 (2021) 40768–40776. 10.1016/j.ijhydene.2021.09.173.
100. C. K. Sumesh, *Int. J. Hydrog. Energy*, 45 (2020) 619–628. 10.1016/j.ijhydene.2019.10.235.
101. N. Alhokbany, T. Ahamad, S. M. Alshehri, J. Ahmed, *Catalysts*, 12 (2022) 530. 10.3390/catal12050530.
102. H. Abdullah, D. H. Kuo, X. Chen, *Int. J. Hydrog. Energy*, (2016) 1 -1 1. 10.1016/j.ijhydene.2016.11.137.
103. B. Patil, B. D. Jadhav, P. V. Bhoir, *Mater. Renew. Sustain. Energy*, 10 (2021) 14. 10.1007/s40243-021-00199-5.
104. Y. Wang, H. Ping, T. Tan, W. Wang, P. Ma, H. Xie, *RSC Adv.*, 9 (2019) 28165. 10.1039/C9RA04975D.
105. I. Ahmad, E., Ahmed, M. Ahmad, *SN Appl. Sci.*, 1 (2019) 327 | 10.1007/s42452-019-0331-9.
106. A. Irshada, A. Ejaza, M. Ahmad, M. S. Akhtar, M. A. Basharat, W. Q. Khan, M. I. Ghauri, A. Ali, M. F. Manzoor, *Mater. Sci. Semicond. Process*, 105 (2020) 104748. 10.1016/j.mssp.2019.104748.
107. Y. J. Yuan, F. Wang, B. Hu, H. W. Lu, Z. T. Yu, Z. G. Zou, *Dalton Trans.*, 44 (2015) 10997. 10.1039/c5dt00906e.
108. B. Archana, K. Manjunath, G. Nagaraju, K. B. Chandrasekhar, N. Kottam, *Int. J. Hydrog. Energy*, xxx (2016) 1–7. 10.1016/j.ijhydene.2016.11.099.
109. A. Wu, L. Jing, J. Wang, Y. Qu, Y. Xie, B. Jiang, C. Tian, H. Fu, *Sci. Rep.*, 5 (2015) 8858. 10.1038/srep08858.

110. Y. C. Chang, S. Y. Syu, Z. Y. Wu, *Mater. Lett.*, 302 (2021) 130435. 10.1016/j.matlet.2021.130435.

111. I. Ahmad, M. S. Akhtar, E. Ahmed, M. Ahmad, M. Y. Naz, *J. Colloid Interface Sci.*, 584 (2021) 182–192. 10.1016/j.jcis.2020.09.116.

112. S. Kumar, A. Kumar, V. N. Rao, A. Kumar, M. V. Shankar, V. Krishnan, *ACS Appl. Energy Mater.*, 8 (2019) 5622–5634. 10.1021/acsaem.9b00790.

113. Y. R. Girish, Udayabhanua, N. M. Byrappa, G. Alnaggar, A. Hezamd, G. Nagaraju, K. Pramoda, K. Byrappa, *J. Hazardous Mater. Adv.*, 9 (2023) 100230. 10.1016/j.hazadv.2023.100230.

114. B. Zhang, Q. Li, D. Wang, J. Wang, B. Jiang, S. Jiao, D. Liu, Z. Zeng, C. Zhao, Y. Liu, Z. Xun, X. Fang, S. Gao, Y. Zhang, L. Zhao, *Nanomaterials*, 11 (2020) 2096. 10.3390/nano10112096.

115. P. C. Nethravathi, D. Suresh, *Inorg. Chem. Commun.*, 134 (2021) 109051. 10.1016/j.inoche.2021.109051.

116. Z. Wu, J. Mei, Q. Liu, S. Wang, W. Li, S. Xing, J. Bai, J. Yang, W. Luo, O. Guselnikova, A. P. O. Mullane, Y. Gu, Y. Yamauchi, T. Liao, Z. Sun, *J. Mater. Chem. A*, 10 (2022) 808–817. 10.1039/D1TA09019D.

117. Z. Wu, T. Liao, S. Wang, J. A. Mudiyanselage, A. S. Micallef, W. Li, A. P. O. Mullane, J. Yang, W. Luo, K. Ostrikov, Y. Gu, Z. Sun, *Nano-Micro Lett.* 14 (2022) 90 10.1007/s40820-022-00832-6.

118. R. Syah, A. Ahmad, A. Davarpanah, M. Elveny, D. Ramdan, M. D. Albaqami, M. Ouladsmane, *Catalysts*, 11 (2021) 1099. 10.3390/catal11091099.

119. F. Fu, H. Shen, W. Xue, Y. Zhen, R. A. Soomro, X. Yang, D. Wang, B. Xu, R. Chi, *J. Catal.*, 375 (2019) 399–409. 10.1016/j.jcat.2019.06.033.

120. S. H. Hsieh, G. J. Lee, C. Y. Chen, J. H. Chen, S. H. Ma, T. L. Horng, K. H. Chen, J. J. Wu, *J. Nanosci. Nanotechnol.*,12 (2012) 5930–5936. 10.1166/jnn.2012.6396.

121. K. Chitrada, R. Gakhar, D. Chidambaram, E. Aston, K. Raja, *J. Electrochem. Soc.*, 163 (2016) 546–558. 10.1149/2.0721607jes

122. P. C. Nethravathi, M. V. Manjula, S. Devaraja, M. Sakar, D. Suresh, *J. Photochem. Photobiol. A: Chem.*, 435 (2023) 114295. 10.1016/j.jphotochem.2022.114295.

123. Y. Li, Y. Liu, Y. Hao, X. Wang, R. Liu, F. Li, *Mater. Des.*, 187 (2020) 108379. 10.1016/j.matdes.2019.108379.

124. M. Xie, Z. Zhang, W. Han, X. Cheng, X. Lia, E. Xie, *J. Mater. Chem. A*, 5 (2017) 10338–10346. 10.1039/C7TA01415E.

125. O. Monfort, L. C. Pop, S. Sfaelou, T. Plecenik, T. Roch, V. Dracopoulos, E. Stathatos, G. Plesch, P. Lianosa, *J. Chem. Eng.*, 286 (2016) 91–97. 10.1016/j.cej.2015.10.043.

126. M. B. Tahir, T. Iqbal, H. Kiran, A. Hasan, *Int. J. Energy Res.*, 43 (2019) 1–8. 10.1002/er.4443.

127. M. Jahan, Z. Liu, K. P. Loh, *Adv. Funct. Mater.*, 43 (2013) 5363–5372. 10.1002/adfm.201300510.

128. Y. P. Wu, W. Zhou, J. Zhao, W. W. Dong, Y. Q. Lan, D. S. Li, C. Sun, X. Bu, *Angew. Chem.*, 129 (2017) 13181–13185.

129. T. Qiu, Z. Liang, W. Guo, S. Gao, C. Qu, H. Tabassum, H. Zhang, B. Zhu, R. Zou, Y. S. Horn, *Nano Energy*, 58 (2019) 1 —10.

130. Molybdenum Disulfide: Understanding Hydrogen Evolution Catalysis Linyou Cao.

131. T. Yang, Y. Bao, W. Xiao, J..Zhou, J. Ding, Y. P. Feng, K. P. Loh, M. Yang, S. J. Wang, *ACS Appl. Mater. Interfaces*, 10 (2018) 22042–22049. 10.1021/acsami.8b03977.

132. C. A. C. Roldan, N. A. Vante, *J. Mex. Chem. Soc*, 63 (2019) 10.29356/jmcs.v63i3.533.

133. T. Wang, L. Liu, Z. Zhu, P. Papakonstantinou, J. Hu, H. Liuc, M. Li, *Energy Environ. Sci.*, 6 (2013) 625–633. 10.1039/C2EE23513G.

134. T. V. Nguyen, M. Tekalgne, T. P. Nguyen, W. Wang, S. H. Hong, J. H. Cho, Q. V. Le, H. W. Jang, S. H. Ahn, S. Y. Kim, *Int. J. Energy Res.*,46, (2022) 11479–11491. 10.1002/er.7896.

135. L. Chen, H. Feng, R. Zhang, S. Wang, X. Zhang, Z. Wei, Y. Zhu, M. Gu, C. Zhao, *ACS Appl. Nano Mater.*, 3 (2020) 6516–6523. 10.1021/acsanm.0c00988.

136. S. M. E. Refaei, P. A. Russo, T. Schultz, N. Koch, N. Pinna, *ChemCatChem*, 13 (2021) 4392–4402. 10.1002/cctc.202100856.

137. D. K. Singh, R. Naidu, S. Sampath, M. Eswaramoorthy, *J. Mater. Chem. A*, 5 (2017) 6025–6031. 10.1039/C6TA11057F.

138. X. Geng, W. Sun, W. Wu, B. Chen, A. A. Hilo, M. Benamara, H. Zhu, F. Watanabe, J. Cui, T. Chen, *Nat. Commun.*, 7 (2016) 10672. 10.1038/ncomms10672.

139. G. Ghanashyama, H. Kyung, G. K. Jeong, *Inorgan. Chim. Acta*, 541 (2022) 121098. 10.1016/j.ica.2022.121098.

140. A. Morozan, L. de. Chimie, B. d. Metaux, H. Johnson, C. Roiron, G. Genay, D. Aldakov, A. Ghedjatti, C. T. Nguyen, P. D. Tran, S. Kinge, V. Artero, *ACS Catal.* 10 (2020) 14336–14348. 10.1021/acscatal.0c03692.

141. C. Sun, P. Wang, H. Wang, C. Xu, J. Zhu, Y. Liang, Y. Su, Y. Jiang, W. Wu, E. Fu, G. Zou, *Nano Res.*, 12,(2019) 1613–1618. 10.1007/s12274-019-2400-1.

142. T. Ling, T. Zhang, B. Ge, L. Han, L..Zheng, F..Lin, Z..Xu, W..Hu, X..Du, K. Davey, S. Z..Qiao,*Adv. Mater.*, 31, (2019) 1807771. 10.1002/adma.201807771.

143. Y. N. Zhu,, X. B. Li, Q. Zhang, F. Peng, *ACS Appl. Mater. Interfaces* 14, (2022) 25592–25600. 10.1021/acsami.2c06698.

144. T. Wang, L. Liu, Z. Zhu, P. Papakonstantinou, J. Hu, H. Liuc, M. Li, *Energy Environ. Sci.*, 6, (2013) 625–633. 10.1039/C2EE23513G.

145. T. V. Nguyen, M. Tekalgne, T. P. Nguyen, W. Wang, S. H. Hong, J. H. Cho, Q. V. Le, H. W. Jang, S. H. Ahn, S. Y. Kim, *Int. J. Energy Res.* (2022) 1–13. 10.1002/er.789.

146. A. Morozan, H. Johnson, C. Roiron, G. Genay, D. Aldakov, A. Ghedjatti, C. T. Nguyen, P. D. Tran, S. Kinge,V. Artero, *ACS Catal.*, 10 (2020) 14336–14348. 10.1021/acscatal.0c03692.

147. M. L. Grutza, A. Rajagopal, C. Streb, P. Kurz, *Sustain. Energy Fuels*, 2 (2018) 1893–1904. 10.1039/C8SE00155C.

148. D. Zhang, F. Wang, W. Zhao, M. Cui, X. Fan, R. Liang, Q. Ou, S. Zhang, *Adv. Sci.*, 9, (2022) 2202445. 10.1002/advs.202202445.

149. S. Lu, M. Hummel, Z. Gu, Y. Gu, Z. Cen, L. Wei, Y. Zhou, C. Zhang, C. Yang, *Int. J. Hydrog. Energy*, 44, (2019) 16144–16153. 10.1016/j.ijhydene.2019.04.191.

150. A. Y. Faid, A. O..Barnett, F. Seland, S..Sunde, *J. Electrochem. Soc.* 166, (2019) F519–F533. 10.1149/2.0821908jes.

151. M. Gong, W. Zhou, M. C. Tsai, J. Zhou, M. Guan, M. C. Lin, B. Zhang, Y. Hu, D. Y. Wang, J. Yang, S. J. Pennycook, B. J. Hwang, H. Dai, *Nat. Commun.* 5, (2014) 4695. 10.1038/ncomms5695.

152. E. C. Lovell, X. Lu, Q. Zhang, J. Scott, R. Amal, *Chem. Commun.*, 56, (2020) 1709–1712. 10.1039/C9CC07486D.

153. S. N. Faisal, E. Haque, N. Noorbehesht, H. Liu, M. M. Islam, L. Shabnam, A. K. Roy, E. Pourazadi, M. S. Islam, A. T. Harrisa, A. I. Minett, *Sustain. Energy Fuels*, 2, (2018) 2081–2089. 10.1039/C8SE00068A.

154. L. Huo, C. Jin, K. Jiang, Q. Bao, Z. Hu, J. Chu, *Adv. Energy Sustain. Res.*, 3, (2022) 2100189. 10.1002/aesr.202100189.

155. Z. H. Ibupoto, A. Tahira, P. Yi Tang, X. Liu, J. R. Morante, M. Fahlman, J. Arbiol, M. Vagin, A. Vomier, *Adv. Funct. Mater.*, 29(2019) 1807562. 10.1002/adfm.201807562.

156. W. Changhao, L. Yahao, W. Xiuli, T. Jiangping, *J. Electron. Mater.*, 50 (2021) 5072–5080. 10.1007/s11664-021-09053-w.
157. A. Y. Faid, A. O. Barnett, F. Seland, S. Sunde, *J. Electrochem. Soc.* 166 (2019) F519–F533. IP address: 106.206.100.158.
158. D. Xiang, B. Zhang, H. Zhang, L. Shen, *Front. Chem.*, 14, (2021). 10.3389/fchem. 2021.773018.
159. H. H. Maghrabi, A. A. Nada, M. F. Bekheet, S. Roualdes, W. Riedel, I. Iatsunskyi, E. Coy, A. Gurlo, M. Bechelan, *J. Colloid Interface Sci.*, 587 (2021) 457–466. 10. 1016/j.jcis.2020.11.103
160. K. L. Zhou, Z. Wang, C. B. Han, X. Ke, C. Wang, Y. Jin, Q. Zhang, J. Liu, H. Wang, H. Yan, *Nat. Commun.* 12, (2021) 3783. 10.1038/s41467-021-24079-8.
161. J. Wang, S. Mao, Z. Liu, Z. Wei, H. Wang, Y. Chen, Y. Wang, *ACS Appl. Mater. Interfaces*, 9,(2017) 7139–7147. 10.1021/acsami.6b15377.
162. A. Karmakar, S. K. Srivastava, *J. Mater. Chem. A*, 7, (2019) 15054. 10.1039/c9ta02 884f.
163. Y. L. Liu, X. Y. Liu, L. Feng, L. X. Shao, S. J. Li, J. Tang, H. Cheng, Z. Chen, R. Huang, H. C. Xu, J. L. Zhuang, *ChemSuschem*,15, (2022) 202102603. 10.1002/cssc. 202102603.
164. H. H. Do, Q. V. Le, T. V. Nguyen, K. A. Huynh, M. A. Tekalgne, V. A. Tran, T. H. Lee, J. H. Cho, M. Shokouhimehr, S. H. Ahn, H. W. Jang, S. Y. Kim, *Int. J. Energy Res.*, 45, (2021) 9638–9647. 10.1002/er.6385.
165. B. Endrődi, S. S. V. Smulders, N. Simi, M. Wildlock, G. Mul, B. T. M. Cornella, *J. Clean. Prod.* 182, (2018) 529–537. 10.1016/j.jclepro.2018.02.071.
166. S. Swathi, Y. Kumara, S. Kumar, G. Ravia, T. Cuong, D. D. Velauthapillaie, *Fuel*, 310, (2022) 122319. 10.1016/j.fuel.2021.122319.
167. L. Zhang, X. Ren, X. Guo, Z. Liu, M. Abdullah, B. Li, L. Chen, X. Sun, *Inorg. Chem*, 57, (2018) 548–552. 10.1021/acs.inorgchem.7b02665.
168. M. S. Anantha, A. Dinesh, M. Kundu, M. Rani, *Mater. Sci. Eng. C*, 284 (2022) 115924. 10.1016/j.mseb.2022.115924.
169. Y. Xu, X. Hao, X. Zhang, T. Wang, Z. Hu, Y. Chen, X. Feng, W. Liu, F. Hao, X. Kong, C. He, S. Ma, B. Xua, *CrystEngComm*, 24, (2022) 3369–3379. 10.1039/D2 CE00209D.
170. D. Ji, C. Liu, Y. Yao, L. Luo, *Nanoscale*,13,9952, (2021). 10.1039/d1nr00069a.
171. Y. Song, Y. Pi, X. Feng, K. Ni, Z. Xu, J. S. Chen, Z. Li, W. Lin, *J. Am. Chem. Soc.*,142, (2020) 6866–6871. 10.1021/jacs.0c00679.

8 Emerging MXene-Derived Photocatalysts for Harvesting Solar Energy into Chemical Energy

Muhammad Tayyab, Umm E Kulsoom, Hummera Rafique, and Ehsan Ullah Mughal
East China University of Science and Technology, P. R. China

University of Gujrat, Pakistan

Summan Aman
University of Gujrat, Pakistan

Asim Mushtaq
Zhejiang Sci-Tech University, China

Laila Noureen
Peking University Shenzhen Graduate School, China

8.1 INTRODUCTION

Over the past few decades, the world's population has increased tremendously as compared to the available energy resources [1]. Due to the rapid consumption of fossil fuel resources, environmental pollution and global climate changes, all the living organisms have been adversely affected. The limited fossil fuel resources could not meet the ever-growing needs of population [2]. There is a dire need of developing efficient, durable and sustainable energy sources to overcome these issues [3]. Therefore, researchers have drawn their attention to produce such renewable energy resources which are sustainable and eco-friendly [4]. For inexhaustible and cleaner energy, solar energy has been regarded as the most promising energy resource [5]. Solar energy can be used in various photovoltaic technologies, photocatalytic water splitting (as shown in Figure 8.1b), nitrogen fixation, CO_2 reduction and degradation of numerous organic pollutants [6].

DOI: 10.1201/9781003364825-8

141

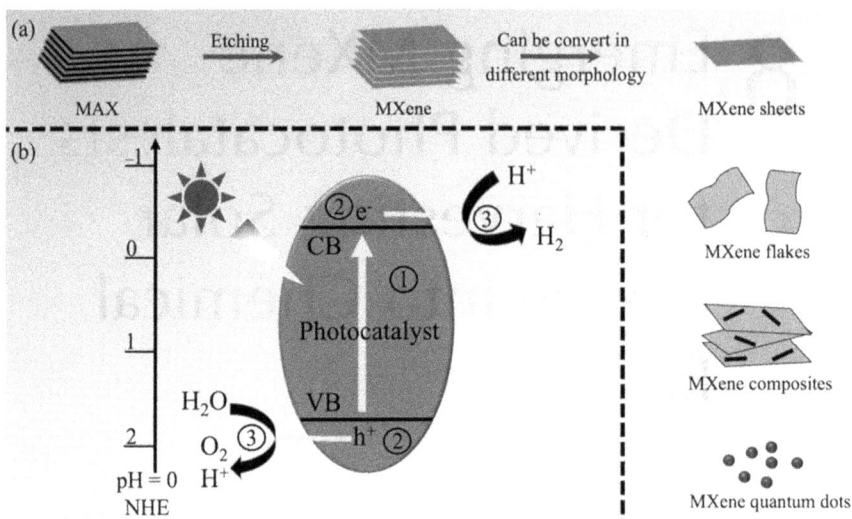

FIGURE 8.1   (a) General method for the synthesis of MXene from MAX and furthermore conversion of MXene into other different morphology composites. (b) A schematic illustration of the water splitting chemical reactions on the surface of the photocatalyst.

Since the discovery of graphene in 2004, two-dimensional (2D) materials have captivated the researchers' attention owing to their outstanding electronic, catalytic and semiconductor properties like MoS_2, $g\text{-}C_3N_4$, WS_2, black phosphorus and silicene [7]. In 2011, the discovery of MXene, a 2D nanomaterial having a general formula $M_{n+1}X_nT_x$ (transition metal carbides/nitrides), produced a gold rush among researchers due to its tremendous photoelectronic properties, unique 2D structure, robust carrier mobility, flexible elemental composition and larger surface area [8]. These unique properties allow MXene to play vital roles in the field of photocatalysis. Up till now, several MXene-based and MXene-derived 0D, 1D and 2D materials have been synthesized and used as efficient photocatalysts as shown in Figure 1.1(a). Surprisingly, they have turned out to be most advantageous and ideal photocatalysts.

MXene are most commonly fabricated by direct HF etching and indirect in-situ HF etching, i.e., by using a mixture of LiF and HCl. Furthermore, to enhance their yield, delamination with different organic intercalants (tetramethylammonium hydroxide, isopropylamine), chemical vapour deposition and high-temperature etching have been employed. The timeline of various etching methods being employed since the first discovery of MXene materials is presented in Figure 8.2 [9]. Owing to their robust characteristics, MXene have been used in other applications like supercapacitors, electrocatalysis, gas sensing and adsorbents. In addition to these applications, MXene have been preferred greatly as precursors or co-catalysts to drive off energy shortage crisis and environmental pollution. The role of MXene as a co-catalyst for photocatalytic reactions is graphically presented in Figure 8.3. $Ti_3C_2T_x$ was reported as the first MXene material fabricated for dye degradation via photocatalysis [10]. This material was initially synthesized by etching the Al layer of Ti_3AlC_2 with HF. Because of the presence of -OH groups or terminal oxygen species

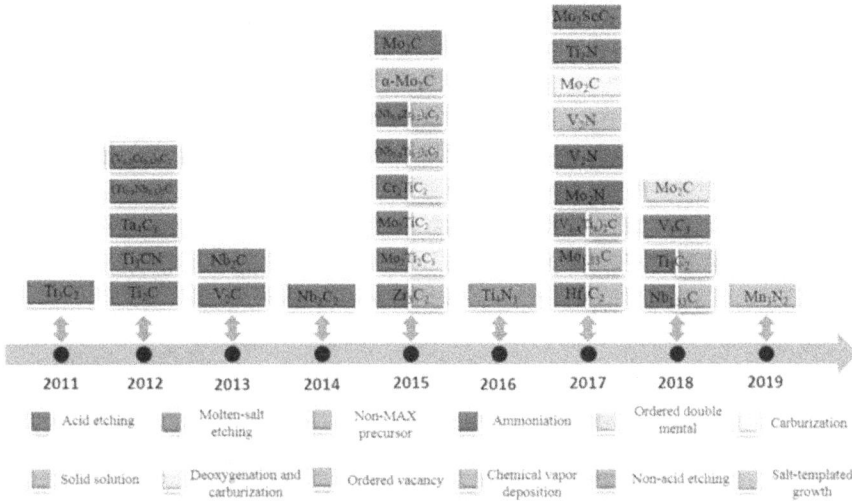

FIGURE 8.2 The timeline of various etching methods employed in MXene journey. Reproduced with permission from Ref. [9]. 2020, Elsevier.

on the surface of MXene, they have comparable conductivity towards transition metal carbides. Moreover, the MXene provide more channels for the ions to move through to enhance their speed, unlike conventional materials. Since then, greater than 20 types of MXene have been fabricated which include TiNbC, Y_2CF_2, Mo_2C, Ti_2CT_x, Nb_2C, $V_4C_3T_x$ and Ti_3CN [11,12].

No doubt, the research works on 2D MXene photocatalysts are promptly progressing in various fields including artificial photosynthesis like H_2 production or H_2O_2 production. MXene perform a multifunctional role, i.e., as a support and cocatalyst in the field of photocatalysis by facilitating charge transfer and separation, acting as a precursor of quantum dots and by providing abundant active sites.

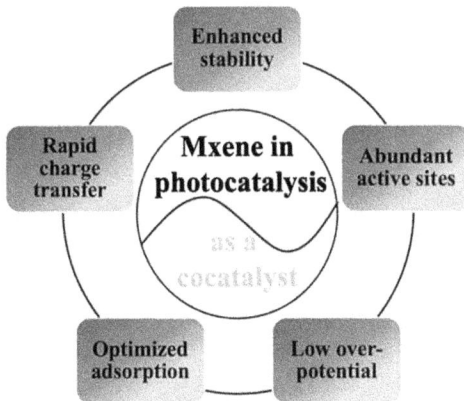

FIGURE 8.3 MXene as a cocatalyst in photocatalysis and its role for photocatalytic reactions.

Besides, MXene can enhance the activity of the catalyst as well as the rate of reaction due to their photothermal conversion characteristics.

8.2 MXENE

8.2.1 MAX PHASE TO MXENE

Generally, MXene are formed by the process of selective etching from MAX phase. A MAX phase comprises an abundance of ternary carbides or nitrides or borides. The "M" layer represents the early transition metal elements like Ti, Cr, Nb, etc. that exist in three forms, i.e., single metal element includes (Ti_2CT_x, $Nb_4C_3T_x$ and $Ti_3C_2T_x$) disordered bimetal solid solution (($Ti,Nb)_4C_3T_x$, $(Ti,V)_3C_2T_x$ and ordered bimetal solution ($Mo_2TiC_2T_x$) [13]. In Figure 8.4, the composed elements of MXene are shown by marked the elements in the periodic table. The schematic illustration of double-M MXene structure is shown in Figure 8.5 [14]. To date, more than 70 MAX phases have already been reported in literature. "A" denotes the IIIA/IVA group elements like Al, Si layers, whereas "X" represents C or N elements. Since the M-A bond is more active than M-X bond, the alumina layer can be removed selectively through etching from precursor phase. MXene are classified into three classes on the basis of the stoichiometry of MAX phase, namely M_2X_1, M_3X_2 and M_4X_3. Their classification is done on the basis of the

H		**M**	Early transition metal elements			**T**	Surface termination										He
Li	Be	**A**	Interleaved elements									**B**	**C**	**N**	**O**	**F**	Ne
Na	Mg	**X**	C and/or N									**Al**	**Si**	P	S	**Cl**	Ar
K	Ca	Sc	Ti	V	Cr	Mn	Fe	Co	Ni	Cu	Zn	Ga	Ge	As	Se	Br	Kr
Rb	Sr	Y	Zr	Nb	Mo	Tc	Ru	Rh	Pd	Ag	Cd	In	Sn	Sb	Te	I	Xe
Cs	Ba	Lu	Hf	Ta	W	Re	Os	Ir	Pt	Au	Hg	Tl	Pb	Bi	Po	At	Rn

FIGURE 8.4   The composed elements of MXene are head-lighted elements in the periodic table.

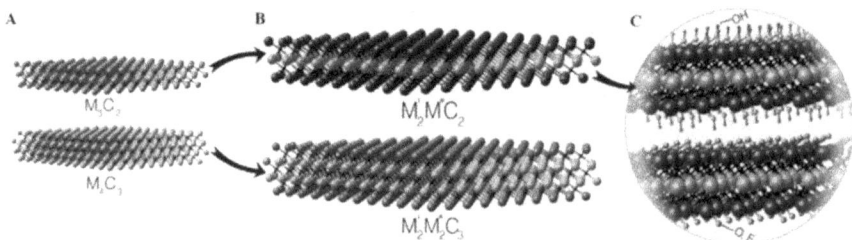

FIGURE 8.5   The schematic representation of double-M MXenes: (a) single-M MXenes, (b) double-M MXene and (c) MXene with surface groups. Reproduced with permission from [14]. 2015, ACS.

composition of metal layer "M". Zhou et al. reported the synthesis of Cr_2CT_x MXene from Cr_2AlC MAX phase via etching process [15].

Since the past few years, the synthesis of composites has been regarded as an efficient way to fabricate effective and durable photocatalysts. Because of the unique properties, MXene have been considered as promising materials to prepare efficient photocatalysts. At present, several MXene-based and MXene-derived photocatalysts have been synthesized. The main difference between these two mentioned photocatalysts is that in the MXene-derived photocatalysts a certain component is due to oxidation of MXene precursor. To prepare these photocatalysts, different synthetic routes have been employed. Among these methods are electrostatic self-assembly, calcination method, solvothermal or hydrothermal method and ultrasonic/mechanical mixing, high energy ball milling and wet chemical oxidation as shown in Figure 8.6.

The common preparation methods for MXene-based photocatalysts and MXene-derived photocatalysts are calcination and hydrothermal techniques. The main difference in their synthesis lies in the fact that MXene-based photocatalysts need controlled reaction conditions in order to prevent the oxidation of MXene. In general, electrostatic self-assembly and ultrasonic/mechanical mixing methods are

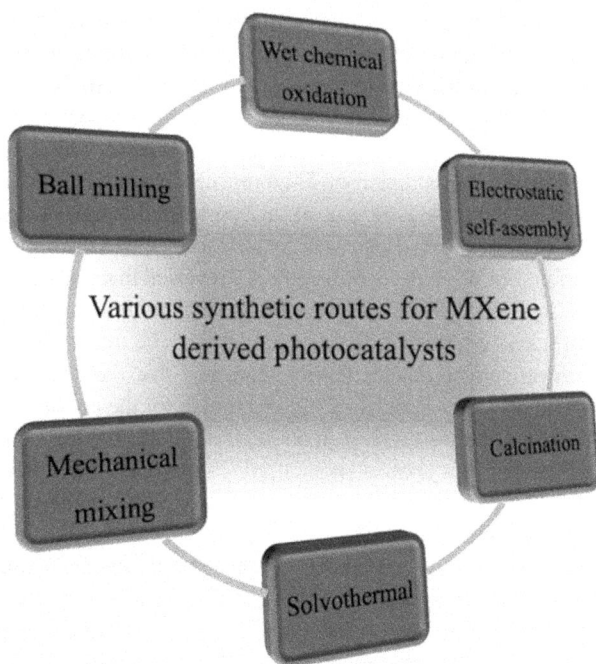

FIGURE 8.6 Various synthetic methods for MXene-derived photocatalysts.

much simpler and more reliable as compared to the calcination and solvothermal techniques while preparing MXene-based photocatalysts, whereas for MXene-derived photocatalysts, a characteristic feature of calcination oxidation imparts the residual presence of the carbon materials that may be used as a co-catalyst ultimately enhancing the visible-light adsorption and inhibiting recombination of charge carriers. We will discuss the emerging MXene-derived photocatalysts in this chapter and how they significantly contribute in converting solar energy into chemical energy.

8.2.2 MXENE-DERIVED PHOTOCATALYSTS

MXene-derived photocatalysts are synthesized by the in-situ oxidation of MXene precursors. Therefore, the most adapted strategies to synthesize these materials are calcination oxidation and hydrothermal routes. In recent years, TiO_2 has been considered as the most promising semiconductor photocatalyst due to its robust photostability, redox potential, eco-friendly nature and low cost. No doubt, it is very advantageous material, but alone TiO_2 still could not overcome some ambiguities such as higher photogenerated carriers' recombination rate and wide bandgap. Therefore, in order to overcome these drawbacks, researchers are trying to expand the light response range and boost the utilization rate of photogenerated carriers using multiple strategies such as co-catalyst loading, heterojunction construction and metal/non-metal doping [16].

Song et al. reported the in-situ construction of heterojunction over 2D Ti_3C_2 MXene precursor by developing intercalation strategy. For this purpose, the deep eutectic solvents (DES) method was adopted to get interlayer expansion of Ti_3C_2 and improve the Ti_3C_2-derived photocatalyst efficiency. Due to the DES method, the photocatalyst Ti_3C_2-DES possessed greater c-lattice parameter and intercalation with water molecules occurred significantly for the oxidation of Ti atoms, hence resulting in the remarkable in-situ growth of TiO_2 crystals [17]. Liu et al. constructed a ternary heterojunction $Ti_3C_2/TiO_2/BiOCl$ through the hydrothermal method to facilitate the separation rate of photoactivated carriers. SEM analysis of bare BOCl revealed that it is made up of thick lamellar, whereas $Ti_3C_2/TiO_2/BiOCl$ looked like a smaller lamellar in which a great number of the particles are adhered. These particles may be assigned as Ti_3C_3/TiO_2. Trapping and electron spin resonance (ESR) confirmed that the construction of heterojunction between TiO_2, BOCl and Ti_3C_2 is favourable for the production of O_2^- and OH radicals. This photocatalyst resulted in an increase in the level of active sites and improved separation efficiency of photoinduced carriers. Furthermore, the results confirmed that the ternary heterojunction exhibited much greater photocatalytic activity as compared to the bare pristine BOCl [18].

Wu et al. synthesized a highly photoactive Z-scheme graphene layered anchored $TiO_2/g-C_3N_4$ (GTOCN) photocatalyst from Ti_3C_2 MXene via a one-step in-situ calcination method. The graphene acted as an electron mediator in the Z-scheme and enhanced the degradation rate much higher than the single TiO_2 and $g-C_3N_4$ photocatalysts [2]. Wang et al. synthesized a novel 2D/3D MXene/CdS nanoflower composite via hydrothermal route. This combination of 2D Ti_3C_2 material with 3D CdS nanoflowers enhanced the charge separation capability, thus improving the performance of CdS. SEM analysis revealed the flower-like morphology of the novel nanocomposite. As compared to the bare CdS, 2D/3D MXene/CdS nanocomposite illustrated longer fluorescence lifetime and lower photoluminescence intensity with smaller impedance [19].

Tayyab et al. reported the synthesis of $In_2S_3/Nb_2O_5/Nb_2C$ Schottky/S-scheme heterojunction by a hydrothermal method. The partial oxidation of Nb_2C MXene onto Nb_2O_5 nanorods along with coupling of In_2S_3 by in-situ chemical anchorage were the prime factors behind the long durability and high efficiency of this ternary photocatalyst. The synergic effect of In_2S_3/Nb_2O_5 heterojunction and redox reaction at multiple sites of the composite facilitated the charge transfer and separation [1]. Besides, Kong et al. synthesized orderly layer-by-layer TiO_2/C superstructures (NPT-TiO_2/C) which were based on the defect engineering of Ti_3C_2 by a nitriding-pretreatment. The assembling of N into Ti_3C_2 not merely stabilized the 2D structure; rather it also enhanced the interlayer spacing by 5.1 Å which proved to be very beneficial for the subsequent intercalation of TiO_2 nanoplates through elevated temperature oxidation. Moreover, the results also confirmed that the intercalation superstructure may definitely enhance the number of active sites and improve the charge separation efficiency [20]. Jiang et al. reported a heterojunction $Nb_2O_5/C/Nb_2C/g$-C_3N_4 in which the Schottky junction and amorphous carbon were generated. This heterojunction was synthesized through the calcination of $Nb_2O_5/C/Nb_2C$ and melamine at an elevated temperature, i.e., 550°C under inert atmosphere. This heterostructure also proved to be fruitful in improving the charge transfer and carrier separation efficiencies [21]. It is a well-known fact that morphology of a photocatalyst plays a significant role in the performance of the catalyst. Likewise, Li et al. synthesized Ti_3C_2-TiO_2 nanoflowers through multi-step processes. Initially, $Na_2Ti_3O_7$-Ti_3C_2 composites were prepared under controlled hydrothermal conditions at temperature 140°C for approximately 12 hours. Subsequently, the as-prepared composites were formed by replacing Na^+ with H^+ in a dilute HCl solution. Finally, the Ti_3C_2-TiO_2 nanoflowers were fabricated after calcination at high temperatures [22].

8.2.3 PROPERTIES OF MXENE

In general, MXenes ($Ti_3C_2T_x$) show excellent conductivity with respect to electric storage performance. This is due to the fact that MXenes have wide spaces between their layers that allow the ions to move through them speedily. In addition, the presence of hydroxyl groups or terminal oxygen species on the surface enhances the conductivity of MXene. These two factors are essential for their use in photocatalysis. Similarly, several surface groups like (-OH, -F and -O) contributed a great number of anchored active sites for base photocatalyst to assemble to form a heterojunction in order to perform the ideal photocatalytic activities. A great number of exposed metal sites are also present on the surface that play the role of active sites in the reactions. The conductivity of the MXene is greatly linked to the morphology, etching and delamination process. For example, the conductivity of $Ti_3C_2T_x$ experienced a remarkable change when the preparation conditions were varied thus proving that the conductivity changes due to the contact and intercalated species present between the sheets [36].

Due to their surface chemical state, MXenes encounter a remarkable influence on their physical state. Because when the surface group -F is replaced by -O, it can be observed that the electrochemical performance is enhanced. As an example, when $Ti_3C_2T_x$ is treated with CH_3OOK and KOH, the number of -O surface groups increases which also increases the electric capacity; ultimately, the number of -F surface groups decreases under inert atmosphere, after which there occurs a great increase in electric capacity. Thus, it is clear that the electronic properties of MXene are associated with the terminal groups and the transition metal elements. The transformation of MXene from metals to semiconductors/topological insulators is due to change in the terminations. Mostly, MXene with surface terminations and naked MXene both show good metallic properties [37]. Figure 8.7 shows a general overview of MXenes in Schottky heterojunctions, all-solid-state Z-scheme hetero-junctions and peculiar properties of noble metals.

The optical properties of MXenes have a strong association with photocatalytic applications. MXenes ($Ti_3C_2T_X$) exhibit excellent transmittance of up to 91.2 % in the UV-Visible region in between the range 300 nm and 500 nm [39]. This transmittance can be attributed to film thickness and intercalation of MXenes layers.

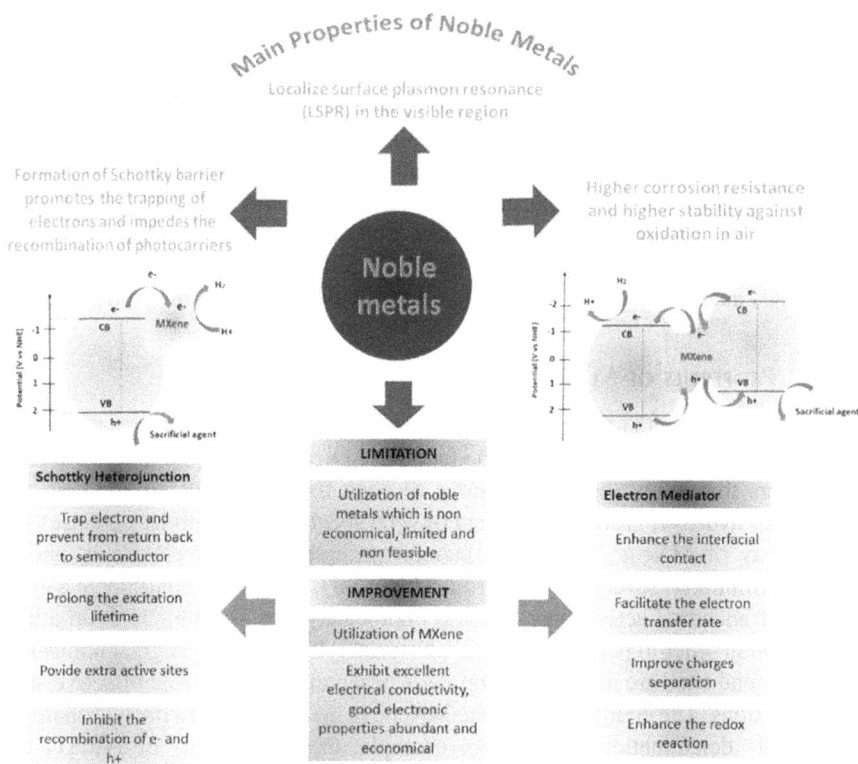

FIGURE 8.7   A general overview of MXenes in Schottky heterojunctions, all-solid-state Z-scheme heterojunctions and peculiar properties of noble metals. Reproduced with permission from [38]. 2021, ACS.

The role of sacrificial agents also signifies the transmittance rate. For instance, $Ti_3C_2T_x$ shows greater transmittance when treated with NH_4HF_2 as compared to DMSO, urea and hydrazines. Also, the precursor phase MAX exhibits least transmittance than MXenes. In particular, the transmittance of Ti_3AlC_2 (precursor phase) was 30% which was much less as compared to MXene ($Ti_3C_2T_x$), i.e., 90%. Lately, research works also revealed that their absorption may be broadened to near-infrared region (NIR). These unique features make MXenes an ideal robust photothermal catalyst.

8.3 APPLICATIONS IN PHOTOCATALYSIS

Fujishima et al. introduced a photoelectrochemical device that can decompose the water into oxygen and hydrogen under light irradiations in 1972. They also suggested that water splitting may occur without voltage being applied under suitable conditions [40]. Basically, photocatalytic reactions comprise three basic steps as shown in Figure 1.1. These steps include the absorption of light and excitation of light. Subsequently, the migration and separation of the photogenerated charge carriers, recombination of photoexcited electrons and holes and the last step include photoinduced carriers undergoing redox reactions. It is important to note that electron excitation takes place from the valence band to conduction band only when the energy of the incident photons is equal to or greater than photocatalyst, thus generating the negatively charged reactive electrons. The process of photocatalysis is widely adopted to carryout chemical reactions. MXene-derived photocatalysts show excellent charge transfer kinetics in hydrogen evolution reactions (HER) because of their high conductivity. Hence, MXenes have been proved to be very favourable series in HER catalysts [41].

8.3.1 PHOTOCATALYTIC HYDROGEN EVOLUTION REACTIONS

Due to the greater energy density, H_2 is considered as an ideal gas to be used as energy source instead of fossil fuels. Also, H_2 is environment friendly as it produces a cleaner combustion product, i.e., H_2O. To date, various methods have been practiced to produce H_2. The most common methods are electrocatalysis, water electrolysis, photocatalysis and coal gasification. Among them, photocatalytic hydrogen evolution is the most promising one due to its sustainability and pollution-free characteristics. Different types of photocatalysts have been tested for hydrogen production including TiO_2, graphene, $BaTiO_3$, metal oxides and metal sulphides. Nevertheless, these photocatalysts show lower light utilization capabilities and faster recombination rate of photogenerated carriers that inhibit their use [42]. Therefore, the fabrication of novel photocatalysts is inevitable.

MXene composites and metal sulphide nanomaterials have been considered as promising materials for hydrogen evolution process. $Ti_3C_2T_x$ MXene is the most widely used MXene in photocatalysis because of several benefits that make this material ideal for photocatalysis. These constitute the hydrophilic surface functional groups that are more conducive to the absorption of H_2O molecules and accelerate the reaction. Second, the Gibbs free energy of MXene-derived photocatalyst on H_2

approaches to zero infinitely which is also conducive to reduction of H^+ ions [43]. Further, theoretical calculations have proved that surface structure functionalization in MXene enhanced the hydrogen evolution active sites on the basal surface through the process of etching.

Shao et al. [44] reported a novel photocatalyst 2D $Ti_2C/g-C_3N_4$ by a calcination method. The efficient synergistic effect between Ti_2C and $g-C_3N_4$ remarkably enhanced water splitting. By loading 0.4 wt % Ti_2C on $g-C_3N_4$, H_2 production rate of 47.5 mmol h^{-1} was achieved, which was 14.4 times than that of pure $g-C_3N_4$. Besides, the morphology of the heterojunction photocatalyst also plays a crucial role in the catalytic activity. The SEM images exhibited that pure $g-C_3N_4$ and heterojunction photocatalyst have the same coral-like morphology, whereas TiCN showed the presence of stacked layers when Ti_2C was loaded on it as shown in Figure 8.8.

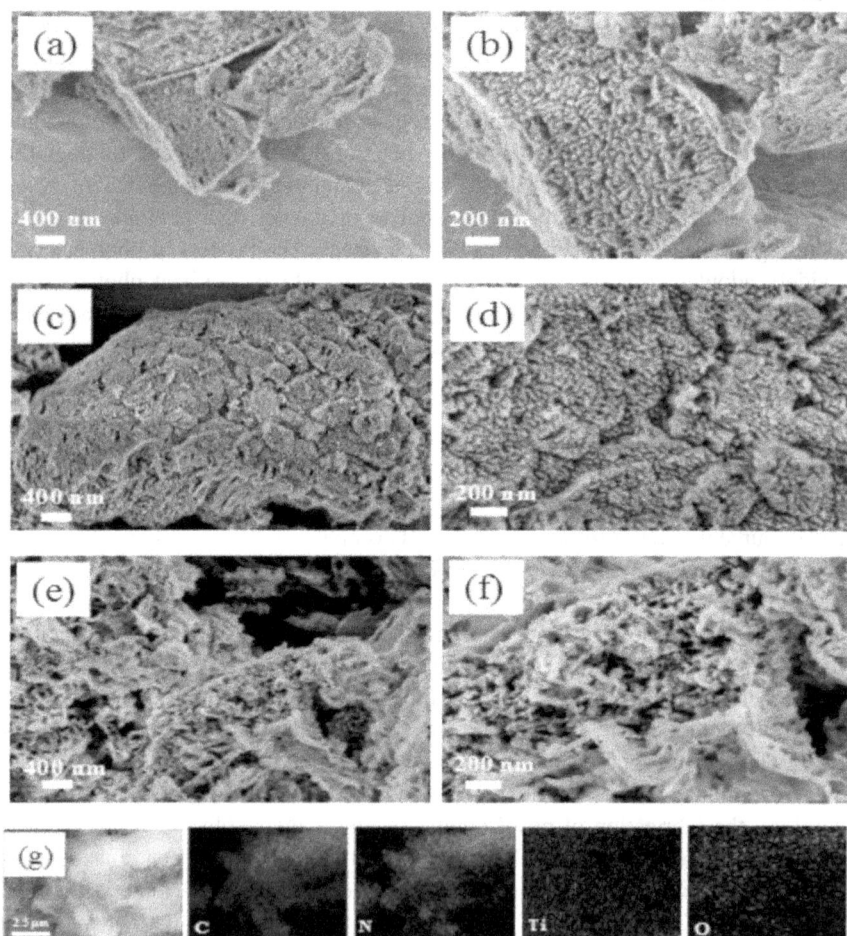

FIGURE 8.8   SEM images of pure $g-C_3N_4$ (a, b), TiCN-0.4 wt % (c, d) and TiCN-1.0 wt % (e, f). (g) EDX elemental mapping of TiCN-0.4 wt %. Reproduced with permission from [44]. 2017, RSC.

Xiao et al. reported an in-situ solvothermal synthesis of 1D/2D CdS/Ti$_3$C$_2$ heterojunction. The Schottky junction and intimate contact between Ti$_3$C$_2$ and CdS showed that the hydrogen production efficiency increased seven-folds as compared to the pristine CdS [30]. MXene-derived photocatalysts both partly and completely oxidized have gained significant importance in the H$_2$ evolution. Peng et al. synthesized Cu/TiO$_2$/Ti$_3$C$_2$ heterostructure through hydrothermal oxidation followed by the photo-deposition method. Results showed that within 300 minutes, the H$_2$ yield 2200–3800 µmol·g^{-1} which was ten-folds greater than TiO$_2$/Ti$_3$C$_2$ under the same time span. Surprisingly, even after the fourth recycling, there was no remarkable decrease in the H$_2$ output. The mechanism studies further revealed that Cu (zero valent metal) behaved as electron mediator, whereas Ti$_3$C$_2$ as a hole mediator which turned out to be improving the transfer and the separation of charges [45] as presented in Figure 8.9(a,b). Tian et al. revealed that Ti$_3$C$_2$/TiO$_2$/UiO-66-NH$_2$ heterojunction exhibited an excellent photocatalytic H$_2$ production due to the synergistic effect of type-II heterojunction and Schottky junctions [46] as shown in Figure 8.9(c). Han et al. reported the fabrication of C-TiO$_2$/g-C$_3$N$_4$ photocatalyst through the calcination method. Results revealed that C-TiO$_2$/g-C$_3$N$_4$ photocatalyst showed a remarkable increase in the H$_2$ production when compared to pure C-TiO$_2$ and g-C$_3$N$_4$. So, this

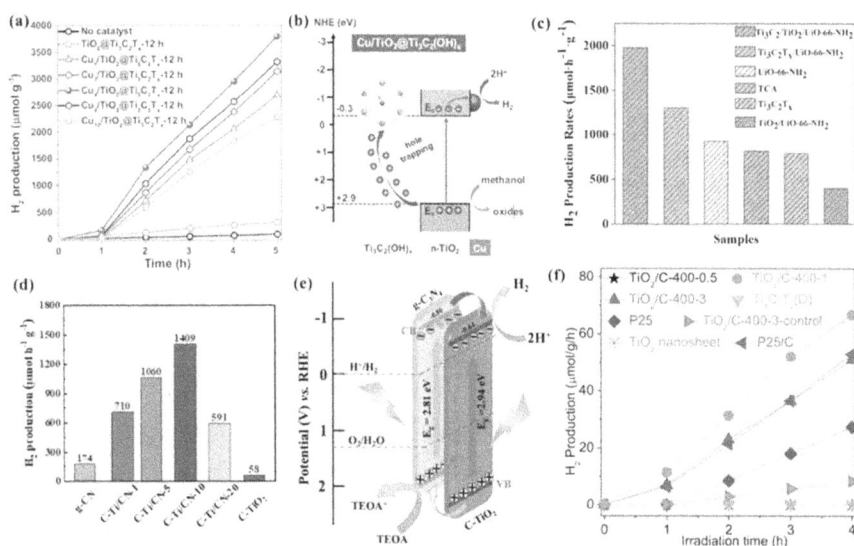

FIGURE 8.9 This figure has been reproduced with permission from [48]. © 2020, Elsevier, which illustrates (a) H$_2$ evolution over Cu$_y$/TiO$_2$/Ti$_3$C$_2$–12 h, (b) and corresponded mechanism of H$_2$ production. Reprinted with permission from [45]. 2018, Elsevier. (c) H$_2$ production over Ti$_3$C$_2$/TiO$_2$/UiO-66-NH$_2$ and other samples. Reprinted with permission from [46]. 2019, Elsevier. (d) H$_2$ production rate (e) H$_2$ generation mechanism over C-TiO$_2$/g-C$_3$N$_4$. Reproduced with permission from [47]. 2020, Elsevier. (f) H$_2$ production over P$_{25}$ and TiO$_2$ composites. Reproduced with permission from [49]. 2020, Elsevier.

significant increase in H_2 production was attributed to the interface heterojunction [47] as shown in Figure 8.9(d,e).

Likewise, Wang et al. reported a highly efficient TiO_2/C composite which was derived from Ti_3C_2. Results concluded that the ordered structure proved to be very fruitful in order to improve the utilization of carriers [49] as shown in Figure 8.9(f). The incorporation of uniformly dispersed C materials resulted in the increase in light absorption range and helped lower the carrier recombination. Kai et al. synthesized a novel CdS/Ti_3C_2 MXene photocatalyst by the in-situ growth method which is considered to be stable and highly efficient. SEM analysis exhibited that the heterostructure possessed regular morphology and size too. When compared to pure CdS, it showed more surface area and higher H_2 evolution rate approximately seven times than that of pure CdS [50]. Some MXene-derived photocatalysts from literature, their synthesis method and photocatalytic activity are shown in Table 8.1.

TABLE 8.1
MXene-Derived Photocatalysts: Synthetic Routes, Morphologies and Their H_2 Evolution Rate

Photocatalyst	Synthesis Method	Morphology	Sacrificial Agent	H_2 Evolution Rate ($\mu mol.g^{-1}.h^{-1}$)	References
Sulfur-doped carbon/ TiO_2	$Ti_3C_2T_x$ oxidation	Nanosheets	Methanol	333	[23]
CdS/MoS_2-MXene	Hydrothermal	Nanosheets	0.25 M Na_2S and 0.35 M Na_2SO_3	9679	[24]
$Mo_xS@TiO_2@Ti_3C$	Two-step hydrothermal approach	Nanosheets	TEOA	10505.8	[25]
C-TiO_2/g-C_3N_4 heterojunction	Calcination	Nanoparticles	TEOA 10%	1146	[26]
$CdLa_2S_4/Ti_3C_2$	Hydrothermal	Nanosheets	0.1 M Na_2S and 0.5 M Na_2SO_3	11182.4	[27]
C-doped g-$C_3N_4@C$, N co-doped TiO_2	Hydrothermal and calcination	Core-shell nanostructures	Methanol 10%	626	[28]
$Ti_3C_2/MoS_2/TiO_2$	$Ti_3C_2T_x$ oxidation	Ti_3C_2 nanosheets with MoS_2 nanoparticles	TEOA	6425.297	[29]
CdS/Ti_3C_2	In-situ solvothermal	Nanosheets	Lactic acid 10%	2407	[30]
Ti_3C_2 nanosheets modified Zr-MOFs	Hydrothermal	Nanosheets	TMAOH 25%	204	[31]
MXene@Au@CdS	Hydrothermal	Nanosheets	0.35 M Na_2S and 0.25 M Na_2SO_3	17070.43	[32]
C-$TiO_2@g$-C_3N_4	Calcination	Hollow nanospheres	Methanol 10%	356	[33]
g-$C_3N_4/Au/C$-TiO_2	Calcination	Hollow nanospheres	TEOA 10%	129	[34]
$Zn_2In_2S_5/Ti_3C_2T_x$	Hydrothermal	Flower-shaped microspheres	0.35 M Na_2S and 0.25 M $Na_2SO_3/$ H_2PtCl6	2596.76	[35]

8.4 CONCLUSION AND FUTURE OUTLOOK

We have focused on the emerging of MXene-derived photocatalysts for harvesting solar energy into chemical energy. MXene-derived photocatalysts proposed novel alternatives to meet the global needs of energy sources and sustainable environment. Due to their unique electrical, optoelectronic properties and presence of hydrophilic groups on the surface, MXene act as a co-catalyst as well as MXene-derived photocatalysts, which favours the induction of Schottky barrier to capture the electrons. To date, hundreds of MXene-derived photocatalysts have been reported. Nevertheless, the efficiency of these photocatalysts is still low due to the fast recombination rate of photoexcited charge carriers and larger bandgap. Attention should be paid to combine the cheap and efficient photocatalyst with MXene in order to achieve better results. In addition, new kinds of transition metal borides have been explored and they have exhibited potential in the photocatalysis. So, more work needs to be done in this direction in the near future.

REFERENCES

1. Tayyab, M., et al., *One-pot in-situ hydrothermal synthesis of ternary $In_2S_3/Nb_2O_5/Nb_2C$ Schottky/S-scheme integrated heterojunction for efficient photocatalytic hydrogen production.* Journal of Colloid and Interface Science, 2022. **628**: p. 500–512.
2. Wu, Z., et al., *MXene Ti_3C_2 derived Z–scheme photocatalyst of graphene layers anchored TiO_2/g–C_3N_4 for visible light photocatalytic degradation of refractory organic pollutants.* Chemical Engineering Journal, 2020. **394**: p. 124921.
3. Liu, Y., et al., *Single-atom Pt loaded zinc vacancies ZnO–ZnS induced type-V electron transport for efficiency photocatalytic H_2 evolution.* Solar Rrl, 2021. **5**(11): p. 2100536.
4. Ye, Z., et al., *Simple one-pot, high-yield synthesis of 2D graphitic carbon nitride nanosheets for photocatalytic hydrogen production.* Dalton Transactions, 2022. **51**(48): p. 18542–18548.
5. Tayyab, M., et al., *Simultaneous hydrogen production with the selective oxidation of benzyl alcohol to benzaldehyde by a noble-metal-free photocatalyst VC/CdS nanowires.* Chinese Journal of Catalysis, 2022. **43**(4): p. 1165–1175.
6. Liu, G., et al., *Direct and efficient reduction of perfluorooctanoic acid using bimetallic catalyst supported on carbon.* Journal of Hazardous Materials, 2021. **412**: p. 125224.
7. Aman, S., et al., *Graphene based nanocomposites: Synthesis, characterization and energy harvesting applications,* in *Advances in Nanocomposite Materials for Environmental and Energy Harvesting Applications.* 2022, Springer. p. 817–857.
8. Kuang, P., et al., *New progress on MXenes-based nanocomposite photocatalysts.* Materials Reports: Energy, 2022. **2**(1): p. 100081.
9. Hong, L.-f., et al., *Recent progress of two-dimensional MXenes in photocatalytic applications: A review.* Materials Today Energy, 2020. **18**: p. 100521.
10. Li, X., et al., *Applications of MXene ($Ti_3C_2T_x$) in photocatalysis: A review.* Materials Advances, 2021. **2**(5): p. 1570–1594.
11. Maeda, K., et al., *Visible-light-induced photocatalytic activity of stacked MXene sheets of Y_2CF_2.* The Journal of Physical Chemistry C, 2020. **124**(27): p. 14640–14645.
12. Khazaei, M., et al., *Recent advances in MXenes: From fundamentals to applications.* Current Opinion in Solid State and Materials Science, 2019. **23**(3): p. 164–178.

13. Persson, I., et al., *Tailoring structure, composition, and energy storage properties of MXenes from selective etching of in-plane, chemically ordered MAX phases*. Small, 2018. **14**(17): p. e1703676.

14. Anasori, B., et al., *Two-dimensional, ordered, double transition metals carbides (MXenes)*. ACS Nano, 2015. **9**(10): p. 9507–9516.

15. Zou, X., et al., *A simple approach to synthesis Cr_2CT_x MXene for efficient hydrogen evolution reaction*. Materials Today Energy, 2021. **20**: p. 100668.

16. Jiang, D., et al., *MXene-Ti_3C_2 assisted one-step synthesis of carbon-supported TiO_2/ Bi_4NbO_8Cl heterostructures for enhanced photocatalytic water decontamination*. Nanophotonics, 2020. **9**(7): p. 2077–2088.

17. Song, H., et al., *Enhanced photocatalytic degradation of perfluorooctanoic acid by Ti_3C_2 MXene-derived heterojunction photocatalyst: Application of intercalation strategy in DESs*. Science of The Total Environment, 2020. **746**: p. 141009.

18. Liu, H., et al., *One-pot hydrothermal synthesis of MXene Ti_3C_2/TiO_2/BiOCl ternary heterojunctions with improved separation of photoactivated carries and photocatalytic behavior toward elimination of contaminants*. Colloids and Surfaces A: Physicochemical and Engineering Aspects, 2020. **603**: p. 125239.

19. Wang, Y., et al., *Ti_3C_2 MXene coupled with CdS nanoflowers as 2D/3D heterostructures for enhanced photocatalytic hydrogen production activity*. International Journal of Hydrogen Energy, 2022. **47**(52): p. 22045–22053.

20. Kong, X., et al., *Orderly layer-by-layered TiO_2/carbon superstructures based on MXene's defect engineering for efficient hydrogen evolution*. Applied Catalysis A: General, 2020. **590**: p. 117341.

21. Jiang, H., et al., *2D MXene-derived Nb_2O_5/C/Nb_2C/gC_3N_4 heterojunctions for efficient nitrogen photofixation*. Catalysis Science & Technology, 2020. **10**(17): p. 5964–5972.

22. Li, Y., et al., *Ti_3C_2 MXene-derived Ti_3C_2/TiO_2 nanoflowers for noble-metal-free photocatalytic overall water splitting*. Applied Materials Today, 2018. **13**: p. 217–227.

23. Yuan, W., et al., *Laminated hybrid junction of sulfur-doped TiO_2 and a carbon substrate derived from Ti_3C_2 MXenes: Toward highly visible light-driven photocatalytic hydrogen evolution*. Advanced Science, 2018. **5**(6): p. 1700870.

24. Chen, R., et al., *Synergetic effect of MoS_2 and MXene on the enhanced H_2 evolution performance of CdS under visible light irradiation*. Applied Surface Science, 2019. **473**: p. 11–19.

25. Li, Y., et al., *Synergetic effect of defects rich MoS_2 and Ti_3C_2 MXene as cocatalysts for enhanced photocatalytic H_2 production activity of TiO_2*. Chemical Engineering Journal, 2020. **383**: p. 123178.

26. Yang, C., et al., *Rational design of carbon-doped TiO_2 modified g-C_3N_4 via in-situ heat treatment for drastically improved photocatalytic hydrogen with excellent photostability*. Nano Energy, 2017. **41**: p. 1–9.

27. Cheng, L., et al., *Boosting the photocatalytic activity of $CdLa_2S_4$ for hydrogen production using Ti_3C_2 MXene as a co-catalyst*. Applied Catalysis B: Environmental, 2020. **267**: p. 118379.

28. Mohamed, M.A., et al., *Revealing the role of kapok fibre as bio-template for In-situ construction of C-doped g-C_3N_4@C, N co-doped TiO_2 core-shell heterojunction photocatalyst and its photocatalytic hydrogen production performance*. Applied Surface Science, 2019. **476**: p. 205–220.

29. Li, Y., et al., *2D/2D/2D heterojunction of Ti_3C_2 MXene/MoS_2 nanosheets/TiO_2 nanosheets with exposed (001) facets toward enhanced photocatalytic hydrogen production activity*. Applied Catalysis B: Environmental, 2019. **246**: p. 12–20.

30. Xiao, R., et al., *In situ fabrication of 1D CdS nanorod/2D Ti$_3$C$_2$ MXene nanosheet Schottky heterojunction toward enhanced photocatalytic hydrogen evolution.* Applied Catalysis B: Environmental, 2020. **268**: p. 118382.

31. Tian, P., et al., *Ti$_3$C$_2$ nanosheets modified Zr-MOFs with Schottky junction for boosting photocatalytic HER performance.* Solar Energy, 2019. **188**: p. 750–759.

32. Yin, J., et al., *Facile preparation of self-assembled MXene@ Au@ CdS nano-composite with enhanced photocatalytic hydrogen production activity.* Science China Materials, 2020. **63**(11): p. 2228–2238.

33. Zou, Y., et al., *In situ synthesis of C-doped TiO$_2$@g-C$_3$N$_4$ core-shell hollow nano-spheres with enhanced visible-light photocatalytic activity for H$_2$ evolution.* Chemical Engineering Journal, 2017. **322**: p. 435–444.

34. Zou, Y., et al., *Fabrication of g-C$_3$N$_4$/Au/C-TiO$_2$ hollow structures as visible-light-driven Z-scheme photocatalysts with enhanced photocatalytic H$_2$ evolution.* ChemCatChem, 2017. **9**(19): p. 3752–3761.

35. Wang, H., et al., *Electrical promotion of spatially photoinduced charge separation via interfacial-built-in quasi-alloying effect in hierarchical Zn$_2$In$_2$S$_5$/Ti$_3$C$_2$ (O, OH)$_x$ hybrids toward efficient photocatalytic hydrogen evolution and environmental remediation.* Applied Catalysis B: Environmental, 2019. **245**: p. 290–301.

36. Pan, H., *Ultra-high electrochemical catalytic activity of MXenes.* Scientific reports, 2016. **6**(1): p. 1–10.

37. Zhao, M.Q., et al., *Flexible MXene/carbon nanotube composite paper with high volumetric capacitance.* Advanced Materials, 2015. **27**(2): p. 339–345.

38. Sherryna, A. and M. Tahir, *Role of Ti$_3$C$_2$ MXene as prominent schottky barriers in driving hydrogen production through photoinduced water splitting: A comprehensive review.* ACS Applied Energy Materials, 2021. **4**(11): p. 11982–12006.

39. Hantanasirisakul, K., et al., *Fabrication of Ti$_3$C$_2$T$_x$ MXene transparent thin films with tunable optoelectronic properties.* Advanced Electronic Materials, 2016. **2**(6): p. 1600050.

40. Fujishima, A. and K. Honda, *Electrochemical photolysis of water at a semiconductor electrode.* Nature, 1972. **238**(5358): p. 37–38.

41. Tayyab, M., et al., *A new breakthrough in photocatalytic hydrogen evolution by amorphous and chalcogenide enriched cocatalysts.* Chemical Engineering Journal, 2022: p. 140601.

42. Wang, L., et al., *Heterogeneous p–n junction CdS/Cu$_2$O nanorod arrays: Synthesis and superior visible-light-driven photoelectrochemical performance for hydrogen evolution.* ACS Applied Materials & Interfaces, 2018. **10**(14): p. 11652–11662.

43. Takanabe, K., *Photocatalytic water splitting: Quantitative approaches toward photocatalyst by design.* ACS Catalysis, 2017. **7**(11): p. 8006–8022.

44. Shao, M., et al., *Synergistic effect of 2D Ti$_2$C and g-C$_3$N$_4$ for efficient photocatalytic hydrogen production.* Journal of Materials Chemistry A, 2017. **5**(32): p. 16748–16756.

45. Peng, C., et al., *High efficiency photocatalytic hydrogen production over ternary Cu/TiO$_2$@ Ti$_3$C$_2$T$_x$ enabled by low-work-function 2D titanium carbide.* Nano Energy, 2018. **53**: p. 97–107.

46. Tian, P., et al., *Enhanced charge transfer for efficient photocatalytic H$_2$ evolution over UiO-66-NH$_2$ with annealed Ti$_3$C$_2$T$_x$ MXenes.* International Journal of Hydrogen Energy, 2019. **44**(2): p. 788–800.

47. Han, X., et al., *Ti$_3$C$_2$ MXene-derived carbon-doped TiO$_2$ coupled with g-C$_3$N$_4$ as the visible-light photocatalysts for photocatalytic H$_2$ generation.* Applied Catalysis B: Environmental, 2020. **265**: p. 118539.

48. Zhong, Q., Y. Li, and G. Zhang, *Two-dimensional MXene-based and MXene-derived photocatalysts: Recent developments and perspectives.* Chemical Engineering Journal, 2021. **409**: p. 128099.

49. Wang, J., et al., *Single 2D MXene precursor-derived TiO₂ nanosheets with a uniform decoration of amorphous carbon for enhancing photocatalytic water splitting.* Applied Catalysis B: Environmental, 2020. **270**: p. 118885.

50. Kai, C.-M., et al., *In situ growth of CdS spherical nanoparticles/Ti₃C₂ MXene nanosheet heterojunction with enhanced photocatalytic hydrogen evolution.* Journal of the Korean Ceramic Society, 2022. **59**(3): p. 302–311.

9 Nanomaterial and Polymer-Based Photocatalysts for Energy Applications

Syed Murtuza Ali
National University of Science and Technology, Sultanate of Oman

Shaik Feroz
Prince Mohammad Bin Fahd University, Kingdom of Saudi Arabia

9.1 INTRODUCTION

Organic contaminants present in aqueous media including dyes, disinfection by-products (DBPs), pharmaceuticals and personal care products (PPCPs), and other emerging contaminants, have been under research lens due to their potential eco-toxicological and adverse health impacts [1,2]. Semiconductor-based photocatalysis has been considered as one of the most promising approaches to treat such organic contaminants in water. Conducting polymers (CPs) and their nanocomposites (CPNs) have attracted much attention due to their tunable electrochemical properties, low costs, and efficacy [3]. The nanocomposites of CPs, with metals and metal oxides, have shown excellent photocatalytic efficiencies for removing organic contaminants, including dyes and PPCPs [4].

Polymers and their composites have shown significant variation in terms of their properties as compared to conventional materials [5–14]. Semiconductor-mediated photocatalysis is a well-established technique for pollutant degradation in aqueous media. In addition, photocatalysis based on metal oxides such as titanium dioxide offers many advantages due to its low cost, high stability, and non-toxic nature (to the environment or humans). It can be supported on various substrates, offer chemical resistance, has high turnover, provide complete mineralization of organic pollutants, has high catalytic activity strong oxidizing power, and is stable against photocorrosion.

Photocatalysts are mostly used for wastewater treatment either in the suspended form or it can be supported over a suitable substrate. It possesses a greater surface area and efficiency when used in powder form. However, the powder form of photocatalyst has certain drawbacks such as low photonic efficiency of suspended

DOI: 10.1201/9781003364825-9

photocatalyst, post-treatment recovery, and problems associated with human health. To address these drawbacks, efforts have been made to incorporate photocatalysts in glass, polymers, and other inert substrates. Immobilization of photocatalysts on polymer substrates offers various advantages: relatively higher utilization of photonic efficiency as compared to powder form, ease of post-treatment catalyst recovery, minimal loss of photocatalyst, and availability of longer contact time with pollutants to be degraded. On the downside, low thermal stability and chemical resistance of polymeric substrates towards photocatalysts itself is one area that needs to be addressed further. It also requires several physical and chemical methods for the incorporation of photocatalysts on the polymer surface. The dissolved organic compounds including aromatic and polyaromatic hydrocarbons in oil-produced water (OPW) are of great concern because of their high resistance to bio-degradation, toxicity to marine biota, and possible carcinogenicity and mutagenicity [15]. Photocatalysis has emerged as a proven technology to treat the recalcitrant trace organic pollutants present in water streams.

Titanium dioxide (TiO_2) is a widely used semiconductor in heterogeneous photocatalysis. TiO_2 photocatalytic activity depends on the crystal structure, phase composition, nanoparticle size distribution, porosity, bandgap, and surface-bound hydroxyl species. TiO_2 exists mainly in three pure solid phases: anatase (A), brookite (B), and rutile (R). The bandgap energy for these three phases varies from 3.00 to 3.21 eV. Charge recombination in pure phase TiO_2 is faster compared to a mixed phase; hence, 80% (A) – 20% (R) TiO_2 is widely used for photocatalysis reactions [16].

Researchers reported the enhancement of photocatalytic activity of TiO_2 by incorporating other semiconductor materials onto the outer surface of titania. Anatase-rutile crystalline structure titania nanoparticles and coated titania with various amounts of tin dioxide (SnO_2) have been prepared by a hydrolysis method. The synthesized particles were characterized using X-ray diffraction analysis (XRD). The characteristic peak and the photocatalytic activity of the synthesized nanoparticles were directly related to the amount of SnO_2 coated over titania nanoparticles [17]. The kinetic investigation of the photocatalytic reactions reveals that the photocatalytic reaction rate constant (kp) enhances with the introduction of SnO_2 nanoparticles in TiO_2 nanoparticles [18]. The photocatalytic degradation efficiency of SnO_2-coated pure anatase titania nanoparticles showed higher removal efficiency of organic pollutant (methyl orange) at lower pH (3) than at pH 11 [19].

Researchers also reported the change in bandgap energy and the photocatalytic efficiency of ZnO nanoparticles incorporated with other semiconductor materials. SnO_2 incorporated on ZnO nanoparticles has been synthesized and characterized. The photocatalytic efficiency was evaluated on photodecomposition of organic pollutant (methylene blue). Statistical studies reveal that main factors have a reasonable effect on the degradation of organic pollutant [20]. Pure zinc stannate (Zn_2SnO_4) nanoparticles were produced by high energy ball milling of ZnO and SnO_2 nanoparticles using a molar ratio of 2:1. The bandgap energy of zinc stannate nanoparticles was around 3.8 eV and the photocatalytic efficiency of hybrid nanoparticles was found to be less than that of pure ZnO nanoparticles [21]. This chapter gives an update on the studies involving efficient use of photocatalysis involving semiconductor-based nanocatalysts, CPs, and their nanocomposites for the treatment of wastewater. Factors influencing the treatment methods of contaminants from wastewater were also discussed.

9.2 PHOTOCATALYTIC MECHANISM

The photocatalytic mechanism was well defined in many research works. When a photocatalyst usually a semiconductor nanomaterial is exposed to sunlight, it will absorb photonic energy. Electrons get excited from the excess energy and promotes to the conduction band from the valance band, thus creating the negative-electron (e^-) and positive-hole (h^+) pair. The positive hole can oxidize the pollutant directly or oxidize water to form ($HO\cdot$) radicals. At the same time, the electron reduces the oxygen adsorbed to the nano photocatalyst, which prevents the combination of electrons and the positive hole. Photocatalytic reaction breaks down the pollutant molecules without any residue. The nano photocatalyst lasts for a longer time and the process don't need any additional chemicals, which makes the operation simple and economical. Most of the traditional nano photocatalysts like TiO_2 and ZnO work under UV band. Hence, great efforts are being made for the development of visible light-responsive nanostructure photocatalysts. In addition to the exploration of novel visible light-responsive nano photocatalysts, the traditional nano photo-catalysts were modified by doping, dye sensitization, or by forming a hetrostructure, coupled with a p-conjugated architecture. Titanium dioxide was introduced as a promising semiconductor nano photocatalyst due to its physical, structural, and optical properties under UV light. The general mechanism of photocatalysis is shown in equations 9.1–9.6.

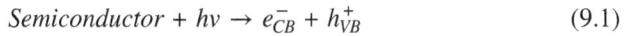

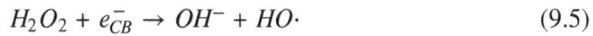

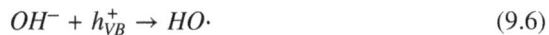

$$Semiconductor + h\nu \rightarrow e_{CB}^- + h_{VB}^+ \tag{9.1}$$

$$H_2O + h_{VB}^+ \rightarrow H^+ + HO\cdot \tag{9.2}$$

$$O_2 + e_{CB}^- \rightarrow O_2\cdot^- \tag{9.3}$$

$$O_2\cdot^- + H_2O \rightarrow 2H_2O_2 \tag{9.4}$$

$$H_2O_2 + e_{CB}^- \rightarrow OH^- + HO\cdot \tag{9.5}$$

$$OH^- + h_{VB}^+ \rightarrow HO\cdot \tag{9.6}$$

Application of nanostructure photocatalysts in energy applications coupled with the removal of pollutants from the environment is discussed in the following sections.

9.3 CASE STUDIES

In our research school, nano photocatalyst (TiO_2) has been immobilized on a low-cost polymer [22]. Composite specimens are synthesized with varying amounts of

TiO_2, viz., 40, 50, and 60 weight%. The composite was used to study the degradation of organics present in OPW. A series of batch experiments were conducted by varying the solution pH, stirring time, dosage, stirring speed, and concentration of TiO_2. Dissolved oxygen (DO), chemical oxygen demand (COD), turbidity, conductivity, and total dissolved solids (TDS) were evaluated to study the performance of the composites. The optimum conditions for the effective treatment of organics present in OPW were determined. The maximum percentage removal efficiency of organics (COD) was achieved between 77% and 88% at optimum operating conditions depending upon the intensity of natural sunlight. The surface morphology of composites using scanning electron microscopy showed that photocatalyst particles aggregated to form particle clusters on the polymer matrix. Thermal gravimetric analysis (TGA) data revealed that the incorporation of TiO_2 into the polymer matrix improved the thermal stability of the composite.

In a research study, $ZnO/TiO_2/H_2O_2$ process in suspension mode under visible light to reduce total organic carbon (TOC) from petroleum wastewater was studied by researchers [23]. Parameters such as treatment period (TP), ZnO concentration, pH, H_2O_2 concentration, and TiO_2 concentration were tested to recognize the optimal circumstances. The efficiency of the treatment improved from 15 percent to 36.5 percent for TOC at pH 5.5 relative to the TiO_2 process. The treatment period (TP) decreased up to 200%. Authors reported that the results in the optimum operational condition met with the effluent standards for wastewater.

Photocatalysis of aqueous organic contaminants using conducting polymers and their nanocomposites (CPNs) was presented in a review by Rahul et al. [24]. Authors discussed different strategies to prepare CPs and CPNs, characterization of CPs, and factors affecting CP-based photocatalysis. This review also highlights the potential, technical challenges, and future directions for CP-based photocatalysis. Highlighting CPs as one of the most promising photocatalytic materials to degrade trace organics, most of the CP-based photocatalysis studies to date have been conducted in a batch-scale mode with dye as a model contaminant. This study highlighted the need for more CP-based photocatalytic research, involving real-world water matrices, UV treatment, stability and reusability of catalysts, continuous flow reactor design, cost-benefit analysis, and pilot-scale testing before CPs could find their application in real-world treatment systems.

Photocatalysis gives an option of utilizing abundantly available light energy source and provides an ecofriendly and inexpensive approach that converts light energy into useful chemical energy and can solve many energy and environmental-related issues. In this direction, several synthetic approaches have been attempted in the last decade to design and develop diverse nanostructured materials that exhibit enhanced performance in visible light photocatalysis [25]. In this context, polymer-based materials have drawn wide attention due to their high spectral response in the visible light associated with excellent photocatalytic performance. In addition, due to their unique advantages of high chemical stability, low cost, and molecularly tunable optoelectronic properties, these materials have recently been proposed as promising alternatives to traditional semiconducting materials for photocatalysis.

The applications of photocatalysis based on visible light were found over conjugated microporous polymers (CMPs) and their functions designed at the molecular level [26]. The oxidation of thiols into disulfides entails proton and electron transfer and requires both Bronsted base and photocatalysis, which could be both combined into CMPs. Using carbazole as a Bronsted base and an electron donor, CMPs were constructed by the authors to implement the synergistic deprotonation and oxidation of thiols into disulfides in ethanol (C_2H_5OH). The bifunctional CMPs activate molecular oxygen (O_2) to superoxide anion ($O_2^{\cdot-}$) and promote the blue light-induced selective oxidation of thiols into symmetrical disulfides with high efficiency in C_2H_5OH. Highly selective formation of unsymmetrical disulfides could be achieved without adding a Bronsted base. Further, feasibility of combining cooperative photocatalysis into CMPs for versatile chemical transformations was reported.

A photocatalyst that is cost-effective, durable, reusable, and efficient for real-time dissipation of environmental pollutants was proposed by authors [27]. Authors suggested a novel approach by homogeneously dispersing CaO/CeO_2 nanocomposites (NCs) onto a 3D network of macro-/meso-porous monolithic polymer templates. The researchers synthesized CaO-doped CeO_2 NCs by a sonochemical-assisted temperature-controlled hydrothermal method. This will help to solve toxicology and recovery issues associated with nanoparticles and their composites. The structural and morphological properties of the synthesized photocatalyst materials have been characterized. The photocatalytic activity was observed under visible light for different stoichiometric ratios of CaO/CeO_2 NCs. The CaO/CeO_2 NCs dispersed porous polymer monolith was reported and it shows better photocatalytic activity than bare CaO/CeO_2 NCs that tend to agglomerate in aqueous solutions which reduces their efficiency. The CaO/CeO_2 NCs dispersed polymer monolith exhibits excellent porosity and surface area for the ultra-fast dissipation of organic pollutants. Considering the growing bacterial resistance due to the uncontrolled discharge/discarding of pharmaceutical medications, authors selected fluoroquinolone-based antimicrobial drug, namely moxifloxacin, as target pollutant. The best results were achieved using the monolithic photocatalyst dispersed with 20:80 CaO/CeO_2 NC, for the complete dissipation of moxifloxacin drug molecules and reported ≥99% degradation, at a pH of 6.0, using 100 mg of photocatalyst to dissipate 15 ppm of moxifloxacin drug, within 2 h of visible light irradiation.

A catalyst having large surface area and low density is preferred in photocatalytic reactions. In this context, hollow spheres (HSs) offer an advantage of large surface area, low density, and high loading capacity; due to this, they are considered as nanoreactors. Li et al. [28] developed new composites with PNVCL on the interior surface of mesoporous TiO_2 HSs (TiO_2-HSs) that were named $PNVCL@TiO_2$-HSs. Grafting of PNVCL on TiO_2-HSs, the light absorption range extended from less than 400 nm to less than 418 nm. Compared with traditional composites (TiO_2-HSs@PNVCL), the new composites have good dispersibility in water at a wide temperature range. Authors suggested that $PNVCL@TiO_2$-HSs can be used as photocatalysts and showed higher catalytic activity in photocatalytic degradation of Rhodamine B (RB) and Acid Red 57 (AR57) in wastewater. ·OH radicals were confirmed to be the main active species and played a dominant role in the degradation.

The presence of micro-pollutants such as triclosan (TCS) present in various water bodies is toxic and harmful, and can pose a serious threat to the aquatic environment. Sun et al. [29] synthesized aluminium acetylacetonate-doped polymeric carbon nitride photocatalysts (PCN-AA) to investigate the degradation properties of TCS under simulated visible light. The results showed that the best ratio material PCN-AA30 (k = 0.0529 min^{-1}) can degrade 99.29% % of TCS in 90 min, which is 2.45 times the degradation of the original polymeric carbon nitride material PCN-AA0 (k = 0.0216 min^{-1}). The degradation process of TCS presented different rules under the changing conditions of catalyst dosage, initial concentration of TCS, pH, common inorganic anions, and natural organic matter in water. The results of radicals quencher experiment showed that $\cdot O_2^-$ has an important role in the photocatalytic degradation in the reaction system. This study also identified ten degradation products of TCS using UPLC-Q-TOF technology and proposed the possible degradation pathways. In addition, the acute biotoxicity of PCN-AA materials was tested by a luminescent bacteria method, indicating that the safety of PCN-AA was relatively high. These results demonstrated that polymeric carbon nitride materials doped with aluminium acetylacetonate are a promising catalyst for the degradation of micro-pollutants in water under visible light.

Crystalline materials such as polyoxometalates (POMs), metals organic frameworks (MOFs), perovskites, and metal oxides are gaining popularity due to their remarkable photocatalytic performances. However, these catalysts are generally produced in powdered form, limiting their reuse and recyclability. A recent solution consisted of associating them with an acrylate polymer via a simple, green, and rapid photopolymerization process under visible light irradiation. The photocomposites exhibited excellent polymer robustness, stability, and malleability, while maintaining remarkable photocatalytic properties. To select the most performant photocatalytic system, authors compared the photocatalytic efficiency of different photocomposites. A comparison with the benchmark catalyst TiO$_2$ was provided for evaluation of their performance. Some of the reported catalysts were significantly better than TiO$_2$. The absorption properties, reusability, and the thermal properties as well as mechanical stabilities were taken into account during the comparison. The MIL-100(Fe)/polymer composite was selected as the most efficient photocatalytic system, which is also reusable during ten successive catalytic cycles with an observed little decrease in the pollutants final degradation percentages starting from the eighth cycle. Moreover, this Fe-based MOF immobilized photocatalyst was characterized by mechanical and thermal stabilities as well as a good absorption property, allowing its application under visible light irradiation.

9.4 SUMMARY

Nano photocatalysts immobilized on polymers show significant advantages in absorbing the photonic energy for the treatment of pollutants in a photocatalysis process. Photocatalytic activity of nano photocatalysts depends on the crystal structure, phase composition, nanoparticle size distribution, porosity, bandgap, and surface-bound hydroxyl species. Nano photocatalysts are mostly applied to wastewater treatment either in the suspended form or it can be supported over a suitable substrate. It shows a greater surface area and efficiency, when used in powder form.

However, powder nanostructure photocatalysts have certain drawbacks such as low photonic efficiency of suspended photocatalysts, post-treatment recovery, and problems associated with human health. To address these drawbacks, efforts have been made to incorporate Nano photocatalysts in glass, polymers, and other inert substrates. Immobilization of nano photocatalysts on polymer substrates offers various advantages: relatively higher utilization of photonic efficiency as compared to powder form, ease of post-treatment catalyst recovery, minimal loss of photocatalyst, and availability of longer contact time with pollutants to be degraded. On the down side, low thermal stability and chemical resistance of polymeric substrates towards photocatalysts itself is one area that needs to be addressed further. It also requires several physical and chemical methods for the incorporation of various types of nanostructure photocatalysts on the polymer surface.

REFERENCES

1. Ding, C., Zhu, Q. Yang, B., Petropoulos, E. Xue, L., Feng, Y., He, S., & Yang, L. Efficient photocatalysis of tetracycline hydrochloride (TC-HCl) from pharmaceutical wastewater using AgCl/ZnO/g-C3N4 composite under visible light: Process and mechanisms. *J Environ Sci* **126**, 249–262 (2023).
2. Antonopoulou, M., Kosma, C., Albanis, T. & Konstantinou, I. An overview of homogeneous and heterogeneous photocatalysis applications for the removal of pharmaceutical compounds from real or synthetic hospital wastewaters under lab or pilot scale. *Sci Total Environ* **765**, 144163 (2021).
3. Shahabuddin, S., Gaur, R., Mukherjee, N., Chandra, P. & Khanam, R. Conducting polymers-based nanocomposites: Innovative materials for waste water treatment and energy storage. *Mater Today Proc* **62**, 6950–6955 (2022).
4. Ahmad, N. *et al.* Visible light-conducting polymer nanocomposites as efficient photocatalysts for the treatment of organic pollutants in wastewater. *J Environ Manage* **295**, 113362 (2021).
5. Syed, M. A., Al-Shukaili, Z. S., Shaik, F. & Mohammed, N. Development and characterization of algae based semi-interpenetrating polymer network composite. *Arab J Sci Eng* **47**, (2022).
6. Syed, M. A., Sawafi, M. A. & Shaik, F. Polyurethane green composites for the treatment of boron present in the oil produced water. *ResearchSquare* Preprint at 10.21203/rs.3.rs-22768 (2020).
7. Syed, M. A. & Syed, A. A. Development of green thermoplastic composites from Centella spent and study of its physicomechanical, tribological, and morphological characteristics. *J Thermoplastic Compos Mater* **29**, 1297–1311 (2016).
8. Syed, M. A. & Syed, A. A. Investigation on physicomechanical and wear properties of new green thermoplastic composites. *Polym Compos* **37**, 2306–2312 (2016).
9. Devi, M. G., Al-kindi, R. S., Chandrasekar, G., Syed, M. A. & Feroz, S. Treatment of textile mill effluent using low molecular weight crab shell chitosan. *Desalin Water Treat* **56**, 1–7 (2015).
10. Syed, M. A. & Syed, A. A. Development of a new inexpensive green thermoplastic composite and evaluation of its physico-mechanical and wear properties. *Mater Des* **36**, 421–427 (2012).
11. Syed, M. A., Akhtar, S., Siddaramaiah & Syed, A. A. Studies on the physicomechanical, thermal, and morphological behaviors of high density polyethylene/coleus spent green composites. *J Appl Polym Sci* **119**, 1889–1895 (2011).

12. Syed, M. A., Ramaraj, B., Akhtar, S. & Syed, A. A. Development of environmentally friendly high-density polyethylene and turmeric spent composites: Physicomechanical, thermal, and morphological studies. *J Appl Polym Sci* **118**, 1204–1210 (2010).

13. Syed, M. A., Siddaramaiah, Syed, R. T. & Syed, A. A. Investigation on physico-mechanical properties, water, thermal and chemical ageing of unsaturated polyester/turmeric spent composites. *Polym – Plastics Technol Eng* **49**, 555–559 (2010).

14. Syed, M. A., Siddaramaiah, Suresha, B. & Syed, A. A. Mechanical and abrasive wear behavior of coleus spent filled unsaturated polyester/polymethyl methacrylate semi in-terpenetrating polymer network composites. *J Compos Mater* **43**, 2387–2400 (2009).

15. Silva, P., Ferraz, N., Perpetuo, E. & Olortiga Asencios, Y. Oil produced water treatment using advanced oxidative processes: Heterogeneous-photocatalysis and photo-fenton. *J Sediment Environ* **4**, 99–107 (2019).

16. Chalastara, K., Guo, F., Elouatik, S. & Demopoulos, G. Tunable composition aqueous-synthesized mixed-phase TiO_2 nanocrystals for photo-assisted water decontamination: Comparison of anatase, brookite and rutile photocatalysts. *Catalysts* **10**, 407 (2020).

17. Abbasi, S. Photocatalytic activity study of coated anatase-rutile titania nanoparticles with nanocrystalline tin dioxide based on the statistical analysis. *Environ Monit Assess* **191**, 206 (2019).

18. Abbasi, S. Investigation of the enhancement and optimization of the photocatalytic activity of modified TiO_2 nanoparticles with SnO_2 nanoparticles using statistical method. *Mater Res Express* **5**, 66302 (2018).

19. Abbasi, S. *et al.* Application of the statistical analysis methodology for photo-degradation of methyl orange using a new nanocomposite containing modified TiO_2 semiconductor with SnO_2. *Int J Environ Anal Chem* **101**, 208–224 (2021).

20. Ghaderi, A., Abbasi, S. & Farahbod, F. Synthesis, characterization and photocatalytic performance of modified ZnO nanoparticles with SnO_2 nanoparticles. *Mater Res Express* **5**, 65908 (2018).

21. Fakhrzad, M., Navidpour, A. H., Tahari, M. & Abbasi, S. Synthesis of Zn_2SnO_4 nanoparticles used for photocatalytic purposes. *Mater Res Express* **6**, 95037 (2019).

22. Syed, M. A., Mauriya, A. K. & Shaik, F. Investigation of epoxy resin/nano-TiO_2 composites in photocatalytic degradation of organics present in oil-produced water. *Int J Environ Anal Chem* **102**, 1–17 (2022).

23. Aljuboury, D., al deen, A. & Shaik, F. Assessment of $TiO_2/ZnO/H_2O_2$ photocatalyst to treat wastewater from oil refinery within visible light circumstances. *S Afr J Chem Eng* **35**, 69–77 (2021).

24. Kumar, R., Travas-Sejdic, J. & Padhye, L. P. Conducting polymers-based photo-catalysis for treatment of organic contaminants in water. *Chem Eng J Adv* **4**, 100047 (2020).

25. Mohanty, S. Polymer-based materials for visible light photocatalysis. in *Nanostructured Materials for Visible Light Photocatalysis* 491–510 (2022) 10.1016/B978-0-12-82301 8-3.00013-0. Arpan Kumar Nayak and Niroj Kumar Sahu (Eds.). Netherlands: Elsevier.

26. Dong, X., Hao, H., Zhang, F. & Lang, X. Combining Brønsted base and photo-catalysis into conjugated microporous polymers: Visible light-induced oxidation of thiols into disulfides with oxygen. *J Colloid Interface Sci* **622**, 1045–1053 (2022).

27. Asu, S. P., Sompalli, N. K., Kuppusamy, S., Mohan, A. M. & Deivasigamani, P. CaO/CeO_2 nanocomposite dispersed macro-/meso-porous polymer monoliths as new gener-ation visible light heterogeneous photocatalysts. *Mater Today Sustain* **19**, 100189 (2022).

28. Li, Z. *et al.* Thermosensitive polymers-TiO_2 hollow spheres composite for photo-catalysis. *Inorg Chem Commun* **146**, 110096 (2022).

29. Sun, D. *et al.* Efficient degradation of triclosan by aluminium acetylacetonate doped polymeric carbon nitride photocatalyst under visible light. *J Environ Chem Eng* **11**, 109186 (2023).

10 The Use of Advanced Nanomaterials for Lithium-Ion Rechargeable Batteries

L. Syam Sundar and Shaik Feroz
Prince Mohammad Bin Fahd University, Kingdom of Saudi Arabia

10.1 INTRODUCTION

Due to their higher voltage (nominal voltage for lithium-ion battery: 3.6 V), higher energy density or specific energy (125 watt-hours per kilogram and per liter), and longer cycle life (>1000 cycles) compared to conventional batteries such as lead-acid [1–3], Ni-Cd, Ni-MH [4–6], and Ag-Zn batteries, lithium rechargeable (or secondary) batteries currently represent the state of the art in small rechargeable batteries. The performance traits of secondary batteries are contrasted in Table 10.1 [7]. Lithium-ion batteries are widely used in cellular phones, laptop computers, and camcorders and they have potential usage in electric vehicles (EVs) and stationary energy storage systems. Huge research is going on in worldwide to improve energy density, cycle life, and safety of lithium-ion batteries.

A typical lithium-ion battery is made up of an electrolyte, cathode, and anode [8]. Li ions from the cathode ($LiCoO_2$) intercalate into the anode's crystal structure (carbon/graphite) during charging. The ions leave the anode upon discharge, turn around, and then re-enter the cathode structure. During discharge, the process is the opposite. Li ion migration in the anode and cathode hosts shouldn't alter or harm the host crystal structure in order to achieve high cycling efficiency and extended cycle life. The intercalation reaction's lithium-ion supply, in particular, comes from the cathode host. Hence, the performance of the entire battery depends on the physical, structural, and electrochemical characteristics of the cathode materials [9].

Energy density and cycle life are crucial performance indicators for Li-ion batteries and are highly reliant on the anode and cathode's capacity and quality. As an anode material, metallic lithium offers great energy density. Yet, due to its high reactivity, this sort of lithium anode, which was utilized in early batteries, presented serious safety issues. When the battery is charged, lithium is electroplated onto the anode, which causes dendrites to develop. The lithium will burn if these expanding dendrites get to the cathode and cause an internal short. Because to this, the metallic

DOI: 10.1201/9781003364825-10

TABLE 10.1

Comparison of the Performance Characteristics of Secondary Batteries

Battery Type	Normal Voltage (V)	Specific Energy Wh/kg	Specific Energy kJ/kg	Volumetric Energy Wh/l	Volumetric Energy kJ/l
Pb-acid	2	35	126	70	252
Ni-Cd	1.2	40	144	100	360
Ni-MH	1.2	90	324	245	882
Ag-Zn	1.5	110	396	220	792
Li-ion	3.6	125	450	44	1584
Li-SPE*	3.1	400	1440	800	2880

Source: Reproduced with permission from [7], D.R. Sado way and A.M. Mayes, *MRS Bulletin* 27, 590 (2002). © 2002, Materials Research Society.

Notes

* Projections based upon thin-film microbattery test results in the laboratory with Li/SPE/VOx; SPE stands for solid polymer electrolyte.

lithium anode has been replaced with a lithium-insertion type anode in more con-temporary lithium secondary batteries [10].

Graphite and other carbonaceous materials, as well as synthetic carbon with a graphite structure, have been employed as anode materials in commercial lithium-ion batteries [11,12]. Researchers have employed nanomaterials as anodes to increase the capacity of lithium batteries, including carbon nanotubes, other nanotubes, inter-metallics and nanocomposites, nanooxides, and nanocrystalline thin films. A nano-composite of Si59 has demonstrated greater cycling performance and an even larger reversible capacity, over 1700 mAh/g at room temperature, compared to single-walled carbon nanotubes, which showed a reversible capacity of 600 mAh/g [13]. The lithium storage capacity of some nanosized transition-metal oxides employed as anodes can reach 700 mAh/g, with 100% capacity retention for up to 100 cycles.

All of these outcomes have pushed industry production of Li-ion rechargeable batteries to a higher level. In February 2005, Sony launched [14] the first tin-based anode battery, and in March of that same year, Toshiba Corporation unveiled [14] a ground-breaking technology that would be used to create new nanoparticles for the negative electrode of lithium-ion batteries. Lithium manganese oxide spinels ($LiMn_2O_4$), layered $LiMO_2$ compounds (such as $LiCoO_2$, $LiNiO_2$, and $LiMnO_2$), and other materials can all be used as cathode materials for lithium-ion batteries. A common cathode material in the manufacture of industrial lithium-ion batteries is $LiCoO_2$ compound. $LiCoO_2$ compound has good electrochemical characteristics such as a long cycle life and a usable capacity of roughly 140 mAh/g; however, there are several issues that need to be resolved.

Because of the high cost and scarce availability of Co, $LiCoO_2$ has a significant disadvantage in terms of pricing. The toxicity of Co is another issue, which has an effect on the environment when the batteries are thrown out. $LiCoO_2$ cannot be used as the cathode material in any large lithium-ion batteries as a result of these two facts. $LiCoO_2$'s cost and toxicity have been reduced through the partial replacement of Co with other metal components such as Ni, Cr, Al, and Mn. Because Ni and Co can substitute for one another to make a solid solution of any percentage without affecting the layer structure, $LiNixCoyO_2$ compounds are the most alluring of them.

In terms of cycle life, capacity, and thermal stability, the electrochemical properties of $LiNixCoyO_2$ compound still fall short of those of $LiCoO_2$ compound. Another desirable substitute for $LiCoO_2$ as the cathode material in lithium-ion batteries is layered $LiMnO_2$ compound. Based on the Mn^{3+}/Mn^{4+} pair, $LiMnO_2$ molecule has a theoretical specific capacity of 285 mAh/g, which is nearly double that of spinel $LiMn_2O_4$. The $LiMnO_2$ compound has two distinct phase configurations. One is m-$LiMnO_2$, which is layered monoclinic $LiMnO_2$ (C2/m space group). The orthorhombic phase o-$LiMnO_2$ is the alternative (Pmnm space group). Under equilibrium conditions, it is challenging to synthesize layered m-$LiMnO_2$.

Aqueous solutions have been used primarily in attempts to create layered $LiMnO_2$ [15,16]. The resultant materials have different stoichiometries than $LiMnO_2$, include protons or water, exhibit poor crystallinity, or lose their structure during cycling. Armstrong and Bruce reported the first productive synthesis of layered $LiMnO_2$ in 1996 [17]. Nevertheless, m-$LiMnO_2$ changes into a spinel phase during the charge/discharge process [18,19]. As a result, m-$LiMnO_2$ cannot be used as a cathode material in lithium-ion batteries. It has been claimed that layering $LiMnO_2$ at high temperatures can be accomplished by doping the structure with other elements. Also, it's possible to anticipate that the doping impact will protect the layer structure against spinel-like transition during charge/discharge cycling.

During more than 200 cycles, the stacked $LiMxMn1xO_2$ cathodes are proven to have a capacity between 150 and 200 mAh/g [20–22]. As manganese oxides are cheap and plentiful, it is anticipated that $LiMxMn1xO_2$ would eventually replace $LiCoO_2$ as the cathode material in lithium-ion batteries. Large-scale rechargeable lithium-ion batteries for EVs and stationary energy storage can also be created using layered $LiMxMn1xO_2$. The commercialization of the layered $LiNixCoyO_2$ and $LiMxMn1xO_2$ compounds is still in its early stages. $LiFePO_4$ is of special interest in the olivine-type $LiMPO_4$ family because it has benefits in terms of environmental friendliness, potential low-cost synthesis, cycling stability, and high temperature capacity.

Even when operating at extremely low current densities, electrochemical experiments on samples prepared at high temperature have revealed that only portion of Li can be removed from the lattice and reversibly cycled. This is because it has a very low electronic conductivity at room temperature. Due to the sluggish diffusivity of lithium ions, this molecule exhibits poor electrochemical characteristics. Nevertheless, LiFePO4 nanocomposites exhibit noticeably enhanced conductivity. Several studies on nanosized cathode materials show that they have better cyclability than electrodes constructed of more common, larger-sized particles. In this regard, the research on conducting polymer films utilized in lithium batteries is also quite intriguing.

Since they offer a great deal of promise for use as power sources in electric and hybrid cars as well as in aerospace applications, all-solid-state rechargeable lithium batteries have undergone substantial research. Solid electrolytes for a solid-state lithium cell must have the following characteristics: strong ionic conductivity, high lithium ion transference number, wide electrochemical stability window, and acceptable thermal and mechanical qualities. A viable all-solid state lithium polymer battery with a nanoceramic polymer electrolyte is widely anticipated to be developed soon. It is important to understand how nanoparticles affect the functionality of lithium batteries and what types of mechanisms the nanomaterials in the lithium battery display.

In this book chapter, we discuss the following topics: (a) the nanomaterials used in anode, cathode, and electrolyte, including carbon nanotubes, two-element nanotubes, intermetallics, nanocomposites, nano-oxides (including nanosized $LiCoO_2$, $LiFePO_4$, $LiMn_2O_4$, and $LiMn_2O_4$), conducting polymer films, and the addition of nanosized metal-oxide powders into the polymer electrolyte. The features of several battery technologies are shown in Table 10.2 [23].

10.2   FUNDAMENTALS OF NANOMATERIALS

The foundation of nanoscience and nanotechnology is nanomaterials. Over the past few years, research and development activity in the vast and interdisciplinary field of nanostructure science and technology has exploded globally. It has the potential to fundamentally alter the processes used to make materials and products, as well as the variety and type of functionalities that may be accessible. It already has a large commercial influence, and this impact will undoubtedly grow in the future.

10.2.1   WHAT ARE NANOMATERIALS?

A group of compounds known as nanoscale materials are those whose minimum dimension is less than or equal to 100 nanometers. Nanometer which is 100,000 times smaller than the diameter of a human hair is one millionth of a millimeter. The unique optical, magnetic, electrical, and other properties that appear at this size in nanomaterials make them interesting. Electronics, medicine, and other fields stand to benefit greatly from these emerging features. Figure 10.1 indicates the difference between the nanometer size with other sizes.

Nanomaterials are incredibly tiny, with each dimension being at least 100 nm or smaller. Nanomaterials can be one dimensional (e.g., surface films), two dimensional (for example, strands or fibers), or three dimensional (e.g., particles). They can be found in solitary, fused, aggregated, or agglomerated forms, and they can have irregular, spherical, or tubular geometries. Nanotubes, dendrimers, quantum dots, and fullerenes are examples of typical nanomaterial kinds. Nanomaterials are used in the field of nanotechnology and exhibit physical chemical properties that are distinct from those of ordinary chemicals (i.e., silver nano, carbon nanotube, fullerene, photocatalyst, carbon nano, silica). Siegel divides nanostructured materials into zero-dimensional, one-dimensional, two-dimensional, and three-dimensional categories.

Nanomaterials are substances with ultra-fine grain sizes (50 nm) or dimensions that are no larger than 50 nm. According to Richard W. Siegel, nanomaterials can be made with modulation dimensionalities of zero (atomic clusters, filaments, and cluster assemblies), one (multilayers), two (ultrafine-grained overlayers or buried layers), or three (nanophase materials made of equiaxed nanometer-sized grains), as shown in Figure 10.2.

TABLE 10.2

Characteristics of Different Battery Technologies [23]

	Life Time Years (Cycles)	Power Density W/kg (kW/m³)	Energy Density Whk/g (kWh/m³)	Discharge Time	Recharge Time	Response Time	Operating Temperature (°C)	Self-Discharge (%day)	Critical Voltage/Cell (V)
Lithium-ion battery	8–15 (500–6000)	230–340 (1300–10000)	100–250 (250–620)	min–h	min–h	20 ms-s	−10 to 50	0.1–0.3	3
Lead-acid battery	3–15 (2000)	75–300 (90–100)	30–50 (75)	min–h	8–16 h	5 ms	−10 to 40	0.1–0.3	1.75
Advanced lead-acid battery	3–15 (3000)	75–300 (90–700)	30–50 (75)	min–h	8–16 h	5 ms	−10 to 40	0.1–0.3	2
Nickel-cadmium battery	15–20 (2500)	150–300 (75–700)	45–80 (<200)	s–h	1 h	ms	−40 to 45	0.2–0.6	1
Zinc bromide flow battery	5–10 (300–1500)	50–150 (1–25)	60–80 (20–25)	s–10 h	4 h	<1 ms	10 to 45	0–1	0.17–0.3
Vanadium redox flow battery	10–20 (13–103)	NA (0.5–2)	75 (20–35)	s–10 h	min	<1 ms	0 to 40	0–10	0.7–0.8

Response time is the time to deliver maximum power output over a specific time, with h = hours; min = minutes; ms = milliseconds; s = seconds; NA = not applicable

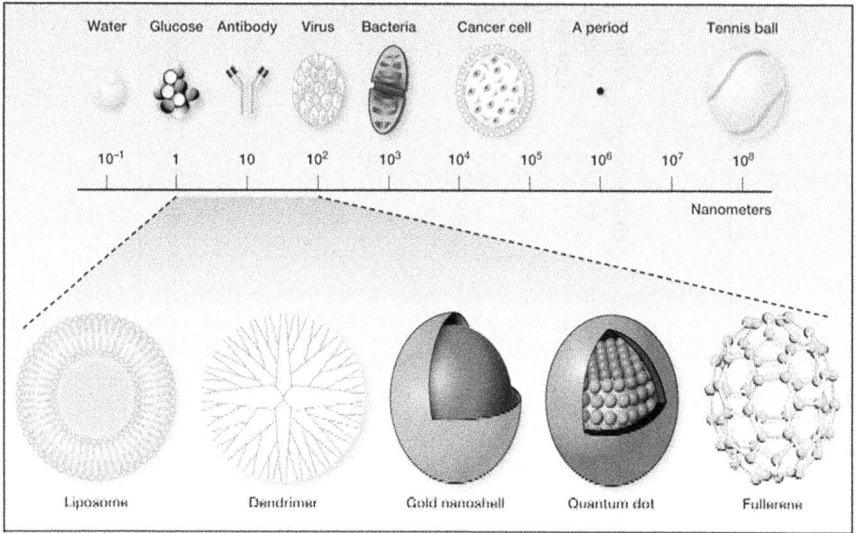

FIGURE 10.1 Indicates the difference between the nanometer size with other sizes.

10 2.2 NANOMATERIAL SYNTHESIS

A nanometer is one billionth of a meter; therefore, the structures we are working with are quite small. This indeed allows us to think in both the 'bottom up' or the 'top down' approaches to synthesize nanomaterials (Figure 10.3), i.e., either to assemble atoms together or to dis-assemble (break, or dissociate) bulk solids into finer pieces until they are constituted of only a few atoms. This domain is a pure example of interdisciplinary work encompassing physics, chemistry, and engineering upto medicine.

10.3 LITHIUM-ION BATTERY COMPONENTS

The cathode, anode, separator, and electrolyte make up a lithium-ion cell. Onto copper and aluminum foil current collectors, respectively, the anode and cathode components are deposited. The separator fits between the anode and the cathode to avoid shorting between the two electrodes while still allowing ion transfer. The electrolyte facilitates the flow of lithium ions between the electrodes. Lithium ions migrate from the anode and insert into the spaces between the cathode crystal layers during the discharge reaction (the process named intercalation). Lithium ions migrate from the positive side of the battery's cathode and insert into the anode during charging. Figure 10.4 depicts the LiB processes and components. The solid-electrolyte interphase (SEI), a passivation layer on the anode that is permeable to lithium ions but not to the electrolyte, is formed when intercalated lithium ions react quickly with the electrolyte solvent during the initial charge [24]. The safety and well-being of LiBs depend on the stability of the SEI. In the blow, the details of the

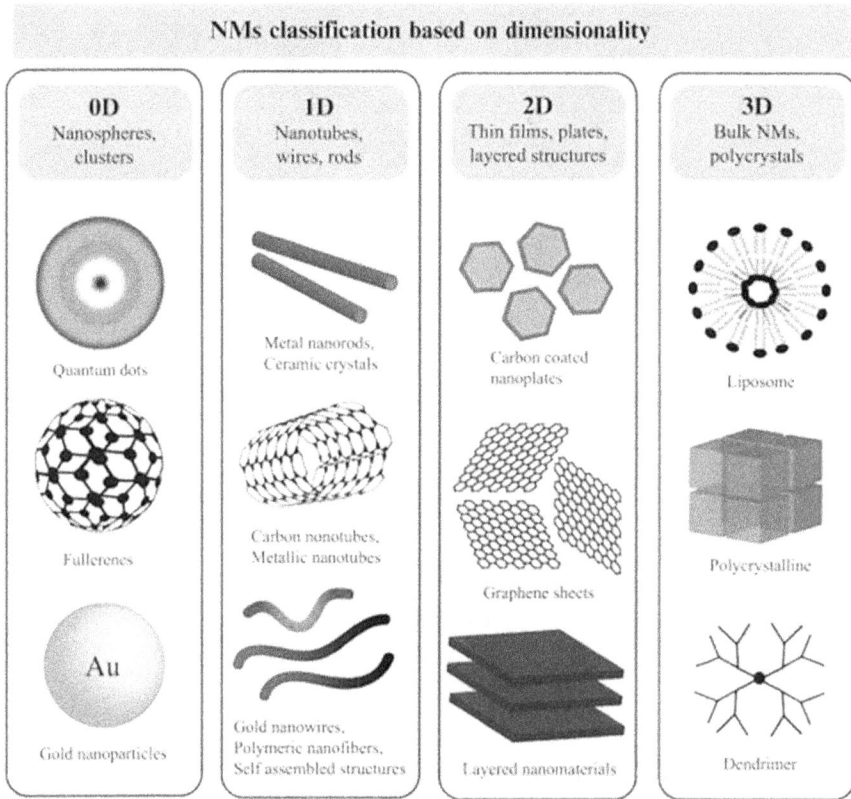

FIGURE 10.2 Classification of nanomaterials (a) 0D spheres and clusters; (b) 1D nanofibers, nanowires, and nanorods; (c) 2D nanofilms, nanoplates, and networks; (d) 3D nanomaterials.

battery are provided. The applications of lithium-ion batteries are mentioned in Figure 10.5. Lithium-ion cell designs and structures shown in Figure 10.6 (a–d) [24] show cylindrical, button, prismatic, and pouch cells, respectively. Figure 10.7 indicates the packaging of typical battery systems. Image is adopted from [25]. Figure 10.8 is the lithium-ion battery pack.

10.3.1 Cathode

The name of the LiB type is derived from the cathode composition. Due to their high capacity for lithium intercalation and compatible chemical and physical properties (e.g., for the reversibility of the intercalation), lithium metal oxides such as lithium cobalt oxide (LCO), nickel cobalt aluminum oxide (NCA), lithium cobalt phosphate (LCP), lithium manganese oxide (LMO), lithium iron phosphate (LFP), and lithium titanium sulfide (LTS) have been used as cathode materials. A variety of cathode materials are shown in Figure 10.9 together with their typical discharge potential and cell specific capacity [27].

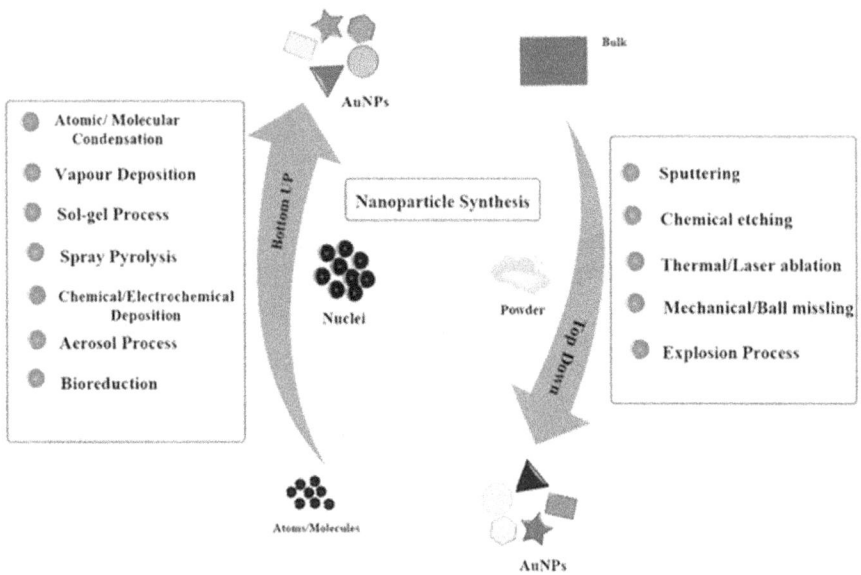

FIGURE 10.3 Top-down and bottom-up process.

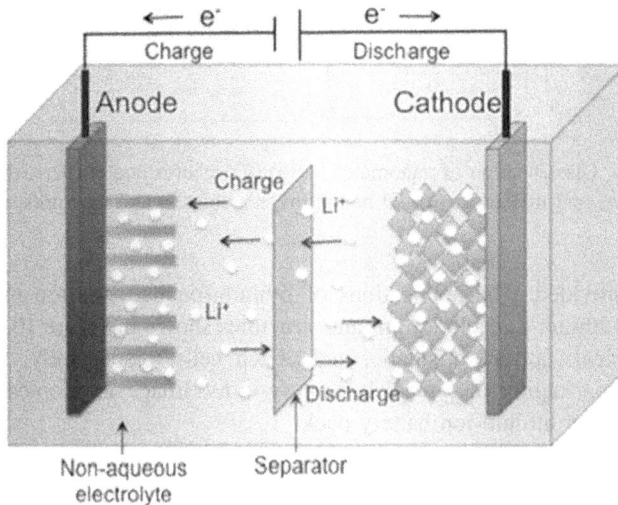

FIGURE 10.4 Principle of the lithium-ion battery (LiB) showing the intercalation of lithium ions (yellow spheres) into anode and cathode matrices upon charge and discharge, respectively [25].

The layered structures of LCO, NCA, and NCM all have high energy densities but lower thermal stabilities, and they all contain pricey cobalt. Despite its modest capacity, LMO features a spinel-like structure, excellent thermal stability, and high voltage. Because of its low cost, high average voltage, and low

FIGURE 10.5   Applications of lithium-ion batteries.

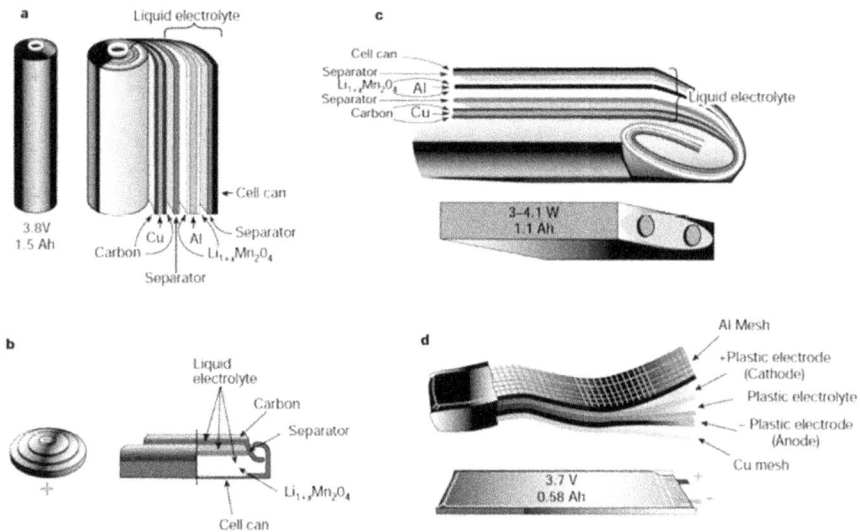

FIGURE 10.6   Lithium-ion cell designs and structures [24]. (a–d) show cylindrical, button, prismatic, and pouch cells, respectively.

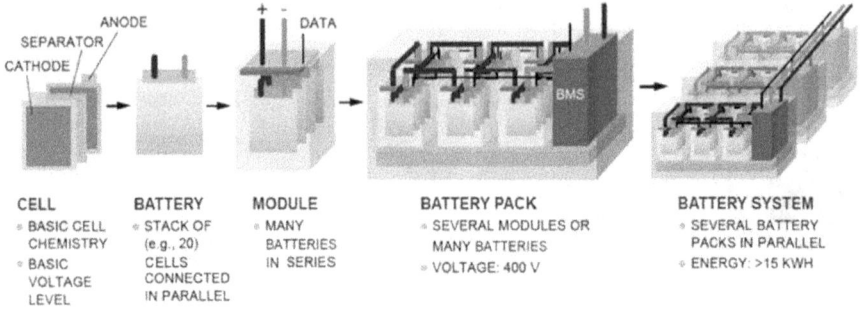

FIGURE 10.7   Packaging of typical battery systems. Image is adopted from [26].

FIGURE 10.8   Lithium-ion battery pack.

sensitivity to thermal runaway compared to NCM and LCO, LFP, which has a stable olivine structure and has gained significant interest, strikes a good compromise between performance and safety. As of now, nickel-, manganese-, and lithium-rich materials, carbon-coated LFP nanospheres, and vanadium pentoxide have all been identified as cathode materials with high capacity and voltage. The specific capacity of the cathode material is increased with higher nickel and lithium levels, while thermal stability is decreased. Synchrotron-based X-ray techniques have been used to systematically characterize the thermal instability of LiBs. These investigation findings offered important knowledge and direction, which is essential for creating more secure batteries with a high energy density.

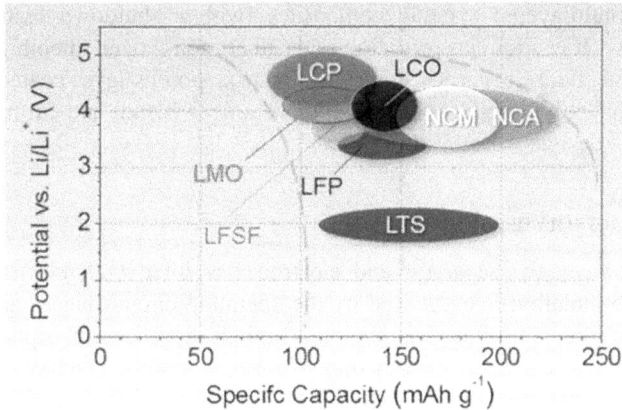

FIGURE 10.9 Average discharge potentials and specific capacity of common cathodes [27].

10.3.2 ANODE

Due to its significant negative potential, graphite is the most often utilized material for LiB anodes. Except for quick charging devices and stationary energy storage, where energy density is less important, other materials such as lithium titanate and silicone operate at lower negative voltages, lowering the energy and power density. In comparison to the graphite anode, lithium ions may move through lithium titanate and silicone more easily, which lowers internal resistance and lessens battery heating. Research have been conducted to find novel materials for anodes with higher voltage and capacity with variety of materials, including silicon, tin, antimony, germanium, silicon oxide, transition metal oxide: MO (where M is cobalt, nickel, copper, or iron), ultra-thin graphene nanosheets, and layered boron-nitrogen-carbon-oxygen material with adjustable composition. Prior to commercialization, features such capacity retention, conductivity, volume expansion, dendritic formation, and stability of the SEI layer should be taken into account in the development of new anode materials.

10.3.3 SEPARATOR

A porous membrane known as the separator separates the cathode and anode in LiBs. A vital part of a lithium-ion cell is the separator, which allows lithium ions to move between the electrodes while preventing electrical short circuits between them. Microporous polyolefin films, such as polyethylene (PE), polypropylene (PP), or laminates of PE and PP, which offer exceptional chemical stability, mechanical qualities, and affordable cost, are the most widely used separator materials for LiBs with organic electrolytes. Because polyolefins have a low melting point (135 C for PE and 165 C for PP), they can be used as a thermal fuse to stop a cell from overheating by losing their porosity and permeability.

With increased mechanical strength and thermal stability, new separator designs have been developed, including ceramic composites (often made of alumina and silica) and multilayered ceramic composites (with a shutdown feature through various phase transition layers). Polyester fiber non-woven membranes, silica/polyvinylidene fluoride porous composite matrix, porous-layer-coated polyimide nanofiber, and polyformaldehyde/cellulose nanofiber blends are further separator materials and designs.

10.3.4 ELECTROLYTE

The area between the separator and electrodes is filled with electrolyte. LiBs' electrolyte formulation is influenced by the operating circumstances and electrode materials. In order to improve cycling, lithium hexafluorophosphate ($LiFP_6$), lithium hexafluoroarsenate monohydrate ($LiAsF_6$), lithium perchlorate ($LiClO_4$), and lithium tetrafluoroborate ($LiBF_4$) are typically added to the flammable carbonate-based organic solvent used to make the typical electrolyte for LiBs.

By utilizing more stable lithium salts, additives, ionic liquid, non-flammable solvents, synthesized aqueous electrolytes, polymer electrolytes, and/or solid-state electrolytes, new electrolytes seek to reduce the fire risk associated with conventional electrolytes. The electrochemistry of LiBs affects their performance, price, and safety. For instance, a lithium-ion cell with a positive electrode made of LCO and a negative electrode made of graphite gives high voltage and energy density but also carries a higher risk of thermal runaway, which could result in cell rupture, venting, electrolyte ignition, and fire.

10.4 NANOMATERIALS USED IN ANODE

10.4.1 CARBON NANOTUBES

Since the discovery of single-walled carbon nanotubes (SWNTs) and multiwalled carbon nanotubes (MWNTs), carbon nanotubes (CNTs) have been of great interest for their diverse and promising applications, including as a candidate anode material for rechargeable Li-ion batteries. Gao et al. [13] reported that SWNTs could have high reversible capacities between 450 and 600 mAh/g, which could be further increased to 1000 mAh/g when SWNTs were ball milled into fractured nanotubes. Carbon nanotubes have a characteristic one-dimensional shape on the nanoscale, and their electrochemical activities are quite high, so that a single CNT can be utilized as a field-effect transistor at room temperature.

When cycling, Claye et al. [28] discovered a constant decline in reversible capacity. After five cycles using flooded cells, SWNTs typically lost 15% of their initial reversible capacity. The electrochemical storage of energy is a central theme in several investigations on carbon MWNTs. Most carbonaceous substances can have some degree of reversible reaction with Li. In order to accommodate Li/Li+, the CNTs have multiple positive sites inside the nanotubes, as well as between the graphene shells for multi-walled nanotubes (MWNTs) and between single-walled nanotubes (SWNTs) within bundles, giving them a very high initial capacity. According to Frackowiak et al. [29], depending on the manufacturing conditions, the MWNTs' reversible capacity ranged from 100 to 400 mAh/g.

According to Chen et al.'s [30] study, nanocomposites of CNTs and Sn2Sb alloys showed better specific capacities than CNTs and increased cyclability in comparison to the original Sn-Sb alloy particles. They explained the improvement as being the result of better Sn2Sb matrix retention with a high dispersion. High first discharge capacities of 570 and 512 mAh/g, high charge/discharge efficiencies of 77.5% and 84.1%, and good cyclability (0.99 loss%/cycle for MWNT/SnNi anode) were all demonstrated by MWNT-Sn and MWNT-SnNi.

Yin et al. [31] mechanically milled Ag-Fe-Sn alloys containing CNTs to create nanostructured Ag-Fe-Sn/C composites. The majority of CNTs were transformed into an amorphous structure that formed a shell around alloy particles during mechanical processing, with the exception of a few remaining CNT segments sticking out from the margins of the composite particles. The milled Ag36 4Fe15 6Sn48/CNT electrode offered 530 mAh/g of discharge capacity in the second cycle and kept 420 mAh/g of rechargeable capacity after 300 cycles. Multi-walled carbon nanotubes made by chemical vapor deposition (CVD) showed good cyclability at moderate current density and a reversible lithium storage capacity of 340 mAh/g.

The charge capacity of slightly graphitized carbon nanotubes was higher (640 mAh/g) than that of well-graphitized carbon nanotubes (282 mAh/g) when carbon nanotubes were created by the catalytic decomposition of acetylene or ethylene at 700°C over iron nanoparticles or iron oxide nanoparticles, but the latter displayed better rate capability and cycle life (with 91.5% and 65.3% of initial capacity remaining after 20 cycles for well graphitized). Freestanding nanotube electrodes will be able to deliver more precise power and energy than any present lithium-ion technology while also being durable, flexible, highly electrically conductive, and strong as steel. So, from both an industrial and a scientific perspective, carbon nanotubes may be a suitable electrode material. Figure 10.10 shows the electrochemical characterization of the MWCNT sample.

10.4.2 OTHER TYPES OF NANOTUBES

Two-element nanotubes have previously been used in lithium rechargeable batteries. In contrast to crystalline WS_2 powders, amorphous WS_3 WS_2 nanotubes produced by sintering it at a high temperature while hydrogen was flowing exhibited stable cyclability across a wide voltage range. Multiwalled TiS_2 nanotubes generated by a low temperature solution method were used to enable lithium intercalation.

Gao et al.'s [13] hydrothermal reaction in NaOH solution produced protonated 10–15 nm diameter TiO_2 nanotubes using rutile particles as the starting material. The anatase nanotubes displayed great initial capacity (170 mAh/g), exceptional high-rate discharge capabilities (640 mA/g), and good reversibility (118 mAh/g after 100 cycles) for electrochemical lithium insertion and extraction. Nickel oxide (NiO) nanotubes were produced utilizing a template processing method. Nickel hydroxide nanotubes were initially produced inside the walls of an anodic aluminum oxide (AAO) template in a two-step chemical process used for the synthesis. After using concentrated NaOH to destroy the template, NiO was created by heating the liberated nanotubes to 350°C while they were still in the air.

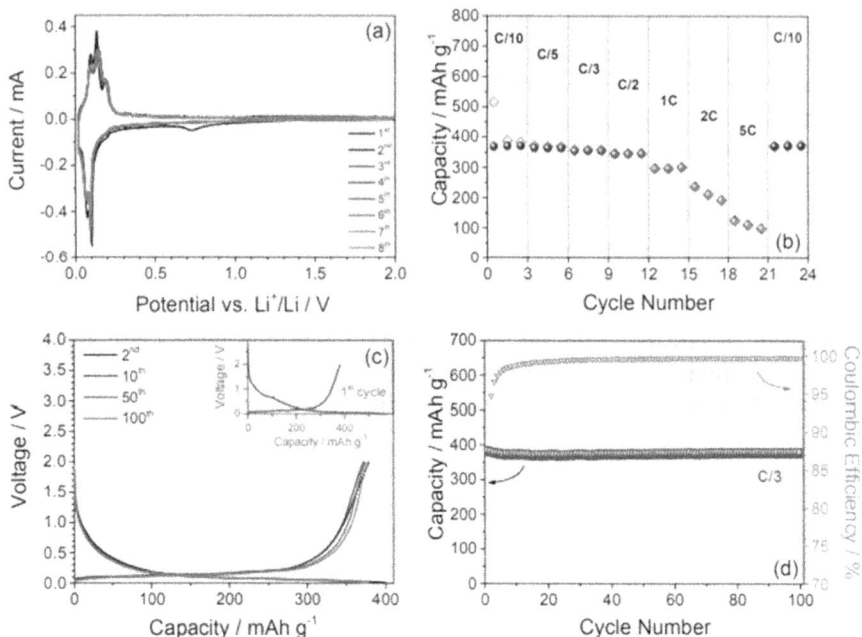

FIGURE 10.10 Electrochemical characterization of the MWCNT sample. (a) Cyclic voltammetry in three-electrode T-cell using lithium counter and reference electrodes at a scan rate of 0.1 mV s−1. (b) Rate capability test in lithium half-cell at C/10, C/5, C/3, C/2, 1 C, 2 C, and 5 C rates (1 C = 372 mAh g−1). (c, d) Galvanostatic cycling test in lithium half-cell at a C/3 rate in terms of (c) voltage profiles and (d) cycling trend (specific capacity and Coulombic efficiency on the left and right y-axes, respectively); the first voltage profile is shown in panel c inset. Lecce et al. [32].

A single nanotube had a length of 60 μm, an outside diameter of 200 nm, and a wall thickness of 20–30 nm. NiO nanotube powder was utilized in Li-ion batteries to test the batteries' ability to store lithium. Preliminary testing demonstrated that the cells exhibited controlled and sustainable lithium diffusion once a SEI is established. Typically, reversible capacities fell around the 300 mAh/g range.

10.4.3 Nano-Oxide Materials

There has been a lot of interest in 3D-transition metal oxides since some nanosized transition-metal oxide anodes may have a lithium storage capacity of approximately 700 mAh/g with 100% capacity retention for up to 100 cycles and high rates. The cell voltage is reduced to 2 V from 4 V when $Li_3CuFe_3O_7$ is used as the negative electrode, and the capacity fading is significant. Wang et al. [33] investigated nanostructured SnO_2, SnO_2-graphite, and Sn-graphite composites created using microemulsion techniques.

A restricted size range of Sn and SnO_2 nanoparticles (7–10 nm) were evenly distributed in graphite for the composites, revealing the material's advantageous role in maintaining reduced particle size. The SnO_2 nanoparticles have a particle size of 12–14 nm. The composites had much higher specific capacities than pure graphite and dramatically improved cyclability as compared to Sn-composites with large particles. SnO_2 powders (20–50 nm) produced utilizing a reverse micelle technique showed a stable reversible lithium storage capacity of about 630 mAh/g over multiple cycles of lithiation and de-lithiation. Initially, the capacity was about 1578 mAh/g. Co_3O_4 (nanosize) produced via chemical deposition exhibited a higher lithium storage capacity (780 mAh/g for the first cycle) than ordinary crystalline Co_3O_4.

10.4.4 NANOCRYSTALLINE THIN FILMS

For sputtered Si thin films, Lee et al. [34] measured the stress during the charge/discharge process. The Li4 4Si phase, created when Li ions are introduced into Si and have a volume three or four times greater than Si, results in microcracking in the electrode. The mesoscale sputtered Si-SnSi-Sn films reported by Beaulieu et al., who give useful insight into the nature of the electrochemical reaction of lithium with sputtered amorphous Si0-Sn0 films. Both the 1 m and 250 nm sputtered Si films were examined by Maranchi et al. The 250 nm films showed reversible capacities of 3500 mAh/g for 30 cycles without any evident evidence of failure, but the 1 _m films showed reversible capacities of 3000 mAh/g for 12 cycles. Si film that was vacuum-deposited onto Ni foil displayed an incredibly high capacity (1700–2200 mAh/g at a 2C discharge rate) and a lengthy cycle life (over 750 cycles at 2C).

Results for sputtered Si-M (M = Cr+Ni, Fe, Mn) thin films with more than 200 compositions were reported by Fleischauer et al. [35]. Si content has a significant impact on capacity, which ranges from nearly pure Si with a capacity of over 3000 mAh/g to nearly nothing at 50–60 at% Si. According to Song et al. [36], pulsed laser deposition produced thinner nanocrystalline films of Mg2Si (30 and 137 nm) that had reversible capacities of 2200 and 890 mAh/g for more than 100 cycles, respectively. However, the thicker films (296 and 380 nm) showed a stabilized capacity of >790 mAh/g with a fade rate of about 0.5–0.6%/cycle.

10.5 NANOMATERIALS USED IN CATHODE

10.5.1 VANADIUM OXIDE (V_2O_5)

Mantoux et al. [37] reported that vanadium oxide films deposited on titanium substrates by CVD are promising candidates as cathode materials in secondary lithium batteries. Reported V2O5 crystallite sizes were 94.2 nm and 65 nm. The reversible capacities reached 250 mAh/g and 410 mAh/g, respectively, close to the theoretical 420 mAh/g. V_2O_5 (nanowire) electrodes prepared by a simple and efficient template synthesis method showed equivalent lithium storage capacity to the thin-film electrode at a low discharge rate (C/20). However, at a rate of 200C, the nanostructured electrode delivered three times the capacity of the thin-film electrode. Above 500C, the nanostructured electrode delivered four times the capacity of the thin-film electrode. Figure 10.11 shows the V–I curve for LiV_3O_8 (Shchelkanova et al. [38]).

All solid-state cell

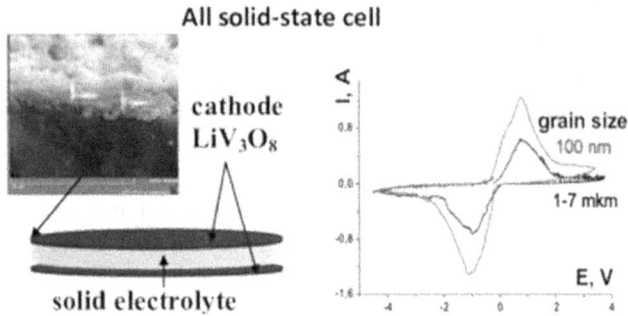

FIGURE 10.11   V–I curve for LiV_3O_8 (Shchelkanova et al. [38]).

10.5.2 α-Fᴇ₂O₃

When α-Fe_2O_3 powders are reacted with metallic lithium, Larcher et al. [39] discovered that the particle size has a significant impact on the reactivity of the powders. When lithium is injected into the corundum structure, the close-packed anionic array of large particles (0.5 m) undergoes an irreversible transition from hexagonal (Fe_2O_3) to disordered cubic stacking ($Li_2Fe_2O_3$). Yet, due to their ability to reversibly and topotactically react with up to one Li per formula unit without phase change, nanoparticles (200) behave significantly differently.

10.5.3   LɪFᴇPO₄

By using solid-state reactions to manufacture both undoped and doped $Li1xMxFePO_4$ (M = Mg, Al, Ti, Nb, or W), it was shown that the electrical conductivity of $LiFePO_4$ is increased by a factor of 108 when regulated cation non-stoichiometry and solid-solution doping by metals are coupled. By heating amorphous $LiFePO_4$, $LiFePO_4$ powders (nanosize) were created, and they nearly reached their full theoretical capacity (170 mAh/g) at the C/10 rate. Sol-gel synthesis was used to create phase-pure $LiMxFe1xPO_4$ (M = Mg, Zr, Ti) compounds, which significantly improved electrical conductivity. At a low rate of C/8, lithium iron phosphates with and without ti-doping showed a consistent discharge capacity of about 160–165 mAh/g, which is quite close to the theoretical capacity. Also, at high charge/discharge rates, the doping effect can greatly improve the electrochemical performance of lithium-iron phosphates.

$LiFePO_4$ particles embedded in amorphous carbon via a spray solution technique or carbon aerogel synthesis produced a carbon-coated $LiFePO_4$ cathode that had a high capacity and stable cyclability. The effect of the carbon coating can considerably increase $LiFePO_4$'s electronic conductivity. At the high discharge rate of 3C and the large discharge rate of 65C, a nanocomposite made of monodispersed nanofibers of $LiFePO_4$ electrode material combined with an electronically conductive carbon matrix that supplied roughly 100% and 36%, respectively, of its theoretical discharge capacity.

Because of the nanofiber shape, which reduces the issue of slow Li^+ transport in the solid state, and the conductive carbon matrix, which overcomes $LiFePO_4$'s naturally low electronic conductivity, this novel nanocomposite electrode exhibits such excellent rate capabilities. Because of this, nanocrystalline $LiFePO_4$ is a viable cathode active material for lithium-ion batteries with high energy density and power. It is particularly ideal as a cathode material for large-sized lithium-ion batteries that will be utilized for stationary storage batteries, hybrid electric cars, and EVs.

10.5.4 $LiMn_2O_4$

The initial capacity of nanoparticles of $LiMn_2O_4$ made by a one-step, intermediate-temperature solid-state reaction was substantially higher (130 mAh/g) than that of the commercial material made by a typical solid-state process (110 mAh/g). The template method was used to create nanostructured $LiMn_2O_4$ electrodes that have $LiMn_2O_4$ nanotubules sticking out from the current collector surface like brush bristles. Reduced wall thickness of the tubules creating the electrode increased rate capability. While cycling, the rate capabilities of electrodes built from the thinnest-walled tubules were extremely high, reaching 109C.

In an all-solid-state battery, LiNi0 5Mn1 5O$_4$ coated with Li_3PO_4 (100 nm) demonstrated the potential to realize both a high energy density and a high-safety battery system. LiCr0 2Mn1 8O$_4$'s capacity fade is discovered to be dependent on the particle size, according to Pascual et al. [40], who observed that larger particles (620 nm and 1560 nm) result in less capacity fade at C and 3C but not at C/24.

10.5.5 $LiMnO_2$

When the synthesis conditions of $LiMnO_2$ (used as a positive electrode material for lithium batteries) were altered, Paterson et al. [41] discovered that a nanodomain structure emerged on cycling. $LiMn_2O_4$ nanoparticles have a significant and advantageous impact on the material's cyclability. In comparison to previous solid solution oxides, the sol-gel-produced Li[Li1/5Ni1/10Co1/5Mn1/2]O$_2$ cathode material (10–15 nm) had a greater discharge capacity, low capacity fading, and improved structural stability.

10.5.6 $LiCoO_2$

According to Kawamura et al. [42], employing cathodes manufactured of $LiCoO_2$ (5 nm) instead of the more common 5 μm $LiCoO_2$ significantly improved the rate capability of Li cells.

10.5.7 CONDUCTING POLYMER FILMS

Electropolymerized conductive polymer films of numerous varieties have been developed and studied recently for use in batteries. Polypyrrole, polythiophene, and polyaniline are three of these polymers that exhibit the highest electroactivity, as well as good reversibility and chemical stability. The battery may have a high energy density and mechanical flexibility by utilizing a polymer cathode. Based on the weight of the electroactive polymers, specific energy values for Li/polypyrrole (PPy) cells with liquid electrolytes have been reported that range from 80 to 390 Whkg-1. The process used to create the polymer electrode has a significant impact on the doping level of polypyrrole and, as a result, on the specific capacity and energy available in the Li/PPy cell.

For instance, Panero et al. [43] observed a specific energy of 151 Whk/g for a Li/ LiClO4-PC/PPy cell with extra electrolyte, a PPy film that was approximately 1 μm thin and doped to a level of y = 0 24. In acetonitrile, the polymer film was electrochemically generated. After perfecting the PPy synthesis, Osaka et al. [44] reported even higher specific energy values for their Li/LiClO4-PC/PPy cells, up to 390 Whk/g (based on the weight of the polymer). Solvent affects polypyrrole's specific capacity. Regarding the characteristics of the resultant polymer, the electrolyte solvent is of utmost significance.

10.5.8 NANOMATERIALS USED IN POLYMER ELECTROLYTES

Conventional polyethylene oxide (PEO)-based electrolytes are the most widely employed of the polymer electrolyte systems reported. While electrolytes based on PEO films exhibit great chemical stability, their room temperature conductivities (107 to 108 S/cm129) are too low for the majority of applications. Thankfully, by adding nanosized ceramic powders as TiO_2, SiO_2, Al_2O_3, MgO, and $BaTiO_3$, the conductivity of PEO-based polymer electrolytes can be increased by orders of magnitudes below 60°C. It has been shown for nanoparticle-dispersed polymer electrolytes (PEO) that the smaller the size of the ceramic particles dispersed, the better the influence on the crystallization kinetics of the PEO polymer chains. This has been done by stirring or high-energy ball milling with a lithium salt ($LiCF_3SO_3$, $LiClO_4$, or $LiPF_6$).

According to Lin et al. [45], the PEO-10% $LiClO_4$–5%TiO_2 (3.7 nm) electrolyte had the best ionic conductivity at 1.40 104 S/cm at 30°C, which was significantly greater than that of the PEO-10% $LiClO_4$ electrolyte. According to Wang et al. [46], 10 wt% TiO_2 (15 nm)-doped PEO-$LiClO_4$ polymer electrolyte had the maximum ionic conductivity, measuring 105 S/cm at 25°C and 103 S/cm at 80°C, and was more than twice as strong as bare PEO-$LiClO_4$ polymer electrolyte both at low and high temperatures. A 10 wt% TiO_2-doped PEO-$LiClO_4$ polymer electrolyte had a lithium-ion transference number (TLi^+) of 0.47, which was twice as high as the naked PEO-$LiClO_4$ polymer electrolyte.

10.6 CONCLUSIONS

The nanoparticles utilized in lithium batteries, their effects on cell performance, their modes of action in the battery, and the nanotechnologies employed to create the nanomaterials have all been covered in this book chapter. Because of the dramatically reduced distance over which lithium ions must diffuse in nanoscale electrodes, the ability of nanoparticles to rapidly absorb and store enormous amounts of lithium ions without endangering the electrode, and the large surface areas, short diffusion lengths, and rapid diffusion rates of nanoparticles along the numerous grain boundaries present in nanometer-scale materials, nanomaterials play a significant role in improving the performance of lithium rechargeable batteries. The high surface area of the nanomaterials will exacerbate the reactivity between the electrode and electrolyte, which will lead to a poor cycle life and a low volumetric energy density.

REFERENCES

1. J. Wang, S. Zhong, H.K. Liu, and S.X. Dou, Beneficial effects of red lead on non-cured plates for lead-acid batteries. *J. Power Sources* 113, 371 (2003).
2. J. Wang, Z.P. Guo, S. Zhong, H.K. Liu, and S.X. Dou, Lead coated glass fibre mesh grids for lead acid batteries. *J. Appl. Electrochem.* 33, 1057 (2003).
3. S. Zhong, J.W. ang, H.K. Liu, S.X. Dou, and M. Skyllas-Kazacos, Influence of silver on electrochemical and corrosion behaviours of Pb–Ca–Sn–Al grid alloys Part I: Potentiodynamic and potentiostatic studies. *J. Appl. Electrochem.* 29, 1 (1998).
4. H.K. Liu, in *Encyclopedia of Nanoscience and Nanotechnology*. Magnesium-Nickel Nanocrystalline and Amorphous Alloys for Batteries, edited by H.S. Nal wa, American Scientific Publishers, Los Angeles (2004), Vol. 4, pp.775–789.
5. B. Luan, H.K. Liu, and S.X. Dou, On the elemental substitutions of titanium-based hydrogen-storage alloy electrodes for rechargeable Ni–MH batteries. *J. Mat. Sci.* 32, 2629 (1997).
6. J. Chen, D.H. Bradhurst, S.X. Dou, and H.K. Liu, Electrode properties of Mg2Ni alloy ball-milled with cobalt powder. *Electrochim. Acta* 44, 353 (1998).
7. D.R. Sadoway and A.M. Mayes, Portable Power: Advanced Rechargeable Lithium Batteries. *MRS Bulletin* 27, 590 (2002).
8. D. Linden and T.B. Reddy, eds., *Handbook of Batteries*, 3rd ed., McGraw Hill, New York (2002).
9. E.C. Gay, D.R. Vissers, F.J. Martino, and K.E. Andersen, EMF Studies of Lithium-Rich Lithium-Aluminum Alloys for High-Energy Secondary Batteries. *J. Electrochem. Soc.* 124, 1160 (1977).
10. B. Scrosati, Lithium Rocking Chair Batteries: An Old Concept? *J. Electrochem. Soc.* 139, 2776 (1992).
11. N. Imanishi, H. Kashiwagi, T. Ichikawa, Y. Takeda, and O. Yamamoto, Charge-Discharge. Characteristics of Mesophase-Pitch-Based Carbon Fibers for Lithium Cells. *J. Electrochem. Soc.* 140, 315 (1993).
12. J.O. Basenhard, M.W. Wagner, and M. Winter, Inorganic film-forming electrolyte additives improving the cycling behaviour of metallic lithium electrodes and the self-discharge of carbon-lithium electrodes. *J. Power Sources* 43–44, 413 (1993).
13. B. Gao, A. Kleinhammes, X.P. Tn, C. Bower, L. Fleming, Y. Wu, and O. Zhou, Electrochemical intercalation of single-walled carbon nanotubes with lithium. *Chem. Phys. Lett.* 307, 153 (1999).
14. http://www.physorg.com/news3061.html
15. M.H. Rossouw, D.C. Liles, and M.M. Thackeray, Synthesis and Structural Characterization of a Novel Layered Lithium Manganese Oxide, $Li_{0.36}Mn_{0.91}O_2$, and Its Lithiated Derivative, $Li_{1.09}Mn_{0.91}O_2$. *J. Solid State Chem.* 104, 464 (1993).
16. F. Leroux, D. Guyomard, and Y.P. Hard, The 2D Rancieite-type manganic acid and its alkali-exchanged derivatives: Part I — Chemical characterization and thermal behavior. *Solid State Ionics* 80, 299 (1995).
17. A.R. Armstrong and P.G. Bruce, Synthesis of layered LiMnO2 as an electrode for rechargeable lithium batteries 18. Lithium Intercalation into Layered $LiMnO_2$. *Nature* 381, 499 (1996).

18. G. Vitins and K. West, Lithium Intercalation into Layered $LiMnO_2$. *J. Electrochem. Soc.* 144, 2587 (1997).

19. G. Ceder and S.K. Mishra, The Stability of Orthorhombic and Monoclinic-Layered $LiMnO_2$. *Electrochem. Solid-State Lett.* 2, 550 (1999).

20. A.R. Armstrong, A.D. Robertson, and P.G. Bruce, Structural transformation on cycling layered $Li(Mn1-yCoy)O_2$ cathode materials. *Electrochim. Acta* 45, 285 (1999).

21. Y.-I. Jang, B. Huang, Y.-M. Chiang, and D.R. Sadoway, Stabilization of $LiMnO_2$ in the α - $NaFeO_2$ Structure Type by $LiAlO_2$ Addition. *Electrochem. Solid-State Lett.* 1, 13 (1998).

22. S. Franger, S. Bach, J.P. Pereira-Ramos, and N. Baffier, Chemistry and Electrochemistry of Low-Temperature Manganese Oxides as Lithium Intercalation Compounds. *J. Electrochem. Soc.* 147, 3226 (2000).

23. K. Cavanagh, J. Ward, S. Behrens, A. Bhatt, E. Ratnam, E. Oliver, and J. Hayward, *Electrical Energy Storage: Technology Overview and Applications*; CSIRO: Canberra, Australia, 2015.

24. J.-M. Tarascon and M. Armand, Issues and challenges facing rechargeable lithium batteries. *Nature* 2001, 414, 359–367.

25. P. Roy and S.K. Srivastava, Nanostructured anode materials for lithium ion batteries. *J. Mater. Chem. A* 2015, 3, 2454–2484.

26. A. Otto, S. Rzepka, T. Mager, B. Michel, C. Lanciotti, T. Günther, and O. Kanoun, Battery management network for fully electrical vehicles featuring smart systems at cell and pack level. In *Advanced Microsystems for Automotive Applications*; Springer: Berlin/Heidelberg, Germany, 2012; pp. 3–14. https://doi.org/10.1007/978-3-642-29673-4_1

27. Nitta, N., Wu, F., Lee, J.T., and Yushin, G. Li-ion battery materials: Present and future. *Mater. Today* 2015, 18, 252–264.

28. A.S. Claye, J.E. Fischer, C.B. Huffman, A.G. Rinzler, and R.E. Smalley, Solid-State Electrochemistry of the Li Single Wall Carbon Nanotube System. *J. Electrochem. Soc.* 147, 2845 (2000).

29. E. Frackowiak, S. Gautier, H. Gaucher, S. Bonnamy, and F. Beguin, Electrochemical storage of lithium in multiwalled carbon nanotubes. *Carbon* 37, 61 (1999).

30. W.X. Chen, J.Y. Lee, and Z. Liu, The nanocomposites of carbon nanotube with Sb and SnSb0.5 as Li-ion battery anodes. *Carbon* 41, 959 (2003).

31. J. Yin, M. Wada, Y. Kitano, S. Tanase, O. Kajita, and T. Sakai, Nanostructured Ag–Fe–Sn/Carbon Nanotubes Composites as Anode Materials for Advanced Lithium-Ion Batteries. *J. Electrochem. Soc.* 152, A1341 (2005).

32. D. Di Lecce, P. Andreotti, M. Boni, G. Gasparro et al. Multiwalled carbon nanotubes anode in lithium-ion battery with $LiCoO_2$, $Li[Ni_{1/3}Co_{1/3}Mn_{1/3}]O_2$, and $LiFe_{1/4}Mn_{1/2}Co_{1/4}PO_4$ cathodes ACS Sustain. Chem. Eng. 2018, 6, 3, 3225–3232

33. Y.W. ang, J.Y. Lee, and B.-H. Chen, Microemulsion Syntheses of Sn and SnO2-Graphite Nanocomposite Anodes for Li-Ion Batteries. *J. Electrochem. Soc.* 151, A563 (2004).

34. S. Lee, J. Lee, S. Chung, and H. Baik, Stress effect on cycle properties of the silicon thin-film anode. *J. Power Sources* 97–98, 191 (2001).

35. M.D. Fleischauer, J.M. Topple, and J.R. Dahna, Combinatorial Investigations of Si-M ($M = Cr + Ni$, Fe, Mn) Thin Film Negative Electrode Materials. *Electrochem. Solid-State Lett.* 8, A137 (2005).

36. S.W. Song, K.A. Striebel, R.P. Reade, G.A. Roberts, and E.J. Cairns, Electrochemical Studies of Nanoncrystalline Mg_2Si Thin Film Electrodes Prepared by Pulsed Laser Deposition. *J. Electrochem. Soc.* 150, A121 (2003).

37. A. Mantoux, H. Groult, E. Balnois, P. Doppelt, and L. Gueroudji, Vanadium Oxide Films Synthesized by CVD and Used as Positive Electrodes in Secondary Lithium Batteries. *J. Electrochem. Soc.* 151, A368 (2004).

38. M.S. Shchelkanova, G.S. Shekhtman, K.V. Druzhinin, A.A. Pankratov, and V.I. Pryakhina, The study of lithium vanadium oxide LiV_3O_8 as an electrode material for all-solid-state lithium-ion batteries with solid electrolyte $Li_{3.4}Si_{0.4}P_{0.6}O_4$, *Electrochim. Acta* 320 (2019) 134570.

39. D. Larcher, C. Masquelier, D. Bonnin, Y. Chabre, V. Masson, J.-B. Leriche, and J.-M. Tarascon, Effect of Particle Size on Lithium Intercalation into αFe_2O_3. *J. Electrochem. Soc.* 150, A133 (2003).

40. L.P. ascual, H. Gadjov, D.K. ovacheva, K. Petrov, P. Herrero, J.K. Amarrilla, R.M. Rojas, and J.M. Rojo, Effect of the Thermal Treatment on the Particle Size and Electrochemical Response of $LiCr_{0.2}Mn_{1.8}O_4$ Spinel. *J. Electrochem. Soc.* 152, A301 (2005).

41. A.J. Paterson, A.R. Armstrong, and P.G. Bruce, Stoichiometric $LiMnO_2$ with a Layered Structure: Charge/Discharge Capacity and the Influence of Grinding. *J. Electrochem. Soc.* 151, A1552 (2004).

42. T. Ka wamura, M. Makidera, S. Okada, K. Koga, N. Miura, and J.I. Yamaki, Effect of nano-size $LiCoO_2$ cathode powders on Li-ion cells. *J. Power Sources* 146, 27 (2005).

43. S. Panero, P. Prosperi, F. Bonino, and B. Scrosati, Characteristics of electrochemically synthesized polymer electrodes in lithium cells—III. Polypyrrole. *Electrochem. Acta* 32, 1007 (1987).

44. T. Osaka, K. Naoi, and S. Ogano, Effect of Polymerization Anion on Electrochemical Properties of Polypyrrole and on Li / LiClO4 / Polypyrrole Battery Performance. *J. Electrochem. Soc.* 135, 1071 (1988).

45. C.W. Lin, C.L. Hung, M. Venkateswarlu, and B.J. Hwang, Influence of TiO2 nano-particles on the transport properties of composite polymer electrolyte for lithium-ion batteries. *J. Power Sources* 146, 397 (2005).

46. G.X. Wang, L. Yang, J.Z. Wang, H.K. Liu, and S.X. Dou, Enhancement of ionic conductivity of PEO based polymer electrolyte by the addition of nanosize ceramic powders. *J. Nanosci. Nanotechnol.* 5, 1135 (2005).

11 Graphene and Graphene-Based Composite Nanomaterials for Rechargeable Lithium-Ion Batteries

L. Syam Sundar, M. Amin Mir, and M. Waqar Ashraf
Prince Mohammad Bin Fahd University, Kingdom of Saudi Arabia

11.1 INTRODUCTION

Energy storage, energy production, and global warming are the three key issues that are of the utmost importance and concern to all nations in the twenty-first century [1]. A lot of work is being done to create a future powered by renewable energy sources while also protecting the environment because of reasons like the growing global population, accelerated economic expansion, rising energy consumption, and imminent climate change [2]. To do this, we must rely less on fossil fuels and use more readily available, environmentally friendly, and renewable energy sources including tidal, solar, wind, and biomaterials [3].

In order to make the most of different energy sources, creative energy storage devices that can store excess energy and use it when necessary to provide a stable power supply for homes and businesses are needed [4]. In this climate, energy storage is certainly a critical issue that must be addressed, and electro-chemical energy storage technologies are unquestionably the key enablers. The electrochemical storage and release of energy are significantly influenced by batteries. Due to their exceptional functionality, rechargeable lithium-ion batteries (LIBs), lithium-sulfur batteries, and lithium-oxygen batteries (LOBs) [5–9] are among the energy storage devices that have recently attracted a lot of attention.

In order to meet the ongoing demand for electronics with improved performance capabilities, such as power consumption devices (such as smart-phones, tablets,

DOI: 10.1201/9781003364825-11

smart watches, and laptops) and electrical hybrid vehicles, fabrication and construction of an efficient material that provides a high specific electrode energy (energy per unit mass) and superior energy density (energy per unit volume) are necessary [10]. These energy storage systems typically include two electrodes and a separator impregnated with electrolyte, which provides the electrons and ions required by the electrochemically active nanomaterials. However, these batteries' energy storage methods, which rely on their electrode structures, are different from one another.

Future energy storage devices will require nanostructured electrode materials with exceptional electrochemical properties, as it has been acknowledged [11]. Nanostructured electrode materials also have advantages over their macro- and micro-sized counterparts in terms of short lengths for electrical ion transportation, enough surface area for electrode/electrolyte interactions, straightforward functionalization for solution chemistry, and novel electrochemistry reactions [12]. It has become abundantly evident in recent years that using synthetic designs with electrode nanomaterials is the only way to significantly improve energy storage devices. As materials for various energy storage technologies, graphene and composites based on it have attracted interest and considerable research [13].

Novoselov et al. [14] developed graphene in 2004, the first name for an advanced aromatic single thick carbon atom layer with a thickness million times smaller than a single hair. Graphene is a hexagonal, two-dimensional (2D) honeycomb lattice structure with chemically sp2 hybridized properties similar to those of graphite precursors. The two strongest sp2 hybridized chemical bonds, the sigma bonds, and the out-of-plane bonds both have electrons that are delocalized in-plane. By creating tenuous connections between the graphene layers and substrates, these bonds cause electron conduction [15–17]. The unique characteristics of graphene make it the perfect material for a variety of applications and have created new possibilities for the advancement of material science research.

In addition to its twofold transparency, graphene has unique thermal and electrical conductivity qualities. Stronger than steel, graphene is still remarkably flexible and light. A revolutionary revolution in modern civilization could be sparked by these remarkable qualities of graphene. The "gold rush" in graphene has been labeled as a result of the steady rise in graphene's popularity in recent years. There are so many industrial uses for graphene that are astounding [18,19].

It is important to note that graphene has many desirable properties, including a large surface area (2630 m^2/g), significantly high Young's modulus (1.0 TPa), breaking strength of 42 N/m, selectivity, extraordinary mechanical strength, electrical conductive superiority ($5 - 6.4 \times 10^6 \mu s/m$), remarkable intrinsic electron mobility ($200000 \ cm^2/\mu s$), exceptional thermodynamic conductivity ($4.8 - 5.3 \times 10^3 W/mk$), optical transmittance activity (~97.7%), material density ($< 1g/cm^3$), stability, catalysis performance and graphene tunable properties have produced substantial interest in the rechargeable battery commercial industry [20–22].

Although graphene has numerous exceptional qualities, it is rarely used in large-scale industrial applications due to the prohibitive cost of creating high-quality graphene. Massive amounts of superior graphene material are currently produced using a number of synthetic methods such as topological metallic growth, topological growth on non-metallic substrates, chemical oxidation of graphite, and micromechanical cleavage [23,24]. Whether functioning as a supporting catalytic material or a non-metal active electrocatalyst, graphene's logical form makes it simpler to adjust an electrocatalytic process in a way that is advantageous and has wonderful electrochemical reaction mechanisms [25,26].

Recently, several theoretical and practical studies on graphene have been conducted. Particularly promising electrocatalytic stability and performance is provided by a supporting material built of graphene [27]. To demonstrate the benefits of graphene, additional alloy materials that serve as additives are generally used. We concentrated on bringing graphene's additional properties to battery electrodes in relation to the previously described. Use different architectural designs, including graphene spheres and cages, to accommodate the volume expansion of composites comprising metal-hosts during the course of sequences. These shapes not only facilitated electronic and ionic transport but also provided an assembly enclosure [28].

In order to create an efficient sulfur-based cathode, Huang et al. [29] created 3D graphene-carbon nanotubes (CNTs). They employed these CNTs and discovered that the electrolyte had fewer soluble species. The function of electrochemical energy storage graphene batteries, with an emphasis on graphene form, has received little attention in the recent reviews of batteries that researchers have written [30–33]. Therefore, it is imperative to carry out a thorough evaluation as soon as possible with a focus on graphene applications. This chapter covers the full spectrum of graphene battery technologies and focuses on theoretical concepts, a newly created hybridization process, and graphene doping that happens in the battery sector.

We concentrate on current advancements in the fictionalization, controlled manufacture, and application of graphene in batteries in this work. The focus is on current research advancements that understand fundamental barriers and product advancements that can be applied in commercialization. In-depth descriptions of the guiding ideas of the contents have been provided. Future research activities that aim to apply graphene in various battery sectors have also been taken into account. Lastly, suggestions are made regarding how graphene can boost electrochemical batteries even more.

11.2 GRAPHENE SYNTHESIS

The graphene structure is shown in Figure 11.1. Synthesizing graphene refers to the process of creating or preparing graphene, which depends on the desired purity, size, forms, and florescence capabilities of the resulting graphene material. Several techniques for producing thin graphitic films have been discovered at the early stage.

In the late 1970s, carbon precipitated as thin graphite coatings on the surfaces of transition metals [34,35]. In 1975, a few sheets of graphite were produced using chemical breakdown techniques on a single platinum crystal, but the material wasn't identified as graphene due to uncharacterized production techniques or perhaps because of its limited prospective applications [36]. They were never tested for their electrical qualities at that time due to a lack of isolation and placement on insulating substrates. However, in the late 1990s, Ruoff [37] and colleagues

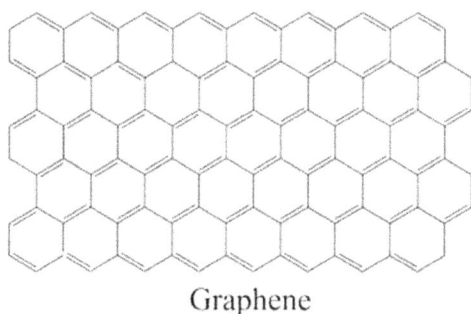

Graphene

FIGURE 11.1   The structure of graphene.

attempted to separate microscopic graphitic flakes on SiO_2 substrates by mechanically rubbing on patterned HOPG (highly oriented pyrolytic graphite) islands.

On the characterization of their electrical properties, there was, however, no information. The electrical properties were subsequently achieved in 2005 by Kim and associates using a comparable technology, and they were published [38]. Nonetheless, graphene research proceeded quite swiftly when Geim and colleagues were acknowledged for their work in isolating graphene on a SiO_2 basis and examining its electrical properties. Once the material was discovered in 2004, many techniques were developed to create very thin, few-layer graphitic films.

Figure 11.2 shows field emission scanning images taken with an electron microscope of graphite, nano-graphite (nGr), multiple-layered graphene nanoplatelets (GNP), and wrinkled generated graphene from GO [39]. Initially from the massively defined counterparts, the thin graphene structure and several substances including hexagonal boron nitride, disulphide, niobium diselenide, and molybdenum were utilized in 2004 [14] and 2005 [40], providing experimental proof of 2D crystals.

However, graphite was first manually exfoliated with scotch tape to produce tiny flakes on the order of few microns, which were then used to make graphene [41,42]. Although the graphene produced by this procedure is of the highest quality, mass manufacturing of graphene requires a fabrication technique that can manufacture wafer-scale graphene. Many techniques have been employed recently for the synthesis of graphene; however, at the present time, exfoliation (mechanical and chemical cleaving), chemical reaction synthesis, and thermal chemical vapor deposition (CVD) approach are the most widely utilized methods. An overview of the processes used to produce graphene is shown in Figure 11.3. The research works by Allen et al. [43], Viculis et al. [44], Park and Rouff [45], and Reina et al. [46] are considered the characterized.

11.2.1   TOP-DOWN GRAPHENE

In a top-down technique, graphite and the derivatives of graphite via graphite oxide (GO) and graphite fluoride are separated from one another or exfoliated to create graphene or graphene sheet modified. Table 11.1 can suggest the involvement of several researchers.

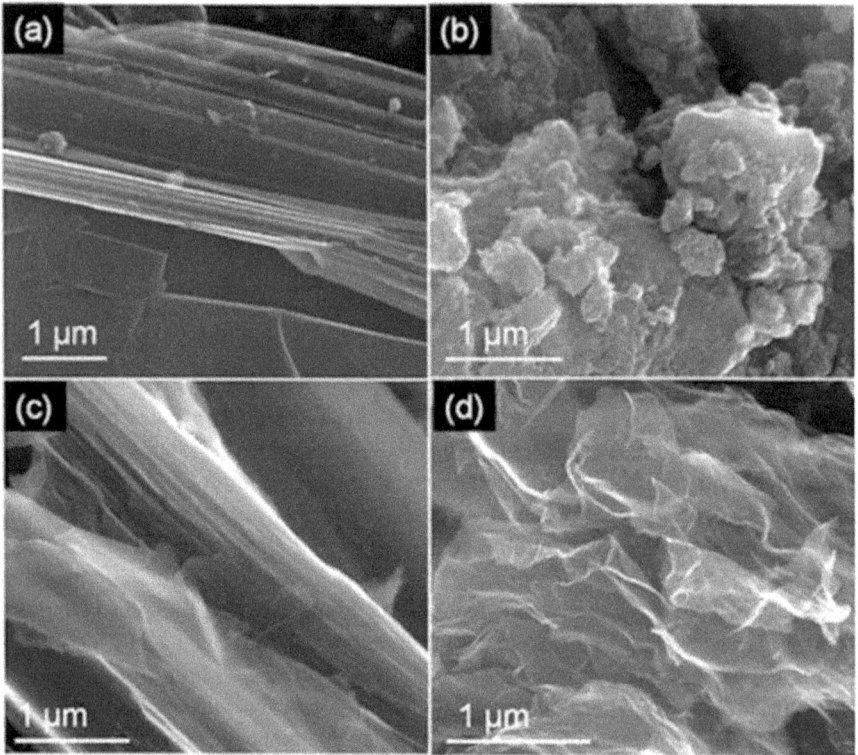

FIGURE 11.2 Electron microscope field emission scanning images of (a) graphite, (b) nano-graphite (nGr), (c) multi-layered graphene nanoplatelets (GNP), and (d) wrinkled derived graphene from GO [39].

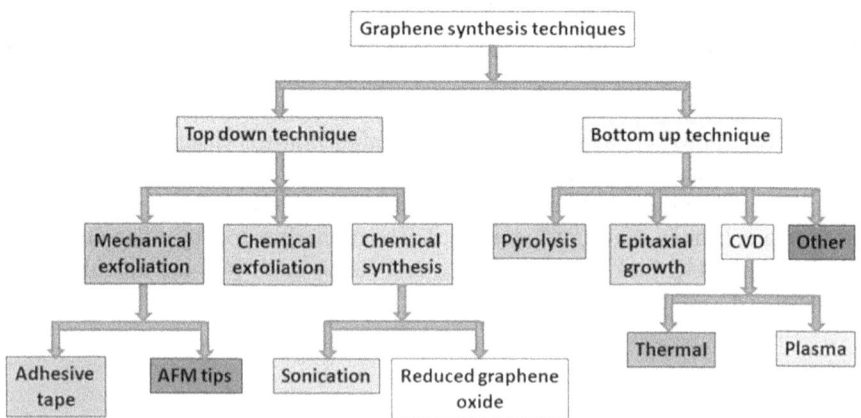

FIGURE 11.3 Graphene synthesis flow chart [47].

TABLE 11.1

Documented Top-Down Graphene Methods

Authors	Methods	Typical Dimensions		Advantages	Disadvantages
Hernandez et al. [48]	Micromechanical exfoliation	Some few layers	μm to cm	Huge size and unmodified graphene sheets	Low scale production
Liu et al. [49]	Sonication of graphite	One layer and multiple layers	μm or sub- μm	Unmodified graphene, inexpensive	Less output: particle separation
Behabtu et al. [50]	Electrochemical exfoliation or functionalization of graphene	One layer and few layers	500–700 nm	It is a single step functionalization and exfoliation; it offers larger electrical conductivity of the functionalized graphene	The cost incurred in the purchase of ionic liquids
Stankovich et al. [51]	Super acid dissolution of graphite	Mostly single layer	300–900 nm	Unmodified graphene; scalable	Utilization of hazardous chlorosulfonic acid; removal of acid involves huge cost

11.2.1.1 Mechanical Exfoliation

The uncommon and yet important method for obtaining graphene flakes in single layers on a particular substrate may be a mechanical process. Hence, this is the first known mechanical processing technique used to produce graphene. The concerned top-down nanotechnology method stresses the materials that make up the layered structure's surface longitudinally or transversely. Weak electrochemical interactive forces allow mono-atomic graphene layers to stack one on top of each other to create graphite. Interlayer separations and interlayer bonding energy are, respectively, 3.35 Å and 2.1 eV/nm^2. To separate layers that are defined as single mono-atomic from graphite mechanically 300 $nN/\mu m^2$, external force is needed. In graphite, stacking of sheets is caused by the overtaking of a partial occupied p orbital that is positioned perpendicular to the graphene plane sheet (showing van der Waals).

Since exfoliation exists as the opposite of pile-up there is weaker bonding and a larger lattice spacing in the right-angle direction than there is in the hexagonal lattice plane and vice versa. In fact, graphitic materials like pyrolytic carbon in graphite, graphite single crystal, or natural graphite could be used to peel off layers to create graphene sheets of various thicknesses [52–56]. A number of tools, including scotch tape, ultrasonication, [57], electric fields, and even transfer printing techniques [58,59], can be used to perform this peeling or exfoliation.

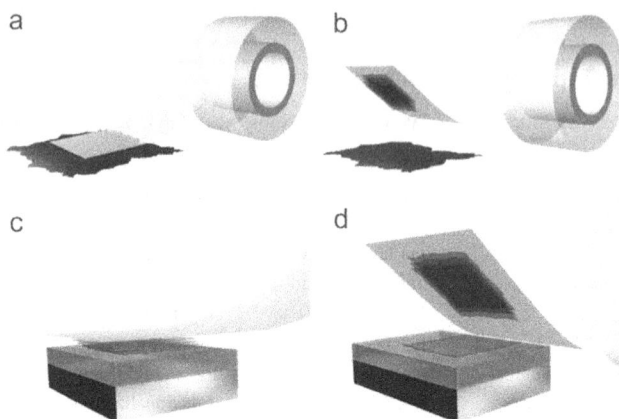

FIGURE 11.4 A mechanical exfoliation procedure step by step: top few layers of adhesive tape are bond to tape when it is placed against a HOPG surface in (a), tape with layered crystal material pressed against a surface of choice in (b), and (d) Tape is peeled off, bottom layer is left on the substrate [60].

The step-by-step process of mechanical exfoliation is mentioned in Figures 11.4 and 11.5 [60]. To increase the percentage of single and few layers of graphene chips, HOPG has been substrate fused using either conventional fixative epoxy resins [61] or SAMs [62]. Additionally, a recent study shows how gold films can be used to transfer print macroscopic graphene designs from patterned HOPG [63]. It is unquestionably the least expensive way to make high-quality graphene. Typically, Raman spectroscopy and optical microscopy, AFM are used to characterize graphene flakes produced by mechanical exfoliation techniques. Exfoliated graphene is subjected to an AFM investigation to determine the number of layers and thickness.

To find a mono-layer chip is purely a matter of chance, and using this method results in poorer yields of mono- or multi-layered graphene with flakes that are dispersed at random on the substrate. Another well-liked technique for finding single-layer graphene is optical microscopy. On top of Si wafers, a 300 nm thick thermally generated SiO_2 layer exhibits a distinctive color contrast depending on the area size of graphene flakes [64]. On graphene obtained through mechanical exfoliation, Raman spectroscopy is also used. The spectroscopic method is the simplest and most accurate way to gauge the thickness of graphene flakes and determine how crystalline they are. This is due to the fact that graphene displays distinctive Raman spectra depending on the quantity of layers present [65–67].

11.2.1.2 Chemical Exfoliation Method

One notable method used to create graphene is chemical synthesis. Colloidal suspensions are produced through a chemical process that modifies graphite and a graphite intercalation component into graphene. The chemical technique has already been used to make several types of graphene paper, graphene derived materials [68–74], graphene polymer composites [75], energy insulation materials [76], and magnetic transparent electrodes [77]. The first production of graphene oxide occurred

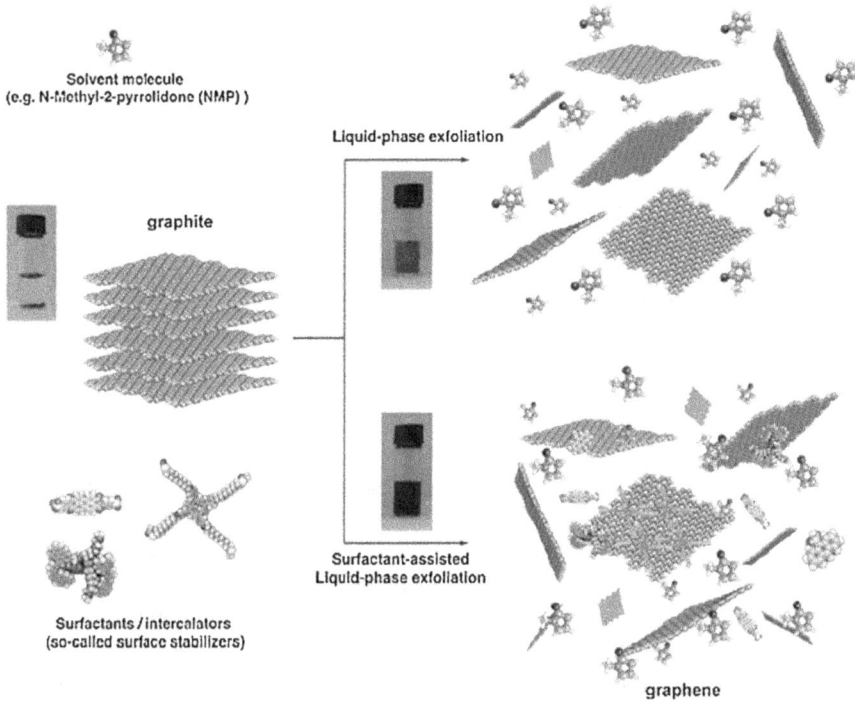

FIGURE 11.5   Schematic of liquid-phase exfoliation (LPE) graphite process in the absence (top-right) and presence (bottom-right) of surfactant [60].

in the year of 1860 by Brodie [78], Hummers [79], and Staudenmaier [80] processes. Exfoliation by chemicals is a two-step procedure. Increases in the interlayer separation by first diminish the between layers of van der Waals forces. Thus, graphene-intercalated compounds (GICs) are created [81]. Furthermore, sonication and fast heating are implemented in the exfoliated layers of graphene.

Ultrasonication [82–86] and density gradient ultracentrifugation [87,88] are used for single-layer graphene oxide (SGO), which has different layer thicknesses. By using the Hummers process, which involves oxidizing graphite using potent oxidizing agents like $KMnO_4$ and $NaNO_3$ in H_2SO_4/H_3PO_4, it is simple to produce graphene oxide (GO) [89]. Single-layer graphene was created via ultrasonication in a DMF/water (9:2) combination. The interlayer space rises from 3.8 to 9.6 A° as a result. Top functional group densities need oxidation, and reduction is required to achieve characteristics akin to graphene. Chemical reduction with hydrazine monohydrate is used to disperse single-layer graphene sheets. For the manufacture of graphene, polycyclic aromatic hydrocarbons (PAHs) [90,91] have been employed.

Utilize a direct precursor that has undergone planarization and cyclodehydrogenation [92] to create a few graphene domains in size. Larger flakes are produced by the precursor of poly-dispersed hyper-branched polyphenylene [92]. The first were created using $FeCl_3$ in an oxidative cyclodehydrogenation reaction [92]. Orthodichloro benzene [93], perfluorinated aromatic solvents [93], and even small-boiling solvents like

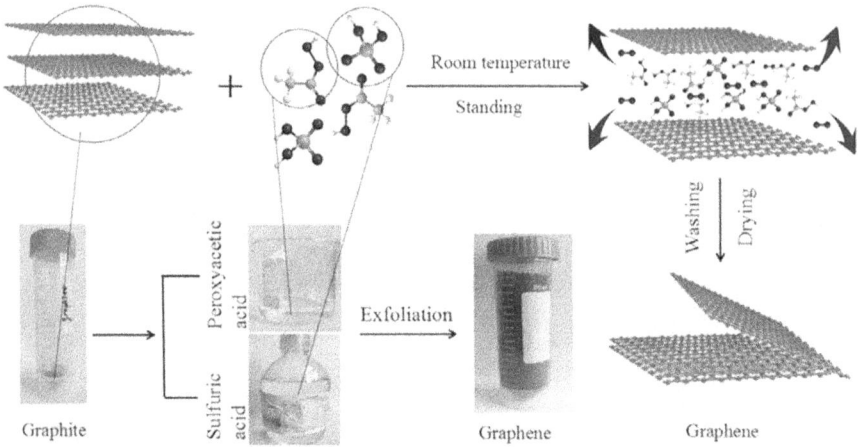

FIGURE 11.6   Schematic of chemical exfoliation mechanism of graphite into graphene [97].

chloroform and isopropanol [94,95] are employed to disperse graphene. Graphene on SiO_2/Si substrates exhibits electrostatic interaction between the Si substrate and HOPG. The HOPG has also been exposed to neodymium-doped yttrium aluminum garnet laser exfoliation in a pulsed order to create FG [96]. Reduced graphene oxide is produced via thermal scrap and graphite oxide (rGO) is reduced. Figure 11.6 illustrates the mechanism involved in the exfoliation of graphite into graphene [97]. Figure 11.7 provides a flow diagram of the rGO process prepared by the oxidation–reduction method.

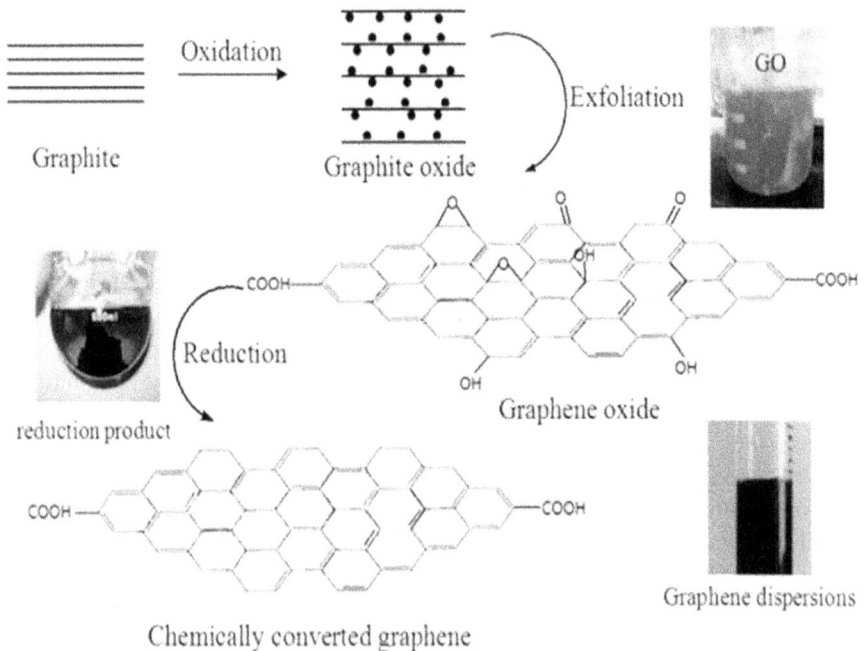

FIGURE 11.7   Process flow diagram of rGO prepared by the oxidation–reduction method [98].

11.2.1.3   Chemical Synthesis

One of the common methods for producing significant graphene amounts is synthetic chemical graphite oxide reduction. According to Brodie [78], Staudenmaier [80], and Hummers [79] methods, concentrated H_2SO_4, HNO_3, and $KMnO_4$ are used for graphite oxidation to produce graphite oxide (GO). The sonication and graphene oxide reduction is another method for the preparation of graphene. A significant amount of sodium boro-hydrate has been used as a reducing agent when H_2 is added across alkenes with nitrogen gas [99]. Phenyl hydrazine [100], hydroxylamine [101], glucose [102], ascorbic acid [103], hydroquinone [104], alkaline solutions [105], and pyrrole [106] are further reducing agents that have been utilized. The chemical reaction that created GO using isocyanates and hydroxyl functional groups is depicted in Figure 11.8 along with information about its FT-IR spectra.

An addition method for graphene production on a big scale is electrochemical reduction [107–109] as reduced graphene oxide monolayer flakes were originally discovered in 1962. To create GO nanoplatelets, the solution of graphite oxide can next be made by sonification. The oxygen functional groups can be reduced by using reducing agenthydrazine, albeit it was discovered that the reduction procedure was ineffective and some oxygen was still present. GO is advantageous because, in contrast to graphite, each of its layers is hydrophilic. To make single- or double-layer graphene, GO is first dipped in water using sonication [110,111] and then it is applied to surfaces by coat spinning or filtration.

After that, graphene films are created by thermally or chemically reducing the graphene oxide using a straightforward, solvothermal reduction process taking one step to create reduced graphene oxide that can be dispersed in an organic solvent [112]. Small organic compounds are used to decorate the suspensions

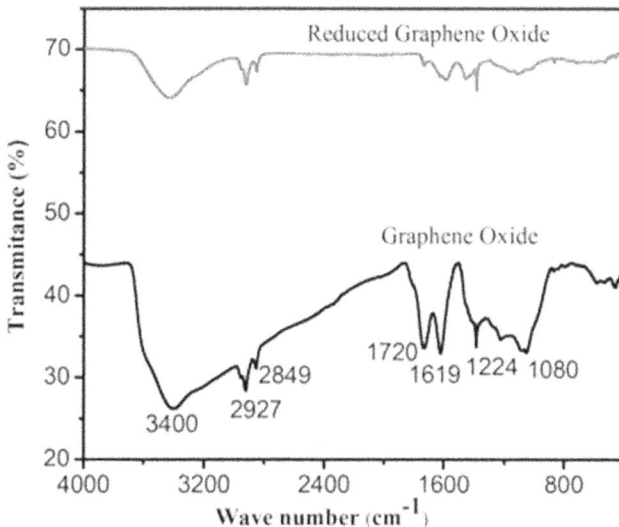

FIGURE 11.8   FTIR of the graphene oxide.

FIGURE 11.9   SEM image of reduced graphene oxide [115].

of colloids of chemically modified graphene (CMG) [72]: 1,2-distearoyl-sn-glycero-3-phosphoethanolamine-N [methoxy (polyethyleneglycol)-5000] [113], poly (m-phenylenevinylene-co-2, 5-dioctoxy-p-phenylenevinylene) (PmPV) [114], and poly (tert-butyl acrylate). Two cross-sectional FE-SEM and TEM images are displayed for the purpose of differentiating between GO and rGO in Figure 11.9 [115].

11.2.2   Bottom-Up Graphene

Table 11.2 provides an overview of the sheet type, width and dimensions, of graphene created via numerous bottom-up techniques, including benefits and drawbacks of each technique.

11.2.2.1   Chemical Vapor Deposition

CVD involves chemical reactions where the precursors, or heated gaseous molecules, are transformed. Substrates are dispersed on thermal precursors that disintegrate elevated temperatures in this CVD technique. On the substrate's surface, there are thin-film made deposits of solid, liquid, crystalline, or gaseous predecessor. High-quality graphene is often deposited via the CVD technique onto a variety of transition-metal bases, including Ni [121], Ru [122], and Ir [123]. Graphene has mostly been grown by CVD on surfaces made of copper [124] and nickel [125]. The first substrate on which large area CVD graphene development was tried was nickel.

TABLE 11.2

Bottom-Up Graphene History

Authors	Methods	Typical Dimensions		Advantages	Disadvantages
		Thickness	**Lateral**		
Zhanget al. [116]	Confined self-assembly	One layer	100's nm	Thickness control	Existence of defects
Li et al. [117]	CVD	Few layer	Very large (cm)	Large size; high quality	Small production scale
Rollings et al. [118]	Arc discharge	Single, bi and few layers	Few 100 nm to a few 1 m	Can produce 10 g/h of graphene	Low yield of graphene; carbonaceous impurities
Hirsch et al. [119]	Epitaxial growth on Si-C	Few layers	Up to cm size	Very large area of pure graphene	Very small scale
Kim et al. [120]	Unzipping of carbon nanotubes	Multiple layers	Few 1 μm long nanoribbons	Size controlled by selection of the starting nanotube	Expensive starting material; oxidized graphene
Novoselov et al. [14]	Reduction of CO	Multiple layers	Sub-lm	Un-oxidized sheets	Contamination with a-Al_2O_3 and a-Al_2S

These initiatives had started as early as 2008 [126]. Thin graphite was created in 1966 by exposing [127] Ni to methane at 900 C with the purpose of serving as an electron microscope sample support. They [128] noticed the creation of FLGs in 1971 by observing the evaporation of C off a graphite rod. The first thermal CVD mono-layer graphitic material deposition on Pt was reported in 1975. Later on, graphite layer development on Ni was reported by Eizenberg and Blakely [129] to examine the catalysts and thermal properties of Ir in the presence of carbon. In 1984, researchers [130] carried a study which led to the first graphene-grown CVD on the surface of Ir [131]. Figure 11.10 provides a complete apparatus setup for CVD of graphene [132]. Figure 11.11 is the schematic diagram of thermal CVD growth of graphene [133].

A new field of graphene-based electronics has been made possible by the de-tailed analysis of graphene's physical and chemical characteristics [134–136]. Camphor (terpenoids, a white, transparent solid with formula as ($C_{10}H_{16}O$)) was used as precedent material in the first effort to synthesize graphene on Ni foil using CVD in 2006. On diverse transition metals, viz., Cu, Ni, Co, Ru, and Au, various hydrocarbons including ethylene, acetylene, benzene and methane were degraded. It was discovered that graphene produced from single crystals employing an ethylene precursor was structurally coherent even at the margins of the Ir step [137].

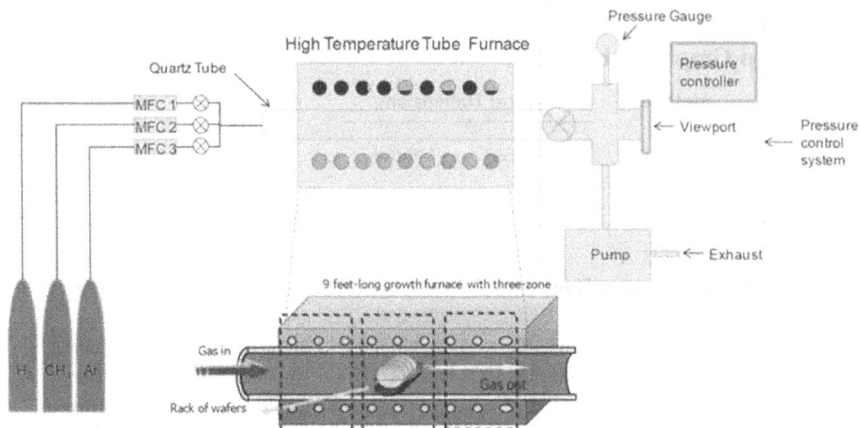

FIGURE 11.10 Complete apparatus setup for chemical vapor deposition of graphene [132].

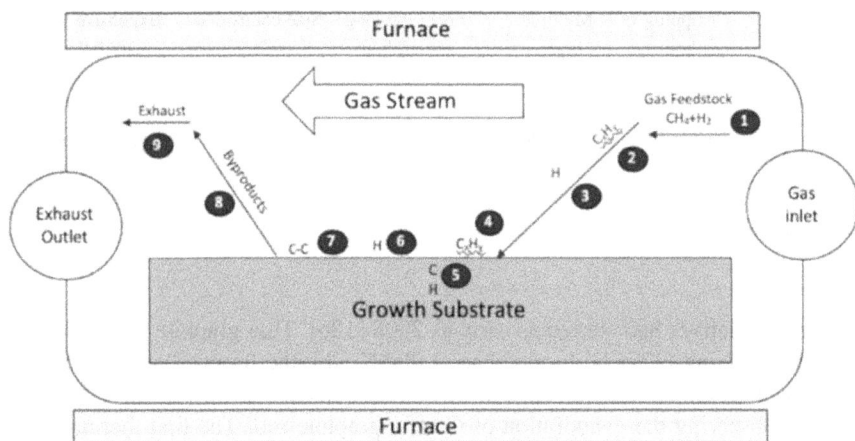

FIGURE 11.11 Schematic diagram of thermal CVD growth of graphene [133].

11.2.2.2 Epitaxial Growth of Graphene

One of the most lauded processes for producing graphene is epitaxial heat growth on (Si-C), single crystalline silicon carbide surface. The word "epitaxy" is known to be Greek in origin; the prefix "epi" implies "upon" or "over" while the word "taxis" implies "arrangement" or "order." Epitaxial growth is a process that results in epitaxial films when a single crystalline layer is deposited on a crystalline single substrate. Single crystalline Si-C substrate fabricates high-crystalline graphene. Depending on the substrate type, there are two general epitaxial processes for growth: homo-epitaxial and hetero-epitaxial development. A homo-epitaxial layer is one that has a film placed on a substrate made of the same material, while a hetero-epitaxial layer has a film substrate made from different substances.

The electrical determination on patterned epitaxial graphene was the first application of Si-C in this context. Si-C has a huge number of semiconductor bandgaps (3 eV) that was used as the substrate for electrical investigations in 2004 [138]. Graphite production on the 6H-SiC (0001) surfaces was originally documented by Bommel et al. in 1975 [139]. In both the Si-C polar planes, graphite was created through a heat treatment that took place between 1000 and 1500 C in extremely high vacuums (1–10 m bar) (0001). After completion of the Si (0001) face of a single 6H-SiC crystal, the de Heer group reported fabricating ultrathin graphite in 2004 [140] and investigated its electrical characteristics [141].

The unique polar Si-C crystal face affects the growth of graphene on Si-C [140,141]. In the C-face rather than the Si-face, graphene develops significantly more quickly [141]. Larger domains of many layered and rotation-wise disordered graphene (200 nm) are created on C-face [142]. Small areas, 30–100 nm in size, are prepared on Si-face by UHV annealing [143]. Evaporation of Si [144,145] materials that have been heated at high T (1000 C) in ultra-high vacuum (UHV) graphitize. Graphene forms on a C-rich 6H3 9 6H3R30 re-preparing the building with respect to the Si-C surface by thermal breakdown of Si-C over 1000 C [146].

A very promising technology for the industrial manufacture and commercialization of graphene for use in electronics has been envisioned as epitaxial graphene growth on Si-C. High-frequency electronic devices, light fluorescent devices, and radiation-resistant devices are all made possible by graphene on Si-C [147]. On a wafer scale, top gated semiconductor made of graphene on Si-C has been created [148]. Additionally, high-frequency transistors with a cut-off frequency of 100 GHz have been demonstrated [149], which is greater than the current-generation Si transistors with a similar gate length [150]. Based on the quantum Hall Effect (QHE), graphene on Si-C has been prepared as a unique resistance standard [151]. Nevertheless, this method is highly costly. Figure 11.12 demonstrates the graphene epitaxial growth mechanism [152].

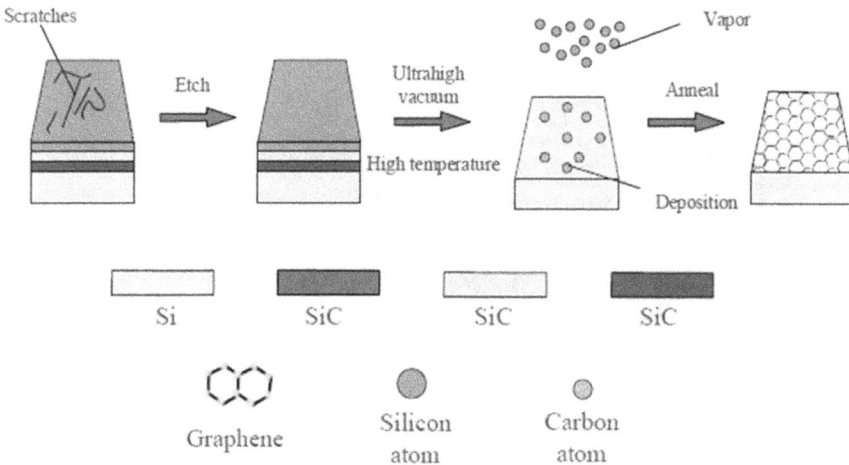

FIGURE 11.12   Epitaxial growth graphene mechanism [152].

11.2.2.3 Unzipping Method

Using chemical and plasma-etched techniques, a CNB is unzipped (CNT). The term "graphene nanoribbon" (GNR) refers to a narrow, long graphene strip with sharp edges. The width of the nanatube affects how an electronic state transitions from metalloid to semiconductor [153]. Depending on whether the initial nanotube has several walls or only one, either multi-layer graphene or one-layered graphene gets produced. The precursor nanotubes' diameter determines the breadth of the nanoribbons that are thus generated. Lithium (Li) with ammonia create multi-walled carbon nanotubes (MWCNTs) (NH3). When growing MWCNTs, dry tetrahydrofuran (THF) and liquid NH3 (99.95%) were utilized to maintain the -77 C dry ice bath temperature [154].

The amount of fully exfoliated nanotubes (60%) and the amount of unexfoliated or partially exfoliated nanotubes (0–5%) of MWCNTs. H_2SO_4, $KMnO_4$, and H_2O_2 were utilized in a sequential process to oxidize the side walls of CNTs [155]. It was stated by researchers that the MWCNT diameter ranged from 40 to 80 nm and then grew to 100 nm. The Si-like substrate was prepared with 3-aminopropyltriethoxysilane before the pristine MWCNT (diameter of 4–18 nm) solution was applied using a controlled unzipping approach. A solution of polymethylmethacrylate (PMMA) [156], researchers developed MWCNTs of high quality with diameters between 6 and 12 nm; the step height of GNRs is between 0.8 and 2 nm. Again, the plasma etching time also affects how many layers are there in GNRs. MWCNTs to GNRs were also unzipped using an electric field. A tungsten electrode was used to provide an electric field to a single MWCNT, and it was noticed that the non-conductive side of the MWCNT began to unravel and produce a graphene nanoribbon. For scalable devices in contemporary electronics, GNRs fabrication results in a high-purity, free defection-controlled preparation. Figure 11.13 is the fully unzipped method for graphene oxide [157] and Figure 11.14 is also a fully unzipped method for graphene oxide [158].

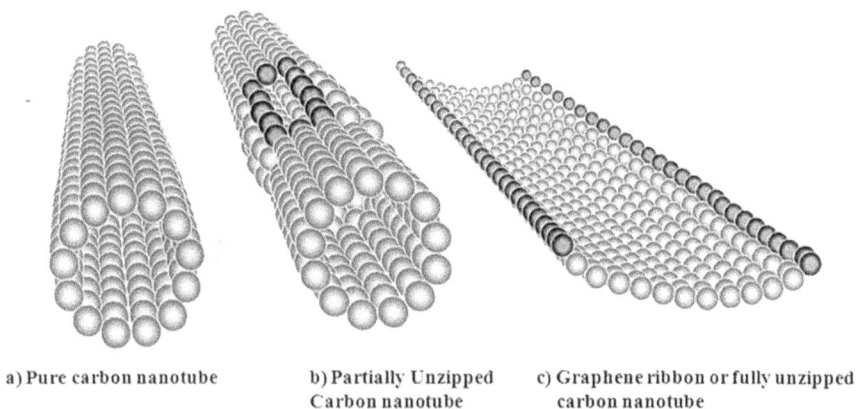

a) Pure carbon nanotube   b) Partially Unzipped   c) Graphene ribbon or fully unzipped
                                     Carbon nanotube   carbon nanotube

FIGURE 11.13 Fully unzipped method for graphene oxide [157].

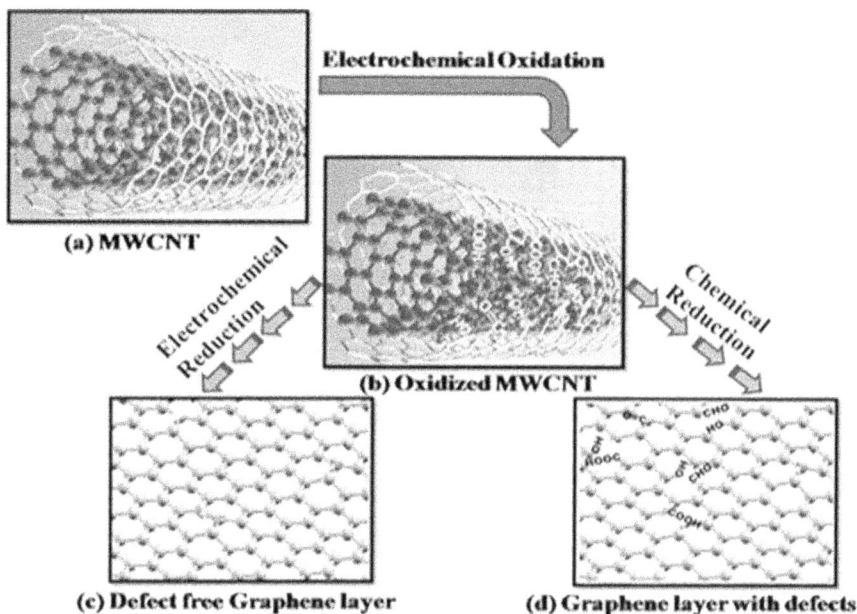

(a) MWCNT

Electrochemical Oxidation

(b) Oxidized MWCNT

Electrochemical Reduction

Chemical Reduction

(c) Defect free Graphene layer

(d) Graphene layer with defects

FIGURE 11.14   Unzipped method for graphene oxide [158].

11.2.2.4   Other Techniques

Other methods of producing graphene include thermal fusion of PAHs [107], the graphite arc discharge [108], PMMA nanofiber electron beam irradiation [109], and inversion of nanodiamond [110]. Arc discharge can be used to create graphene with two to three layers and flake sizes between 100 and 200 nm [111]. The nanosheets of graphene with 110–195 nm diameter typically have two layers produced by arc discharge during a quick heating procedure. Elevated current (over 100 A), elevated voltage (above 50 V), and elevated hydrogen pressures are ideal circumstances for getting the graphene in the inner walls (more than 200 Torr).

The starting air pressure has a significant impact on the formation of the graphene layer [112]. As an arc discharge method, He and NH_3 atmosphere are also utilized [72]. To obtain various numbers of graphene sheets in the He environment, gas pressure and currents were considered. Ethylene gas source was employed in the molecular beam deposition process to deposit on a nickel substrate. Depending on the cooling rate, large area and extremely pure quality of graphene layers are prepared.

11.3   GRAPHENE MATERIAL FOR BATTERY APPLICATIONS

Currently, the LIBs are highly utilized type of energy storage materials. The LIBs have several significant advantages over other battery types, including a high amount of energy storage, high Coulombic forces, energy efficiency, no memory effect, a huge cycle life, all designed with flexibility. As a result, LIBs are preferred for a variety of products, including smart phones, laptop computers, and power banks. Among the currently commercially available batteries, LIBs are also thought

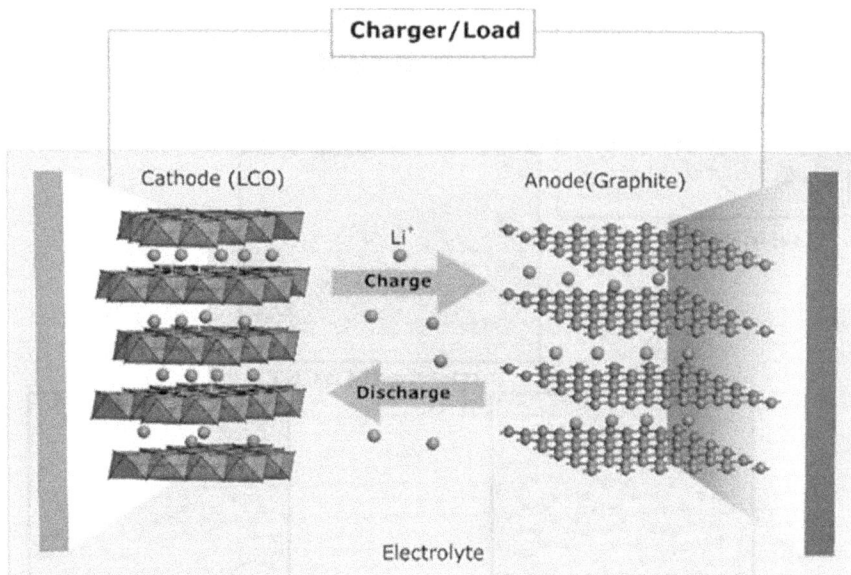

FIGURE 11.15   The operating principles of LIBs [164].

of as a promising source of power for electric vehicles (EVs) and hybrid electric vehicles (HEVs). As a result, significant resources have been invested in research to develop LIBs with improved performance in a more economical manner.

Since they have significantly better energy densities, LIBs have volumetric and gravimetric properties compared to other commercial batteries. The LIBs are considered essential in the rapidly expanding EV market as well as for portable power supply. Today, energy storage lithium compounds have the cathodes and anodes used in secondary batteries, better known as LIBs. Figure 11.15 [98,159–164] depicts the LIBs' operational principles.

The LIBs are frequently referred as "rocking chair batteries" since they oscillate backward and forward between the electrodes when the battery is being charged or depleted. The positive electrode layered metal oxides include Li-Ni-Co-MgO (LiNi$_1$/3Co$_1$/3Mn$_1$/3O$_2$ or NCM), Li-CoO$_2$ Li-cobalt oxide (LiCoO$_2$ or LCO), and lithium nickel cobalt aluminum oxide (LiNiCoAlO2 or NCA), as well as lithium manganese oxide (LiMnO$_2$ or LMO), spinel metal oxide materials, and olivine lithium iron phosphate (LiFePO$_4$ or LFP) materials. Graphite and lithium titanate oxide (Li$_4$Ti$_5$O$_{12}$ or LTO) both fall under the category of insertion materials and are frequently used as the negative electrodes in commercial batteries. Research is now being done on alloys including Ge, Si, Sn, and metal oxide conversion materials.

11.4   GRAPHENE-BASED ANODES

According to the reaction mechanism, LIB anode materials are divided into a few categories: insertion (carbon-containing materials, lithium titanate, etc.), alloys such as (Si, Sn, etc.), and conversions (e.g. transition metal oxides, sulfides, nitrides, and so on). Carbonaceous materials, such as hard carbon, soft carbon, graphite, and meso-carbon microbeads (MCMB), all are categorized as insertion type materials and are the most often utilized anode materials in commercial cells; graphite interlayers can have Li intercalated into them up to stoichiometry LiC_6. The applied graphite use as a LIB anode was mentioned in the previous section as a SEI layer that possessed stability which restricts electrode from reacting with electrolyte, while speed Li+ transfer is essential and allowed. This is because the electrochemical Li-ion potential for insertion reactions into graphite is out of scope for the electrolyte window.

Other carbonaceous materials, such as hard and soft carbons, may also exhibit Li-ion intercalation at electrochemical charges similar to those of the Li metal electrode, requiring creation of a SEI layer [165]. Li-ions are consumed during SEI development and become a SEI layer and additional becoming irreversible. Higher specific surface area carbon materials are likely to consume more Li ions in the first few cycles to form SEI layer. This appears to have a larger initial irreversible capacity because the Li consumption for irreversible formation of SEI layer is dependent upon the surface area of the carbon [166]. Little fissures on the SEI layer developed during charging and discharging the graphite anode can be repaired by the electrolyte's reaction with a freshly exposed graphite surface, which will create a new SEI layer. Nevertheless, long cycling results in thicken SEI and increased electrode impedance [167], as shown in Figure 11.16. Due to its high charge/discharge potential that prevents the formation of a SEI layer, graphite anodes have a shorter cycle life than lithium titanate $Li_4Ti_5O_{12}$ (LTO) [167].

Li plating and dendrite formation, other problems brought on by the charge/discharge potential are extremely dangerous, especially in large-scale, high-power batteries [165]. In materials made of disordered carbon, the Li-ion storing process is more intricate. The disordered carbon materials frequently contain some micro- or nano-crystallites, which are composed of a minimal number of GE layers held due to the van der Waals force. On dispersive graphene layers, electrolyte/solvent co-intercalation and other side reactions result in the disordered carbon capacitive behavior, which is characterized by elevated capacity, elevated rate performance, and low Coulombic proficiency [169]. Structurally flawed disordered carbons exhibit increased polarization due to elevated charge transfer resistance, which results in high internal power loss. Mesocarbon microbeads (MCMB) and other chemically synthesized carbon products, including vapor grown carbon fibers (VGCFs), mesophase-pitch-based carbon fibers (MCFs), and massive artificial graphites (MAGs), have low specific capacities (i.e. 372 mAh/g), which becomes problematic for HEV, PHEV, and PEV applications. Materials that can be alloyed with Li include substances like Si, Sn, or SnO_x. They frequently exhibit potentials that necessitate SEI production, similar to carbon materials, and have large theoretical capacity values, typically between 1000 and 4000 mAh/g.

11.4.1 ACTIVE GRAPHENE ANODE MATERIAL

Graphene in 2D has attracted much interest because of the extraordinary properties it possesses such as high charge carrier mobility, high conductivity, mechanical

Fading mechanisms of graphite electrodes in any "good" solutions

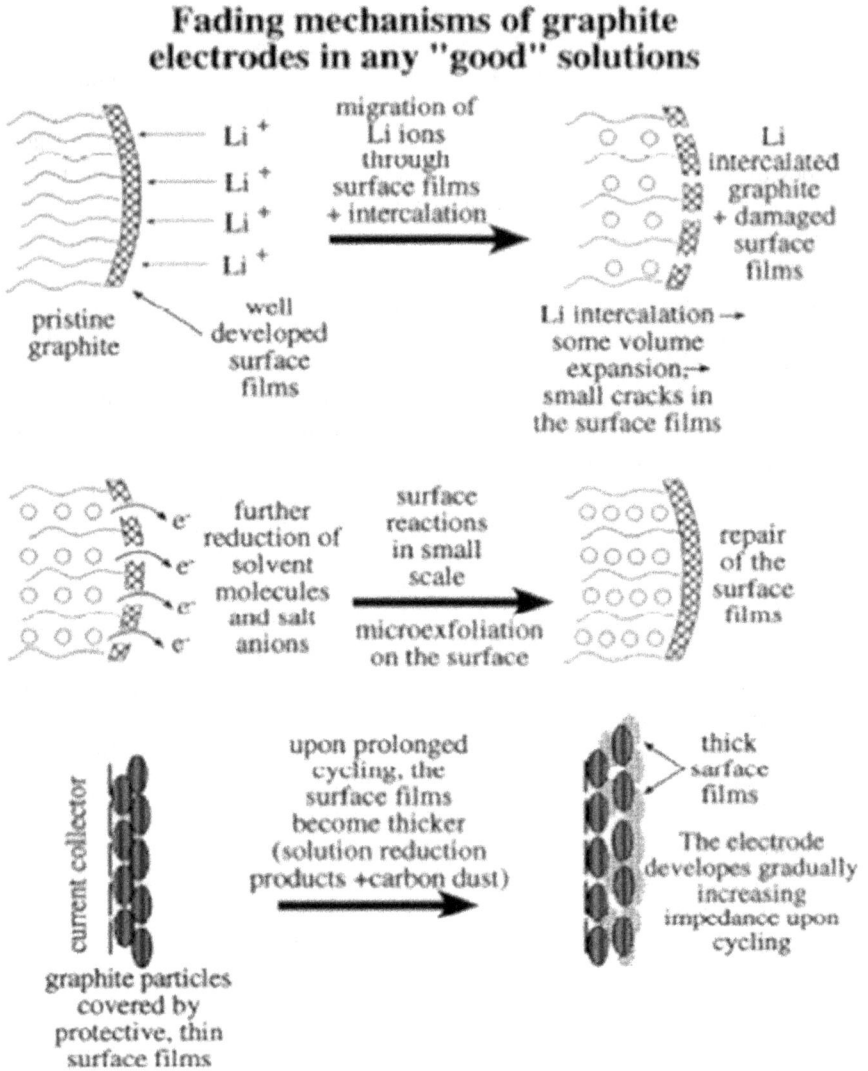

FIGURE 11.16 Development of SEI upon cycling [168].

strength, and high surface area. Graphene is the perfect electrode for LIBs due to these characteristics. Graphene has a theoretical capacity of 744 mAh/g based on Li_2C_6 stoichiometry, which permits Li ions to link on to both single-layer sides of the graphene sheet. However, the LiC_6 phase cannot form on single-layer graphene because of the weaker Li and carbon binding energies and significant Coulombic repulsive forces of the Li atoms on the sidewalls of graphene, as shown by Raman spectroscopy *in-situ* on CVD single and layered graphene. There is low surface coverage (5%Li) for a single sheet of graphene, and this corresponds to an extremely low Li capacity of just LiC_2O [170].

Defects can significantly improve Li-ion adsorption and diffusion as well as decrease aggregation, according to recent theoretical research on flaws in graphene materials [171]. This further improves the electrochemical performance. Exfoliation of pure graphite in solution allows for the vast production of chemically produced graphene compounds like G-O and rGO. In order to improve conductivity and restore sp^2 bonding regions, GO is often reduced to rGO. GO is a very unique material with disturbed sp^2 bonding. In comparison to those on the graphene basal plane, rGO provides significant edge and defect sites with substantially greater binding energies and low Li ion diffusion batteries.

The quick electron transit to the current collector is made possible by the high conductivity of rGO. As of right now, the best rGO anodes have specific capacities of about 1200 mAh g^{-1} discharged fifth at 100 mA/g in Li half-cells [172] and about 100 mAh/g at 29 mA/g, when a fully combined battery is present [173].

11.5 GRAPHENE-ACTIVE COMPOSITE MATERIALS

11.5.1 INSERTION MATERIAL

The anode material utilized in commercial batteries is $Li_4Ti_5O_{12}$ (LTO). In applications where safety and stability are more important, LTO is a contender that also has an extended cycle life. When combined with its potential platform at 1.5 V against Li/Li+, it is a strong anode. This is due to the reaction, which is known as a zero-strain insertion reaction, which involves lithium ions and electron extraction insertion from the solid matrices. The low electronic conductivity (1013 S/cm) of LTO is a disadvantage; however, it is commonly compensated for by composites including carbon materials, especially graphene materials [174].

Shi et al. [174] used a straightforward mixing technique that produced LTO-rGO composite. In comparison to 170 mAh/g at 1 Cycle, the hybrid materials delivered a specific capacity of 122 mAh/g at a charge/discharge rate of 30 Cycles. Moreover, this material has remarkable cycling behavior, with a discharge loss in capacity less than 6% after 300 cycles at 20°C. LTO/rGO was created by Ding et al. [175] and offers a better capacity of 120 mAh/g with a high current density of 1600 mA/g. The LTO nanoparticles anchored on rGO sheets was found to be less in size than those made without rGO. The substantial pseudo-capacitive effect on LTO/rGO can be attributed to the enhanced high-rate performance since current-voltage analysis shows that reactions that are surface-confined predominate at higher frequency.

With its extended cycle life, low cost, and relatively high Li insertion/extraction voltage, TiO_2 has numerous benefits that are similar to those of LTO. Researchers want to increase TiO_2's natural electrical conductivity (10–12 S/cm) by using graphene-based compounds.

Using $TiCl_4$ as a precursor, Tao et al. [176] created a composite of TiO_2 nanoparticles and rGO and discovered that TiO_2 produced with G-O (TiO_2-GNS) exhibited an anatase phase, whereas pure TiO_2 produced the rutile phase only. TiO_2-GNS has enhanced electrochemical performance because anatase is often thought to be the phase that is more electrochemically active. The best TiO_2-GNS offers 198 mAh/g reversible capacity after 50 cycles at a density of 50 mA/g, compared to TiO_2, i.e., 96 mAh/g.

The anatase TiO_2 nanospindles were created and disseminated across functional G-O by Qiu et al. [177] using a spontaneous self-assembly technique.

The composite was further reduced and annealed in NH_3 gas to nitridate the surface of the TiO_2. The anatase@oxynitride/titanium nitride-GS ($TiO_2@TiO_xN_y/$ TiN-GS) nanocomposite hybrid electrode displayed a capacity of 166 mAh/g at 1 Cycle compared with 145 mAh/g for pure TiO_2 nanospindles, and 150 mAh/g for anatase oxynitride/titanium nitride($TiO_2@TiO_xN_y/TiN$) without rGO. A significantly low polarization was also obtained. Compared to the 260 mV for pure TiO_2 nanospindles, the polarization of $SP-20@TiO_xN_y/TiN$ was 220 mV, and that of $TiO2@TiO_xN_y/TiN$-GS was 170 mV.

Hu et al. [178] used a binder-free rGO-TiO_2 flexible paper as an electrode. He was able to attain a capacitance of 157 mAh/g after 100 cycles at a 200 mA/g density. When the density rose to 2 A/g, specific capacity remained at a constant 122 mAh/g. In each of the aforementioned cases, the standard voltage had 1.0 V break-off voltage versus what Li/Li+ was used. The composite was discharged at a voltage that is roughly Li/Li+ and is typically less than the voltage cut-off of graphene anodes which is referred to as opening the voltage window. This research emphasizes benefits and weaknesses of graphene-based anodes by viewing them as active Li+ insertion materials.

Sol-gel synthesis was employed by Xiang et al. [179] to create an LTO/rGO composite containing a graphene content of 7.8 wt.%. The LTO/rGO composite had a reversible capacity of 174 and 146 mAh/g at 0.2C and 10C, respectively. When measured in a regular potential window and compared to the reference sample capacity at 0.2C, the capacity was significantly lower at 10C (131 mAh/g). The discharge curve can be divided into 2.5–1.0 V and 1.0–0 V regions in the enlarged voltage window. LTO has a total specific capacity of 270 mAh/g, of which 87.5 mAh/g at 0 V related to the change of phase from $Li_7Ti_5O_{12}$ to $Li_{8.5}Ti_5O_{12}$ and 160 mAh/g at 1.0–2.5 V to the shift from spinel to rock-salt structure.

The memory of Li and irreversible SEI development on rGO sheets are responsible for the LTO/rGO composite's elevated capacity (270 mAh/g) in the 0–1.0 V range. The LTO/rGO composite charge curve outside the primary platform slopes at 1.5 V; this too is a key property of charging graphene, in contrast to pristine LTO sample's vertical charging curve.

TiO_2 nanoparticles anchored on rGO sheets were studied by Cai et al. [180] at two distinct window potential of 1.0–3.0 V and 0.01–3.0 V and under comparable conditions. TiO_2/rGO exhibits improved specific charge capacity (499 mAh/g) in the wider voltage window (0.01–3 V) and is comparable to pure TiO_2 with a voltage window of 1.0–3 V (200 mAh/g). When measured from 0.01 to 3 V, only 45.0% Coulombic efficiency of TiO_2/rGO detected; hence, this is much lower than the Coulombic efficiency of bare TiO_2, which is 71.8%. These results suggest extreme SEI production and irreversible Li ion insertion when lower cut-off voltage was discharged.

11.5.2 ALLOY ANODE MATERIAL

Due to their elevated specific capacity, alloy materials are a potential for anode materials. Yet, cycling's significant volume growth drastically shortens their cycle life. In

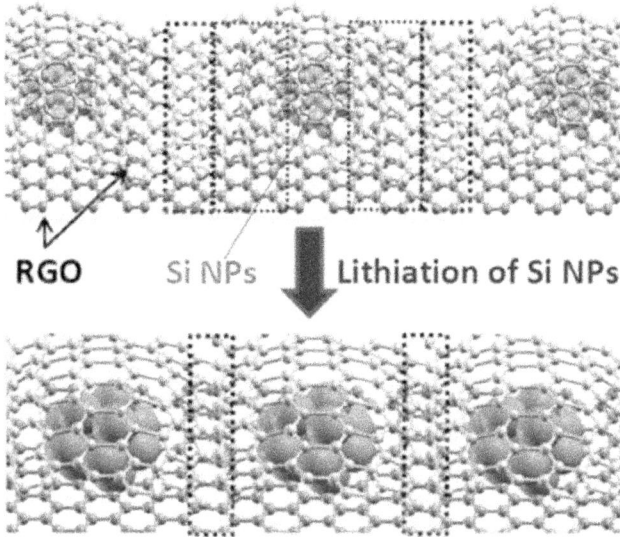

FIGURE 11.17   Graphene paper buffering Si volume change providing a conductive cycle network [181].

order to achieve both mechanical integration and electrical conductivity, graphene materials are utilized in composite materials with active alloy materials.

Due to high capacity (theoretical capacity of 4200 mAh/g) and low cost, Si is a widely researched alloying-type anode material. However, Si anodes are poor electrical conductors with high volume expansion of 300% following lithiation. Much research has used G-O to create binderless, free-standing electrodes.

The Si nanoparticle-in-graphene paper model underwent significant research, as seen in Figure 11.17. In this structure, a conductive network consisting of rG-O sheets connects islands of Si nanoparticles. Van de Waals forces cause regions furthest from the Si nanoparticles to stack up again, improving mechanical stability, whereas rG-O sheets may deform easily to fit the Si nanoparticles volume change [181].

By evaporating a steady aqueous dispersion of silicon nanoparticle/G-O and collecting the thin film that developed at the air/liquid interface, Tang et al. [182] were able to create a Si nanoparticle/GO film. Hydrazine was then used to decrease the GO, creating a Si/porous rG-O composite film. The free-standing electrode can reach a 1000 mAh/g capacity at 500 mAh/g. Moreover, it has a capacity of nearly 2100 mAh/g initially at 50 mAh/g and 1261 mAh/g steady after 70 Cycles. Jiang et al. [183,184] used a filtering technique to create Si/GO paper and used thermal reduction to create Si/G paper. At a current density of 100 mA/g, the Si/G paper has a capacity of 1500 mAh/g and is stable over 100 cycles.

11.5.3 ANODE CONVERSION MATERIAL

Lithiation conversion reactions are processes where lithium completely reduces active substances to create a metal and compound that contains lithium. Several transition metal oxides, phosphides, sulfides, and nitrides are components of this type of material. An amorphous organic matrix and a Li-containing discharge product were combined to generate an anode-electrolyte inter-phase layer for Li-ion battery anodes, which was then broken down during charging. The material must be nanosized in order for this kinetically constrained process to occur since phase change in nanomaterials is frequently heterogeneous [180]. Large overpotentials are typically needed by conversion electrodes to move the reaction. Even though the cause of this specific potential is not entirely known [185], the design of the electrodes has a significant impact on polarization [186]. In conversion anode materials, graphene materials were introduced primarily to inhibit aggregation, accommodate volume variation, and give excellent electrical conductivity.

After 130 cycles, Yang et al.'s [187] electrostatic co-assembly of graphene (rGO)-encapsulated metal oxide (GE-Co_3O_4) produced a high-capacity device with a capacity of 1000 mAh/g. According to Su et al. [188] a2D carbon-coated graphene (rGO)/metal oxides nanosheets G@MO@C were produced by hydrolyzing metal salt on functionalized G-O, polymerizing phenol and formaldehyde to create a phenol-formaldehyde (PF) coating and carbonizing. The G@MO@C carbon shells preserve mechanical stability and electrical conductivity by buffering the volume change of MO nanoparticles and have high electrical conductivity.

After 100 cycles, the G@SnO_2@C and G@Fe_3O_4@C electrodes produced 800 and 920 mAh/g capacities at a rate of 200 mA/g, respectively. The mesoporous nanostructure Fe_3O_4 that Luo et al. [189] produced on a 3D graphene foam substrate material using the ALD has a high capacity of 785 mAh/g at 1C rate and maintained with insignificant decay up to 500 cycles. At a capacity of 200 mAh/g, the performance rate was measured for densities up to 60C. Sun et al. [190] created a hybrid anode MnO/rGO with a reversible capacity as high as 2014 mAh/g and a charge/discharge current of 200 mAh/g. The capacity rises throughout the course of the first 150 cycles steadily. Platform appears at about 2.1 V; this illustrates that Mn is accessible at higher oxidation states. During cycling, the MnO nanosheets might disintegrate into ultrafine manganese oxide nanoparticles that could be sandwiched between the rGO layers to create a new platform with higher capacity. An analysis by TEM made this point clear.

Due to graphene materials' predominant conductor rather than active material role, materials that have greater charge/discharge voltages are not likely to experience major issues with first-cycle efficiency. This is because parameters (number of layers, density of defects, functional groups, etc.) of the graphene utilized differ in these studies. Hence, there is no significant association between graphene material quantities, first-cycle efficiency, and capacitor-like behavior. It is challenging to relate graphene content and electrochemical behavior since nanosized active materials might possibly be a factor in the aforementioned shortcomings.

11.6 GRAPHENE CATHODE MATERIALS

As previously indicated, metal oxides that are layered like LCO, $LiNiO_2$ or LNO and NCM, olivine transition-metal phosphates like LFP and LVP, and spinel LMO are the most typical cathode for LIBs materials. These cathode materials have acceptable Li-ion diffusion coefficients but have poor conductivity for rapid reactions. For example, electrical conductivities of pure LFP, LCO, and LMO are 10^{-3}, 10^{-5}, and 10^{-9} S/cm, respectively [191,192]. They frequently have lower rate capabilities because of their poor conductivity, which makes it more difficult to fast charge or discharge a battery. Carbon black is used as a conducting component in the production of commercial LIBs to enhance their electrochemical characteristics.

High conductivity, chemical stability, and graphene mechanical strength have all been shown to make it a good addition for enhancing cathode's electrochemical performance. Because of graphene's planar, flexible sheet-like structure and "liberated" electrons, which can travel more freely, conducting networks can be formed with materials that are electrochemically active. There is a decrease in internal resistance of the batteries and an increase in the output power due to the passage of electrons through graphene current collectors and the electrochemical activity of the nanoparticles. Strong mechanical qualities of graphene sheets help to increase cycle stability.

11.6.1 GRAPHENE-OLIVINE CATHODE MATERIAL COMPOSITES

11.6.1.1 Graphene-LiFePO$_4$ Composites

Often known material is LiFePO$_4$(LFP), an olivine-structured cathode material that has undergone extensive research and used in large-scale battery pack productions due to its low costs, environmental friendliness, and safety features. LFP has a theoretical capacity of 170 mAh/g but negligible intrinsic electrical conductance (10^{-9} S/cm). Furthermore, because Li-ion diffusion channels inside LFP particles are 1D channels, obstruction is frequently brought on by the flaws. To address these issues, carbon coating and nanosizing have been utilized.

The introduction of rGO as an additive to fabricate a graphene-based conducting network employing the "plane-to-point" conducting mode was announced by Su et al. In order to achieve the same conductive carbon black super-P(SP) results used in conventional electrode manufacture, rGO (abbreviated as GN) was added to the conductance material and distributed among particles, as illustrated in Figure 11.18. rGO with the active material have a contact mode of plane-to-point, which provides enhanced interaction between the active material and conductive filler, as opposed to the point-to-point model between active materials and SP. Rapid electron transit inside the plane is made possible by rGO, and modest amounts of rGO additives can create an efficient conductive network. In the EIS testing, it was illustrated that samples with 2 wt% rGO had lower charge transfer resistance and greater charge/discharge execution (138 mAh/g) than samples with 20% carbon black (122 mAh/g).

LiFePO$_4$ nanoparticles anchored on G-O that is decreased using various techniques were created by Yang et al. [194]. Reduced graphene LiFePO$_4$/hydrazine and LiFePO$_4$/thermal reduced graphene had discharge capacities of 166 and 86 mAh/g in the 50th cycle, respectively, when charged at 0.1C. The former's polarization value of 31.4 mV was 47.4 mV lower than the latter. In order for graphene materials to work as intended in Li-ion batteries, the reduction process of graphene is equally crucial.

The structured LiFePO$_4$/rGO (referred to as LFP@GNs by the authors) "platelet-on-sheet" was created by Wang et al. [195] using the solvothermal process. According to Figure 11.19, the electrostatic interaction between G-O and Li ions helps to ensure that Li$_3$PO$_4$ is distributed evenly on GO.

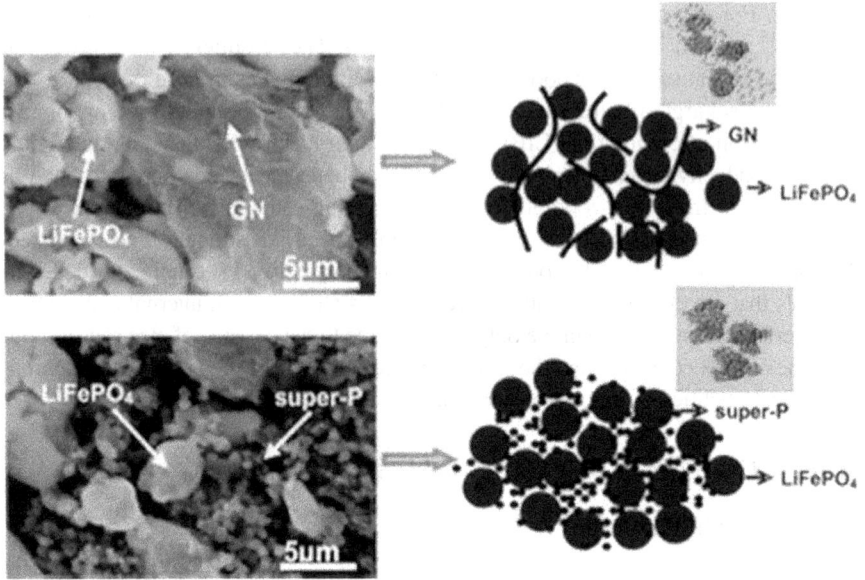

FIGURE 11.18 Graphical illustrations of the point-to-point mode for SP and plane-to-point mode for GN showing conducting mechanisms of GN and SP as conductive additives in LiFePO$_4$ [193].

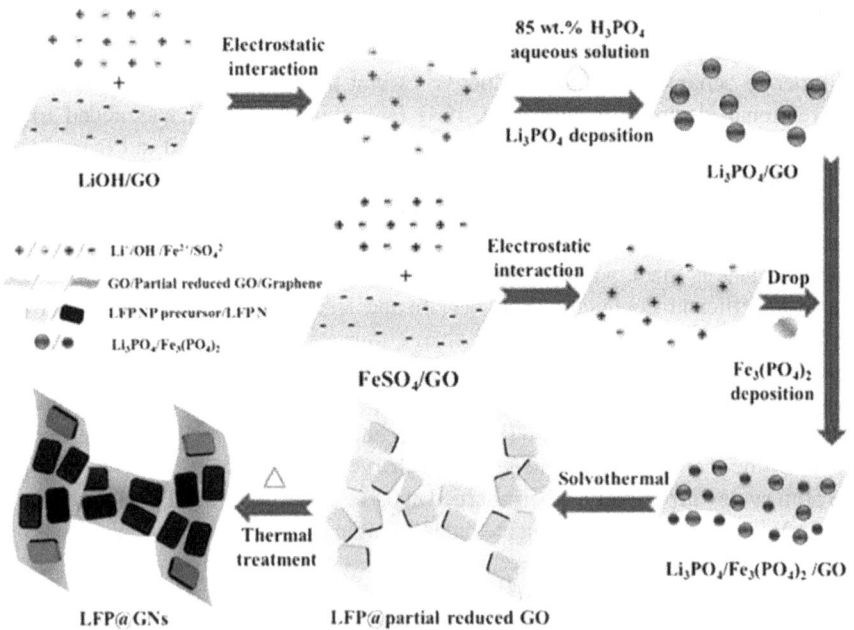

FIGURE 11.19 Schematic preparation process of LFP@GN [195].

FeSO4 was used in place of LiOH during the same procedure to first generate $Fe_3(PO_4)_2$ on GO sheets. LiFePO4 was produced by the hydrothermal treatment of $Li_3PO_4/Fe_3(PO_4)_2/GO$, and GO was partially decreased. The facets exposure of LiFePO4, LiFePO4LiFePO4 nanoplatelets are formed on graphene oxide sheets in-situ. The graphene oxide sheets were subsequently partially reduced to cross-link to create a 3D porous network.

11.6.1.2   Graphene-Li$_3$V$_2$(PO$_4$)$_3$ Composites

The amount of energy that can be isolated from a cell is inversely proportional to the square of the cell voltage. The comparatively low voltage plateau of LFP (3.4 V vs. Li/Li+) significantly reduces the energy that can be extracted from cells when cells are assembled. Major research has been conducted on materials such as LiCoPO4 (4.8 V vs. Li/Li+), LiNiPO4 (5.2 V vs. Li/Li+), and Li_3V_2 (PO$_4$)$_3$ (3.8 V vs. Li/Li+) in attempts to generate LFP safe high voltage cathodes. The LVP has high theoretical capacitance of 197 mAh/g, and is favored as a phosphate high-voltage cathode due to its accessibility, fast ionic diffusion, and safety. The monoclinic crystal structure, high operating voltage, and rate capacity of LVP allow it to be considered as a possible cathode material. Graphene aids to control LVP particle agglomeration during the heat treatment process, in addition to the heterogeneous nucleation site to control the growth of LVP nanoparticles and improve low electronic conductivity (10^{-8} S/cm).

Yang et al. [196] used chitosan, sucrose, and nanoporous G-O as carbon bases to construct LVP/C composite utilizing sol-gel and solid-state techniques. They found the LVP/C produced using the sol-gel method had a larger capacity than when produced by the solid-state technique (187 and 112 mAh/g at 0.1 and 10C) (157 and 108 mAh/g at 0.1 and 10C). The sol-gel approach allows for enhanced precursor mixing, which leads to improved dispersion. Using a freeze-drying technique, Cheng et al. [197] produced nano-LVP/rG-O with a capacity of 105.7 mAh/g at 20C. Researchers discovered that liquid N_2 can more effectively synthesize graphene materials with porous structures, as well as quickly generate solid G-O and vanadium salt precursors. Moreover, segregation is avoided and smaller LVP particles are made.

LVP @rGO and LVP @amorphous carbons were both synthesized by Li et al. [198], and their relative capacities at 5C were 141.3 mAh/g and 118.9 mAh/g. LVP/rGO and LVP/rmGO were both synthesized by Pei et al. [199], the latter of which used G-O (rmGO) modified with CTAB and sucrose for carbon precursors. The voltage ranges of 3.04.3 V and 3.04.8 V have larger discharge capacities for LVP/rmGO, although the latter voltage range in particular delivered 167 mAh/g at 10C as opposed to 118 mAh/g for LVP/rGO. In order to create LVP-rGO composites, Rajagopalan et al. [200] described a microwave solvothermal method followed by a freeze-drying technique. Specific capacities of 192 and 65 mAh/g at 0.5C and 5C, respectively, were delivered. Between 1000 and 5000 cycles, there was no discernible capacity degradation.

11.6.1.3   Graphene-Li$_2$MSiO$_4$ Composites

Iso-structural orthosilicates Li_2MSiO_4 (M = Mn and Fe) and their corresponding solid solutions have generated a lot of investigative interest due to their large theoretical capacities (332 mAh/g for Li_2FeSiO_4) and low costs because of the availability of Fe and Si in abundance on earth. They are severely constrained in their actual performance by their low specific conductivity, which is close to that of a general insulator (5×10^{-16} S/cm for Li_2MnSiO_4 and 6×10^{-14} S/cm for Li_2FeSiO_4 at room temperature). Therefore, for these materials to function better, nanoscale particles and electrically conductive networks are needed, which encourages research into Li_2MSiO_4-graphene composite materials [201].

Li_2FeSiO_4 nanorods and rGO combine to create a 2D hybrid material with high capacity at 300 mAh/g at 0.1C and 134 mAh/g at 12C. Moreover, it retains 95% of its capacity after 240 cycles. According to EIS studies, the Li_2FeSiO_4@rGO cathode's charge transfer resistance is less than one-third that of Li_2FeSiO_4 before the cycling process and after 10 cycles [202]. The same investigative team created graphene-coated 3D-hierarchical Li_2FeSiO_4 that also contained secondary nano-petals (G@3D-HFLFS), capable of discharging 327.2 mA per square meter per hour at 0.1°C and 100.5 mA per square meter per hour at 20°C [203].

11.6.2 Graphene-Spinel Cathode Material Composites

The spinel structure of $LiMn_2O_4$ (LMO) features tetrahedral sites occupied by lithium and manganese occupying the octahedral sites. Instead of occurring on planes, Li intercalates and de-intercalates in 3D network of channels. $LiMn_2O_4$ has a poor charge capacity and experiences phase changes during cycling; particularly when Mn^{2+} is dissolved into the electrolyte, the life cycle is greatly reduced. Other transition metals including Co, Ni, and Fe may be added; this is thought to help with electrochemical performance.

By combining rGO with previously synthesized LMO material, Xu et al. [204] were able to produce a capacity of 107 mAh/g at 50C, this is much greater than 80 mAh g^{-1} for bare sample. Furthermore, the self-assembly mechanism is effective for uniformly securing LMO nanoparticles on rGO nanosheets (LMO-GNS). Nearing LMO theoretical capacity, the LMO-GNS delivered a capacity of 146 mAh g^{-1}. At 100 C, the LMO-GNS retains 74% of its capacity. The charge-discharge behavior was even more similar to that of a capacitor [205,206].

Similar findings were made by Bak et al. [155], who discovered that LMO/rG-O behaved in a capacitive-like manner. By using CV analysis at potential scan speeds of 1 and 10 mV s^{-1}, respectively, they discovered that the surface reaction contributed 59% and 71% of the total stored charge.

11.6.3 Graphene-Layered Cathode Material Composites

$LiCoO_2$ has a layered structure, and it is considered as a cathode material that is employed in commercial applications most frequently. Nevertheless, because of toxicity, safety issues, and elevated cobalt cost, researchers are working to develop layer-structured solid solutions that are less expensive, less poisonous, safer, with greater capacities by replacing cobalt with nickel or manganese. Their efforts have led to consider Li [CoxNiyMn(1-x-y)]O$_2$ (NCM) as one of the outcomes. It is advantageous to utilize graphene materials to increase these materials' electronic conductivity because they generally have electronic conductivity lower than that of LCO.

Additionally, these materials tend to need air or oxygen during calcination and are susceptible to reduction by carbon at high temperatures. The synthesis of graphene and layered cathodes composite materials is classically processed differently with cathode materials that require inert atmosphere. Jiang et al. [207] used rGO to replace 50% of conductive carbon black while creating the reaction slurry for NCM cathode. The rGO-free NCM cathode has essentially minimal capacity at 6C, whereas the rGO-added electrode generates 110 mAh/g at 6C and 55 mA h/g at 20C. The capacity of the NCM and rGO composite generated using ball milling is 153 mAh/g at 5C, which is higher than that of NCM without rGO (138 mAh/g).

The NCM/rGO composite's charge transfer resistance is one-fifth less than that of pristine NCM [208]. The rGO sheets are wrapped around NCM particles in the composite material known as rGO/NCM, which was created by spray drying and then heat treatment at 300 °C. At a rate of 5C, the capacities of rGO/NCM and NCM are 88.5 and 6 mAh/g, respectively. However, the former charge-transfer resistance is only one-fifth that of the latter [209].

11.6.4 GRAPHENE IN CATHODE AS ACTIVE MATERIAL

Recent studies have shown that in addition to offering superior double-layer capacitance and electrical conductivity as previously reported, graphene materials are used as additives. Furthermore, rG-O in LIB cathode contributes to an active role in the redox processes. The high surface area of graphene materials and the capacity to reversibly store Li+ in functional groups with oxygen such as carbonyl, carboxylic, and ester groups are well known [210]. The capacity of partially reduced G-O is created by slightly lowering G-O to 250 mA/g, according to Kim et al. [211]. The free-standing G-O cathode paper-based model has a maximum capacity of 360.1 mA/g [212]. The rGO with a significant pyrrolic-N doping has a greater LIB cathode capacity, according to Xiong et al. [213].

It is however noted that these cathodes lack characteristic charge/discharge plateau and behave more like capacitors, making them unsuitable for application. Yet, research into G-O and rG-O as LIB cathodes may be able to clarify some of the phenomena that have been discovered while studying composite cathode materials made of graphene-based materials. When LFP is above the theoretical capacity limit, LFP surface is modified with 2 wt% electrochemically exfoliated rGO and can achieve a specific capacity of 208 mAh/g. As a result of the inherent ability of electrochemically exfoliated rGO to store Li ions, scientists have studied the CV of this material and have discovered that extra capacity was attributed to the reversible reduction-oxidation reaction between the Li ions from the electrolyte and the rGO flakes [214].

Sheets from rGO boost the overall capacity and reduce electroactive cathode material size, which makes it easier for Li+ ions to move during charging and discharging. It also makes it easier for electrons to transfer. It was found that RGO might contribute up to 16% of LVP/rGO composite's total capacity at a low rate of 0.075C. As graphene sheets restack, the content of rGO rise, reaching a peak at 10 wt% of the overall capacity before steadily declining.

11.7 CONCLUSION

The pores offer greater access to the electrolyte and contribute to the mechanical resilience, which is essential for materials with high volume expansion during cycling. They also increase electrical conductivity, crystal size, and orientation management. The key benefits that graphene materials can provide for composite materials are those mentioned above. Graphene-based materials are receiving a lot of interest, and LIBs have paved the way for the creation of numerous composite materials with advantageous characteristics like improved rate performance, higher capacity, and longer cycle lifetimes. These advancements make it possible to build LIBs with more power, better energy densities, and longer life cycle.

Notwithstanding these benefits, graphene-based materials have several disadvantages, including low volumetric energy density and tap density, capacitor-like behavior, and poorly defined charging and discharging platforms. However, other nanomaterials besides graphene-based and graphene-based composite materials have the same disadvantages. It is challenging to establish a strong link since graphene-based material characteristics frequently differ between studies. The low first-cycle efficiency of the graphene composite anode materials has received the most attention. Several techniques have been used to improve the acceptability of high irreversible capacity anode materials for whole-cell applications.

REFERENCES

1. D. Larcher, J.M. Tarascon, Towards greener and more sustainable batteries for electrical energy storage, *Nat. Chem.* 7 (2015) 19–29.
2. G.L. Soloveichik, Flow batteries: current status and trends, *Chem. Rev.* 115 (2015) 11533–11558.
3. Z.W. Seh, Y. Sun, Q. Zhang, Y. Cui, Designing high-energy lithium-sulfur batteries, *Chem. Soc. Rev.* 45 (2016) 5605–5634.
4. Y. Sun, N. Liu, Y. Cui, Promises and challenges of nanomaterials for lithium-based rechargeable batteries, *Nat. Energy* 1 (2016) 16071.
5. Y. Tang, Y. Zhang, W. Li, B. Ma, X. Chen, Rational material design for ultrafast rechargeable lithium-ion batteries, *Chem. Soc. Rev.* 44 (2015) 5926–5940.
6. J. Cai, C. Wu, Y. Zhu, P.K. Shen, K. Zhang, Hierarchical porous acetylene black/$ZnFe_2O_4$@carbon hybrid materials with high capacity and robust cycling performance for Li-ion batteries, *Electrochim. Acta* 187 (2016) 584–592.
7. Y. Zhu, H. Zheng, J. Cen, A. Ali, X. Chen, P.K. Shen, High performance lithium-sulfur batteries based on CoP nanoparticle-embedded nitrogen-doped carbon nanotube hollow polyhedra, *J. Electroanal. Chem.* 885 (2021) 114996.
8. K. Liu, Y. Fan, A. Ali, P.K. Shen, A flexible and conductive MXene-coated fabric integrated with in situ sulfur loaded MXene nanosheets for long-life rechargeable Li–S batteries, *Nanoscale* 13 (2021) 2963–2971.
9. D. Aurbach, B.D. McCloskey, L.F. Nazar, P.G. Bruce, Advances in understanding mechanisms underpinning lithium–air batteries, *Nat. Energy* 1 (2016), 16128.
10. G. Liang, F. Mo, X. Ji, C. Zhi, Non-metallic charge carriers for aqueous batteries, *Nat. Rev. Mater.* 6 (2021) 109–123.
11. A.R. Dehghani-Sanij, E. Tharumalingam, M.B. Dusseault, R. Fraser, Study of energy storage systems and environmental challenges of batteries, *Renew. Sustain. Energy Rev.* 104 (2019) 192–208.

12. L. Wang, Y. Han, X. Feng, J. Zhou, P. Qi, B. Wang, Metal–organic frameworks for energy storage: batteries and supercapacitors, *Coord. Chem. Rev.* 307 (2016) 361–381.
13. G. Li, B. Huang, Z. Pan, X. Su, Z. Shao, L. An, Advances in three-dimensional graphene-based materials: configurations, preparation and application in secondary metal (Li, Na, K, Mg, Al)-ion batteries, *Energy Environ. Sci.* 12 (2019) 2030–2053.
14. K.S. Novoselov, A.K. Geim, S.V. Morozov, D. Jiang, Y. Zhang, S.V. Dubonos, I.V. Grigorieva, A.A. Firsov, Electric field effect in atomically thin carbon films, *Science* 306(5696) (2004) 666–669.
15. P.K. Shen Ali, Nonprecious metal's graphene-supported electrocatalysts for hydrogen evolution reaction: fundamentals to applications, *Carbon Energy* 2 (2020) 99–121.
16. P.K. Shen Ali, Recent progress in graphene-based nanostructured electrocatalysts for overall water splitting, *Electrochem. Energy Rev.* 3 (2020) 370–394.
17. P.K. Shen Ali, Recent advances in graphene-based platinum and palladium electrocatalysts for the methanol oxidation reaction, *J. Mater. Chem.* 7 (2019) 22189–22217.
18. W. Hooch Antink, Y. Choi, K.-D. Seong, J.M. Kim, Y. Piao, Recent progress in porous graphene and reduced graphene oxide-based nanomaterials for electrochemical energy storage devices, *Adv. Mater. Interfaces* 5 (2018) 1701212.
19. N. Mahmood, C. Zhang, H. Yin, Y. Hou, Graphene-based nanocomposites for energy storage and conversion in lithium batteries, supercapacitors and fuel cells, *J. Mater. Chem.* 2 (2014) 15–32.
20. L. Ji, P. Meduri, V. Agubra, X. Xiao, M. Alcoutlabi, Graphene-based nanocomposites for energy storage, *Adv. Energy Mater.* 6 (2016) 1502159.
21. P.W. Albers, V. Leich, A.J. Ramirez-Cuesta, Y. Cheng, J. Ho€nig, S.F. Parker, The characterisation of commercial 2D carbons: graphene, graphene oxide and reduced graphene oxide, *Mater. Adv.* 3 (2022) 2810–2826.
22. H. Huang, H. Shi, P. Das, J. Qin, Y. Li, X. Wang, F. Su, P. Wen, S. Li, P. Lu, F. Liu, Y. Li, Y. Zhang, Y. Wang, Z.S. Wu, H.M. Cheng, The chemistry and promising applications of graphene and porous graphene materials, *Adv. Funct. Mater.* 30 (2020) 1909035.
23. S. Hossain, A.M. Abdalla, S.B.H. Suhaili, I. Kamal, S.P.S. Shaikh, M.K. Dawood, A.K. Azad, Nanostructured graphene materials utilization in fuel cells and batteries: a review, *J. Energy Storage* 29 (2020) 101386.
24. X. Yu, H. Cheng, M. Zhang, Y. Zhao, L. Qu, G. Shi, Graphene-based smart materials, *Nat. Rev. Mater.* 2 (2017) 17046.
25. D.G. Papageorgiou, I.A. Kinloch, R.J. Young, Mechanical properties of graphene and graphene-based nanocomposites, *Prog. Mater. Sci.* 90 (2017) 75–127.
26. P.T. Yin, S. Shah, M. Chhowalla, K.-B. Lee, Design, synthesis, and characterization of graphene–nanoparticle hybrid materials for bioapplications, *Chem. Rev.* 115 (2015) 2483–2531.
27. Y. Li, H. Zhang, Y. Chen, Z. Shi, X. Cao, Z. Guo, P.K. Shen, Nitrogen-doped carbon-encapsulated SnO_2@Sn nanoparticles uniformly grafted on three-dimensional graphene-like networks as anode for high-performance lithium-ion batteries, *ACS Appl. Mater. Interfaces* 8 (2016) 197–207.
28. M. Yang, Y. Liu, T. Fan, D. Zhang, Metal-graphene interfaces in epitaxial and bulk systems: a review, *Prog. Mater. Sci.* 110 (2020) 100652.
29. S. Huang, L. Zhang, J. Wang, J. Zhu, P.K. Shen, In situ carbon nanotube clusters grown from three-dimensional porous graphene networks as efficient sulfur hosts for high-rate ultra-stable Li–S batteries, *Nano Res.* 11 (2018) 1731–1743.
30. X. Guo, S. Zheng, G. Zhang, X. Xiao, X. Li, Y. Xu, H. Xue, H. Pang, Nanostructured graphene-based materials for flexible energy storage, *Energy Storage Mater.* 9 (2017) 150–169.

31. Z. Chen, X. An, L. Dai, Y. Xu, Holey graphene-based nanocomposites for efficient electrochemical energy storage, *Nano Energy* 73 (2020), 104762.
32. K. Chen, Q. Wang, Z. Niu, J. Chen, Graphene-based materials for flexible energy storage devices, *J. Energy Chem.* 27 (2018) 12–24.
33. T. Sattar, Current review on synthesis, composites and multifunctional properties of graphene, *Top. Curr. Chem.* 377 (2019) 10.
34. E. Eizenberg, J.M. Blakely, Carbon monolayer phase condensation on Ni (111), *Surf. Sci.* 82(1–2) (1979) 228–236.
35. M. Eizenberg, J.M. Blakely, Carbon interaction with nickel surfaces: monolayer formation and structural stability, *J. Chem. Phys* 71(8) (1979) 3467.
36. A Lang LEED study of the deposition of carbon on platinum crystal surfaces. *Surf. Sci.* 53(1) (1975) 317–329.
37. X.K. Lu, M.F. Yu, H. Huang, R.S. Ruoff, Tailoring graphite with the goal of achieving single sheets, *Nanotechnology* 10(3) (1999) 269–272.
38. Y.B. Zhang, J.P. Small, W.V. Pontius, P. Kim, Fabrication and electric-field dependent transport measurements of mesoscopic graphite devices, *Appl. Phys. Lett.* 86 (2005) 073104.
39. S. Kaniyoor, A Ramaprabhu, Raman spectroscopic investigation of graphite oxide derived graphene, *AIP Adv* 2 (2012) 032183.
40. K.S. Novoselov, D. Jiang, F. Schedin, T.J. Booth, V.V. Khotkevich, S.V. Morozov, S.V., A.K. Geim, Two-dimensional atomic crystals. *PNAS* 102(3) (2005) 10451–10453.
41. A.A. Balandin, S. Ghosh, W.Z. Bao, I. Calizo, D. Teweldebrhan, F. Miao, C.N. Lau, Superior thermal conductivity of single-layer graphene. *Nano. Lett.* 8(3) (2008) 902–907.
42. H.P. Boehm, R. Setton, E. Stumpp, Nomenclature and terminology of graphite intercalation compounds (IUPAC Recommendations 1994). *Pure Appl. Chem.* 66(9) (1994) 1893–1901.
43. M.J. Allen, V.C. Tung, R. B. Kaner, Honeycomb carbon: a review of graphene. *Chem. Rev.* 110(1) (2010) 132–145.
44. L.M. Viculis, J.J. Mack, R.B. Kaner, A chemical route to carbon nanoscrolls. *Science* 299(5611) (2003) 1361.
45. S. Park, R.S. Ruoff, Chemical methods for the production of graphenes. *Nat. Nanotechnol.* 4 (2009) 217–224.
46. A. Reina, X.T. Jia, J. Ho, D. Nezich, H. Son, V. Bulovic, S. Mildred Dresselhaus, J. Kong, Large area, few-layer graphenefilms on arbitrary substrates by chemical vapor deposition. *Nano Lett.* 9(1) (2009) 30–35.
47. K. Geim, K.S. Novoselov, The rise of graphene. *Nat. Mater.* 6(3) (2007) 183–191. doi:10.1038/nmat1849
48. Y. Hernandez, V. Nicolosi, M. Lotya, F.M. Blighe, Z. Sun, S. De, I.T. McGovern, B. Holland, M. Byrne, Y.K. Gun'Ko, J.J. Boland, P. Niraj, G. Duesberg, S. Krishnamurthy, R. Goodhue, J. Hutchison, V. Scardaci, A.C. Ferrari, J.N. Coleman, High-yield production of graphene by liquid-phase exfoliationof graphite. *Nat. Nanotechnol.* 3 (2008) 563–568. doi:10.1038/nnano.2008.215
49. N. Liu, F. Luo, H. Wu, Y. Liu, C. Zhang, J. Chen, One-stepionic-liquid-assisted electrochemical synthesis of ionic-liquidfunctionalizedgraphene sheets directly from graphite. *J. Adv. Funct. Mater.* 18(10) (2008) 1518–1525. doi:10.1002/adfm.200700797
50. N. Behabtu, J.R. Lomeda, M.J. Green, A.L. Higginbotham, A. Sinitskii, D.V. Kosynkin, D. Tsentalovich, A.N.G. Parra-Vasquez, J. Schmidt, E. Kesselman, Y. Cohen, Y. Talmon, J.M. Tour, M. Pasquali, Spontaneous high-concentration dispersionsand liquid crystals of grapheme. *Nat. Nanotechnol.* 5 (2010) 406–411. doi:10.1038/nnano.2010.86

51. S. Stankovich, R.D. Piner, S.T. Nguyen, R.S. Ruoff, Synthesis and exfoliation of isocyanate-treated graphene oxide nanoplatelets. *Carbon* 44(15) (2006) 3342–3347. doi:10.1016/j.carbon.2006.06.004

52. H. Hiura, T.W. Ebbesen, J. Fujita, K. Tanigaki, T. Takada, Role of sp3 defect structures in graphite and carbon nanotubes. *Nature* 367 (1994) 148–151.

53. T.W. Ebbesen, H. Hiura, Graphene in 3-dimensions: towards graphite origami. *Adv. Mater.* 7(6) (1995) 582–586.

54. T.M. Bernhardt, B. Kaiser, K. Rademann, Formation of superperiodic patterns on highly oriented pyrolytic graphite by manipulation of nanosized graphite sheets with the STM tip. *Surf. Sci.* 408(1–3) (1998) 86–94.

55. X. Lu, M. Yu, H. Huang, R.S. Ruoff, Tailoring graphite with the goal of achieving single sheets. *Nanotechnology* 10(3) (1999) 269.

56. H.V. Roy, C. Kallinger, K. Sattler, Manipulation of graphitic sheets using a tunneling microscope. *J. Appl. Phys.* 83 (1998) 4695.

57. L.J. Ci, L. Song, D. Jariwala, A.L. Elias, W. Gao, M. Terrones, P.M. Ajayan, Graphene shape control by multistage cutting and transfer. *Adv. Mater.* 21(44) (2009) 4487–4491.

58. X. Liang, Z. Fu, S.Y. Chou, Graphene transistors fabricated via transfer-printing in device active-areas on large wafer. *Nano Lett.* 7(12) (2007) 3840–3844.

59. J.-H. Chen, M. Ishigami, C. Jang, D.R. Hines, M.S. Fuhrer, E.D. Williams, Printed graphene circuits. *Adv. Mater.* 19(21) (2007) 3623–3627.

60. K. Parvez, S. Yang, X. Feng, K. Müllen, Exfoliation of graphene via wet chemical routes, *Synthetic Metals* 210 Part A (2015) 123–132.

61. H. Vincent, N. Bendiab, N. Rosman, T. Ebbesen, C. Delacour, V Bouchiatand, Large and flat graphene flakes produced by epoxy bonding and reverse exfoliation of highly oriented pyrolytic graphite. *Nanotechnology* 19(45) (2008) 455601.

62. L.-H. Liu, M. Yan, Simple method for the covalent immobilization of graphene. *Nano Lett.* 9(9) (2009) 3375–3378.

63. L. Song, L. Ci, W. Gao, P.M. Ajayan, Transfer printing of graphene using gold film. *ACS Nano* 3(6) (2009) 1353–1356.

64. Z.H. Ni, H.M. Wang, J. Kasim, H.M. Fan, T. Yu, Y.H. Wu, Y.P. Feng, Z.X. Shen, Graphene thickness determination using reflection and contrast spectroscopy. *Nano Lett.* 7(9) (2007) 2758–2763.

65. A.C. Ferrari, J.C. Meyer, V. Scardaci, C. Casiraghi, M. Lazzeri, F. Mauri, S. Piscanec, D. Jiang, K.S. Novoselov, S. Roth, A.K. Geim, Raman spectrum of graphene and graphene layers. *Phys. Rev. Lett.* 97(18) (2006) 187401.

66. A.C. Ferrari, Raman spectroscopy of graphene and graphite: disorder, electron–phonon coupling, doping and nonadiabatic effects. *Solid State Commun.* 143(1–2) (2007) 47–57.

67. Z. Ni, Y. Wang, T. Yu, Z. Shen, Raman spectroscopy and imaging of graphene. *Nano Res.* 1(4) (2008) 273–291.

68. C. Casiraghi, A. Hartschuh, E. Lidorikis, H. Qian, H. Harutyunyan, T. Gokus, K.S. Novoselov, A.C. Ferrari, Rayleigh imaging of graphene and graphene layers. *Nano Lett.* 7(9) (2007) 2711–2717.

69. D.A. Dikin, S. Stankovich, E.J. Zimney, R.D. Piner, G.H.B. Dommett, G. Evmenenko, S.T. Nguyen, R.S. Ruoff, Preparation and characterization of graphene oxide paper. *Nature* 448 (2007) 457–460.

70. S. Park, K.-S. Lee, G. Bozoklu, W. Cai, S.T. Nguyen, R.S. Ruoff, Graphene oxide papers modified by divalent ions—enhancing mechanical properties via chemical cross-linking. *ACS Nano* 2(3) (2008) 572–578.

71. D. Li, M.B. Muller, S. Gilje, R.B. Kaner, G.G. Wallace, Processable aqueous dispersions of graphene nanosheets. *Nat. Nanotechnol.* 3 (2008) 101–105.

72. Y. Xu, H. Bai, G. Lu, C. Li, G. Shi, Flexible graphene films via the filtration of water-soluble noncovalent functionalized graphene sheets. *J. Am. Chem. Soc.* 130(18) (2008) 5856–5857.

73. S. Park, J.H. An, R.D. Piner, I. Jung, D.X. Yang, A. Velamakanni, S.T. Nguyen, R.S. Ruoff, Aqueous suspension and characterization of chemically modified graphene sheets. *Chem. Mater.* 20(21) (2008) 6592–6594.

74. H. Chen, M.B. Muller, K.J. Gilmore, G.G. Wallace, D. Li, Mechanically strong, electrically conductive, and biocompatible graphene paper. *Adv. Mater.* 20(18) (2008) 3557–3561.

75. S. Stankovich, D.A. Dikin, G.H.B. Dommett, K.M. Kohlhaas, E.J. Zimney, E.A. Stach, R.D. Piner, S.T. Nguyen, R.S. Ruoff, Graphene-based composite materials. *Nature* 442 (2006) 282–286.

76. M.D. Stoller, S.J. Park, Y.W. Zhu, J.H. An, R.S. Ruoff, Graphene-based ultracapacitors. *Nano Lett.* 8(10) (2008) 3498–3502.

77. X. Wang, L. Zhi, K. Mullen, Transparent, conductive graphene electrodes for dye-sensitized solar cells. *Nano Lett.* 8(1) (2008) 323–327.

78. B.C. Brodie, Sur le poidsatomique du graphite. *Ann. Chim. Phys.* 59 (1860) 466–472.

79. W.S. Hummers, R.E. Offeman, Preparation of graphitic oxide. *J. Am. Chem. Soc.* 80(6) (1958) 1339.

80. L. Staudenmaier, VerfahrenzurDarstellung der Graphitsä̈ure *Eur. J. Inorg. Chem.* 31(2) (1898) 1481–1487.

81. Y.H. Wu, T. Yu, Z.X. Shen, Two-dimensional carbon nanostructures: fundamental properties, synthesis, characterization, and potential applications. *J. Appl. Phys.* 108 (2010) 071301.

82. D.C. Marcano, D.V. Kosynkin, J.M. Berlin, A. Sinitskii, Z. Sun, A. Slesarev, L.B. Alemany, W. Lu, J.M. Tour, Improved synthesis of graphene oxide. *ACS Nano* 4(8) (2010) 4806–4814.

83. S. Park, J. An, I. Jung, R.D. Piner, S.J. An, X. Li, A. Velamakanni, R.S. Ruoff, Colloidal suspensions of highly reduced graphene oxide in a wide variety of organic solvents. *Nano Lett.* 9(4) (2009) 1593–1597.

84. M.J. Allen, V.C. Tung, R.B. Kaner, Honeycomb carbon: a review of graphene. *Chem. Rev.* 110(1) (2009) 132–145.

85. V.C. Tung, M.J. Allen, Y. Yang, R.B. Kaner, High-throughput solution processing of large-scale graphene. *Nat. Nanotechnol.* 4 (2009) 25–29.

86. J.I. Paredes, S. Villar-Rodil, A. Marti'nez-Alonso, J.M.D. Tasco'n, "Graphene oxide dispersions in organic solvents". *Langmuir* 24(19) (2008) 10560–10564.

87. A.A. Green, M.C. Hersam, Emerging methods for producing monodisperse graphene dispersions. *J. Phys. Chem. Lett.* 1(2) (2009) 544–549.

88. A.A. Green, M.C. Hersam, Solution phase production of graphene with controlled thickness via density differentiation. *Nano Lett.* 9(12) (2009) 4031–4036.

89. J. Wu, W. Pisula, K. Mu¨llen, Graphenes as potential material for electronics. *Chem. Rev.* 107(3) (2007) 718–747.

90. J. Cai, P. Ruffieux, R. Jaafar, M. Bieri, T. Braun, S. Blankenburg, M. Matthias, A.P. Seitsonen, S. Moussa, X. Feng, K. Mu¨llen, R. Fasel, Atomically precise bottom-up fabrication of graphene nanoribbons. *Nature* 466 (2010) 470–473.

91. X. Yan, X. Cui, L. Binsong, L. Liang-shi, Large, solution-processable graphene quantum dots as light absorbers for photovoltaics. *Nano Lett.* 10(5) (2010) 1869–1873.

92. L. Zhi, K.A. Mullen, A bottom-up approach from molecular nanographenes to unconventional carbon materials. *J. Mater. Chem.* 18 (2008) 1472–1484.

93. C.E. Hamilton, J.R. Lomeda, Z. Sun, J.M. Tour, A.R. Barron, High-yield organic dispersions of unfunctionalized graphene. *Nano Lett.* 9(10) (2009) 3460–3462.

94. A. ONeill, U. Khan, P.N. Nirmalraj, J. Boland, J.N. Coleman, Graphene dispersion and exfoliation in low boiling point solvents. *J. Phys. Chem. C* 115(13) (2011) 5422–5428.

95. Y. Hernandez, M. Lotya, D. Rickard, S.D. Bergin, J.N. Coleman, Measurement of multicomponent solubility parameters for graphene facilitates solvent discovery. *Langmuir* 26(5) (2009) 3208–3213.

96. M. Qian, Y.S. Zhoul, Y. Gao, J.B. Parkl, T. Feng, S.M. Huang, Z. Sun, L. Jiang, Y.F. Lul, Formation of graphene sheets through laser exfoliation of highly ordered pyrolytic graphite. *Appl. Phys. Lett.* 98 (2011) 173108.

97. M. Liu, X. Zhang, W. Wu, T. Liu, Y. Liu, B. Guo, R. Zhang, One-step chemical exfoliation of graphite to ~100% few-layer graphene with high quality and large size at ambient temperature, *Chem. Eng. J.* 355 (2019) 181–185.

98. Z. Hou, Y. Zhou, G. Li, S. Wang, M. Wang, X. Hu, S. Li, Reduction of graphene oxide and its effect on square resistance of reduced graphene oxide films, 36 (2015) 1681–1687.

99. H.-J. Shin, K.K. Kim, A. Benayad, S.-M. Yoon, H.K. Park, Efficient reduction of graphite oxide by sodium borohydride and its effect on electrical conductance. *Adv. Funct. Mater.* 19(12) (2009) 1987–1992.

100. V.H. Pham, T.V. Cuong, T.-D. Nguyen-Phan, H.D. Pham, E.J. Kim, S.H. Hur, E.W. Shin, S. Kim, J.S. Chung, One-step synthesis of superior dispersion of chemically converted graphene in organic solvents. *Chem. Commun.* 46 (2010) 4375–4377.

101. X. Zhou, J. Zhang, H. Wu, H. Yang, J. Zhang, S. Guo, Reducing graphene oxide via hydroxylamine: a simple and efficient route to graphene. *J. Phys. Chem. C* 115(24) (2011) 11957–11961.

102. C. Zhu, S. Guo, Y. Fang, S. Dong, Reducing sugar: new functional molecules for the green synthesis of graphene nanosheets. *ACS Nano* 4(4) (2010) 2429–2437.

103. J. Zhang, H. Yang, G. Shen, P. Cheng, J. Zhang, S. Guo, Reduction of graphene oxide viaL-ascorbic acid. *Chem. Commun.* 46 (2010) 1112–1114.

104. G. Wang, J. Yang, J. Park, X. Gou, B. Wang, H. Liu, J. Yao, Facile synthesis and characterization of graphene nanosheets. *J. Phys. Chem.* 112(22) (2008) 8192–8195.

105. X. Fan, W. Peng, Y. Li, X. Li, S. Wang, G. Zhang, F. Zhang, Deoxygenation of exfoliated graphite oxide under alkaline conditions: a green route to graphene preparation. *Adv. Mater.* 20(23) (2008) 4490–4493.

106. C.A. Amarnath, C.E. Hong, N.H. Kim, B.-C. Ku, T. Kuila, J.H. Lee, Efficient synthesis of graphene sheets using pyrrole as a reducing agent. *Carbon* 49(11) (2011) 3497–3502.

107. H.-L. Guo, X.-F. Wang, Q.-Y. Qian, F.-B. Wang, X.-H. Xia, A green approach to the synthesis of graphene nanosheets. *ACS Nano* 3(9) (2009) 2653–2659.

108. R.S. Sundaram, C. G´omez-Navarro, K. Balasubramanian, M. Burghard, K. Kern, Electrochemical modification of graphene. *Adv. Mater.* 20(16) (2008) 3050–3053.

109. O.C. Compton, B. Jain, D.A. Dikin, A. Abouimrane, K. Amine, S.T. Nguyen, Chemically active reduced graphene oxide with tunable C/O ratios. *ACS Nano* 5(6) (2011) 4380–4391.

110. M.J. McAllister, J.L. Li, D.H. Adamson, H.C. Schniepp, A.A. Abdala, J. Liu, M.H. Alonso, D.L. Milius, R. Car, K. Robert, R.K. Prud'homme, I.A. Aksay, Single sheet functionalized graphene by oxidation and thermal expansion of graphite. *Chem. Mater.* 19(18) (2007) 4396–4404.

111. J.I. Parades, S. Villar-Rodil, A. Mart´ınez-Alonso, J.M.D. Tasc´on, Graphene oxide dispersions in organic solvents. *Langmuir* 24(19) (2008) 10560–10564.

112. S. Dubin, S. Gilje, K. Wang, V.C. Tung, K. Cha, Hall, A one-step, solvothermal reduction method for producing reduced graphene. *ACS Nano* 4(7) (2010) 3845–3852.

113. X.L. Li, X.R. Wang, L. Zhang, S.W. Lee, H.J. Dai, Chemically derived, ultrasmooth graphene nanoribbon semiconductors. *Science* 319(5867) (2008) 1229–1232.

114. X.L. Li, G.Y. Zhang, X.D. Bai, X. Sun, X. Wang, E. Wang, H. Dai, Highly conducting graphene sheets and Langmuir- Blodgett films. *Nat. Nanotechnol.* 3(9) (2008) 538–542.

115. A.S. Ali Shalaby, D. Nihtianova, P. Markov, A. Staneva, Structural analysis of reduced graphene oxide by transmission electron microscopy, *Bulg. Chem. Commun.* 47(1) (2015) 291–295

116. W. Zhang, J. Cui, C.-A. Tao, Y. Wu, Z. Li, L. Ma, Y. Wen, G. Li, A Strategy for producing pure single-layer graphene sheets based on a confined self-assembly approach. *Angew. Chem. Int. Ed.* 48(32) (2009) 5864–5868.

117. X.S. Li, W.W. Cai, J.H. An, S. Kim, J. Nah, D.X. Yang, R. Piner, A. Velamakanni, I. Jung, E. Tutuc, S.K. Banerjee, L. Colombo, R.S. Ruoff, Large-area synthesis of high-quality and uniform graphene films on copper foils. *Science* 324(5932) (2009) 1312–1314.

118. E. Rollings, G.-H. Gweon, S.Y. Zhou, B.S. Mun, J.L. McChesney, B.S. Hussain, A.V. Fedorov, P.N. First, P.N. First, W.A. de Heer, A. Lanzar, Synthesis and characterization of atomically thin graphite films on a silicon carbide substrate. *J. Phys. Chem. Solids* 67(9-10) (2006) 2172–2177.

119. A. Hirsch, Unzipping carbon nanotubes: a peeling method for the formation of graphene nanoribbons. *Angew. Chem. Int. Ed.* 48(36) (2009) 6594–6596.

120. C.-D. Kim, B.-K. Min, W.-S. Jung, Preparation of graphene sheets by the reduction of carbon monoxide. *Carbon* 47(6) (2009) 1610–1612.

121. K.S. Kim, Y. Zhao, H. Jang, S.Y. Lee, J.M. Kim, K.S. Kim, J.-H. Ahn, P. Kim, J.-Y. Choi, B.H. Hong, Large-scale pattern growth of graphene films for stretchable transparent electrodes. *Nature* 457 (2009) 706–710.

122. P.W. Sutter, J.-I. Flege, E.A. Sutter, Epitaxial graphene on ruthenium. *Nat. Mater.* 7 (2008) 406–411.

123. J. Coraux, A.T. N'Diaye, C. Busse, T. Michely, Structural coherency of graphene on Ir(111)". *Nano Lett.* 8(2) (2008) 565–570.

124. D. Wei, Y. Liu, Y. Wang, H. Zhang, L. Huang, G. Yu, Synthesis of N-doped graphene by chemical vapor deposition and its electrical properties. *Nano Lett.* 9(5) (2009) 1752–1758.

125. Q. Yu, J. Lian, S. Siriponglert, H. Li, Y.P. Chen, S-.S. Pei, Graphene segregated on Ni surfaces and transferred to insulators. *Appl. Phys. Lett.* 93(11) (2008) 113103.

126. A.N. Obraztsov, Obraztsova, A.A. Zolotukhin, Chemical vapor deposition of thin graphite films of nanometer thickness. *Carbon* 45(10) (2007) 2017–2021.

127. A.E. Karu, M. Beer, Pyrolytic formation of highly crystalline graphite films. *J. Appl. Phys.* 37 (1966) 2179.

128. J. Perdereau, G.E. Rhead, LEED studies of adsorption on vicinal copper surfaces. *Surf. Sci.* 24(2) (1971) 555–571.

129. M. Eizenberg, J.M. Blakely, Carbon monolayer phase condensation on Ni(111). *Surf. Sci.* 82(1–2) (1979) 228–236.

130. N.A. Kholin, E.V. Rut'kov, A.Y. Tontegode, The nature of the adsorption bond between graphite islands and iridium surface. *Surf. Sci.* 139(1) (1984) 155–172.

131. N.R. Gall, E.V. Rut'kov, A.Y. Tontegode, Intercalation of nickel atoms under two-dimensional graphene film on (111)Ir. *Carbon* 38(5) (2000) 663–667.

132. L.G. De Arco, Y. Zhang, C. Zhou, Large scale graphene by chemical vapor deposition: synthesis, characterization and applications, *Graphene – Synthesis, Characterization, Properties and Applications*, (2011), 161–184. Editor: Jian Ru Gong, United Kingdom: IntechOpen.

133. M. Saeed, Y. Alshammari, S.A. Majeed, E. Al-Nasrallah, Chemical vapour deposition of graphene—synthesis, characterisation, and applications: a review, *Molecules* 25 (2020) 3856.

134. I.M. Katsnelson, Graphene: carbon in two dimensions. *Mater. Today* 10(1–2) (2007) 20–27.

135. A.K. Geim, P. Kim, Carbon wonderland. *Sci. Am.* 298(4) (2008) 90–97.

136. D.R. Dreyer, S. Park, C.W. Bielawski, R.S. Ruoff, The chemistry of graphene oxide. *Chem. Soc. Rev.* 39(1) (2010) 228–240.

137. P.R. Somani, S.P. Somani, M. Umeno, Planar nano-graphenes from camphor by CVD. *Chem. Phys. Lett.* 430(1–3) (2006) 56–59.

138. C. Berger, Z.M. Song, T.B. Li, X.B. Li, W.A. De Heer, Ultrathin epitaxial graphite: 2D electron gas properties and a route toward graphene-based nanoelectronics. *J. Phys. Chem. B* 108(52) (2004) 19912–19916.

139. A.J. Van Bommel, J.E. Crombeen, A. Van Tooren, LEED and Auger electron observations of the SiC(0001) surface. *Surf. Sci.* 48(2) (1975) 463–472.

140. W. De Heer, The development of epitaxial graphene for 21st century electronics. *MRS bulletin* 36 (2011) 633.

141. W.A. Heera, C. Bergera, M. Ruana, Sprinklea, E. Conrada, Large area and structured epitaxial graphene produced by confinement controlled sublimation of silicon carbide. *PNAS* 108 (2011) 16900.

142. J. Hass, R. Fengl, T. Li, X. Li, Z. Zong, C. Berger, Highly ordered graphene for two dimensional electronics. *Appl. Phys. Lett.* 89 (2006) 143106.

143. J. Hass, J.E. Milla´n-Otoya, P.N. First, E.H. Conrad, Interface structure of epitaxial graphene grown on 4H-SiC(0001). *Phys. Rev. B* 78 (2008) 205424.

144. I. Forbeaux, J.-M. Themlin, A. Charrier, F. Thibaudau, J.-M. Debever, Solid-state graphitization mechanisms of silicon carbide 6H–SiC polar faces. *Appl. Surf. Sci.* 162 (2000) 406.

145. J. Hass, C.A. Jeffrey, R. Feng, T. Li, X. Li, Z. Song, C. Berger, W.A. de Heer, P.N. First, E.H. Conrad, Highly-ordered graphene for two dimensional electronics. *J. Appl. Phys.* 92 (2002) 2479.

146. K.V. Emtsev, F. Speck, Th Seyller, L. Ley, J.D. Riley, Interaction, growth, and ordering of epitaxial graphene on SiC{0001} surfaces: a comparative photoelectron spectroscopy study. *Phys. Rev. B* 77, 155303 (2008).

147. R.F. Davis, et al., Thin film deposition and microelectronic and optoelectronic. *Proc. IEEE* 79 (1991) 677.

148. J. Kedzierski, P.-L. Hsu, P. Healey, P. Wyatt, C. Keast, M. Sprinkle, C. Berger, W. D. Heer, Epitaxial graphene transistors on SiC substrates. *IEEE Trans. Electron. Devices* 55 (2008) 2078.

149. Y.-M. Lin, C. Dimitrakopoulos, K. A. Jenkins, D. B. Farmer, H.- Y. Chiu, A. Grill, Ph. Avouris, 100-GHz Transistors from Wafer-Scale Epitaxial Graphene, *Science* 327 (2010) 662.

150. F. Schwierz, Graphene transistors. *Nat. Nanotechnol.* 5 (2010) 487.

151. K.S. Novoselov, A.K. Geim, S.V. Morozov, D. Jiang, M.I. Katsnelson, I.V. Grigorieva, A.A. Firsov, Two-dimensional gas of massless Dirac fermions in graphene. *Nature* 438 (2005) 197.

152. H. Tan, D. Wang, Y. Guo, Thermal growth of graphene: a review, *Coatings* 8 (2018) 40; doi:10.3390/coatings8010040

153. Z.H. Chen, Y.M. Lin, M.J. Rooks, P. Avouris, Graphene nano-ribbon electronics. *Phys. E-Low-Dimens. Syst. Nanostruct.* 40(2) (2007) 228–232. doi:10.1016/j.physe.2007.06.020.

154. A.G. Cano-Marquez, F.J. Rodriguez-Macias, J. Campos-Delgado, G.C. Espinosa-Gonza´lez, F. Trista´n-Lo´pez, D. Ramı´rez- Gonza´lez, A.D. Cullen, J.D. Smith, M. Terrones, I.Y. Vega- Cantu´, Ex-MWNTs: graphene sheets and ribbons produced by lithium intercalation and exfoliation of carbon nanotubes. *Nano Lett.* 9(4) (2009) 1527–1533. doi:10.1021/nl803585s

155. L.Y. Jiao, X.R. Wang, G. Diankov, H.L. Wang, H.J. Dai, Facile synthesis of high-quality graphene nanoribbons. *Nat. Nanotechnol.* 5(5) (2010) 321–325.

156. L.Y. Jiao, L. Zhang, X.R. Wang, G. Diankov, H.J. Dai, Narrow graphene nanoribbons from carbon nanotubes. *Nature* 458(7240) (2009) 877–880.

157. P. Sivaraman, Sarada P. Mishra, Darshna D. Potphode, Avinash P. Thakur, K. Shashidhara, Asit B. Samui, Arup R. Bhattacharyya, A supercapacitor based on longitudinal unzipping of multi-walled carbon nanotubes for high temperature application, *RSC Adv.* 5 (2015) 83546–83557.

158. D.B. Shinde, J. Debgupta, A. Kushwaha, M. Aslam, V.K. Pillai, Electrochemical unzipping of multi-walled carbon nanotubes for facile synthesis of high-quality graphene nanoribbons, *J. Am. Chem. Soc.* 133(12) (2011) 4168–4171.

159. N.R. Wilson, P.A. Pandey, R. Beanland, R.J. Young, I.A. Kinloch et al., Graphene oxide: structural analysis and application as a highly transparent support for electron microscopy, *ACS Nano* 3 (9) (2009) 2547–2556.

160. M. Pelaez-Fernandez, A. Bermejo, A.M. Benito, W.K. Maser, R. Arenal, Detailed thermal reduction analyses of graphene oxide via in-situ TEM/EELS studies, *Carbon* 178 (2021) 477–487.

161. F.T. Johra, J.-W. Lee, W.-G. Jung, Facile and safe graphene, *J. Ind. Eng. Chem.* 20 (2014) 2883–2887.

162. V. Scardaci, G. Compagnini, Raman spectroscopy investigation of graphene oxide reduction by laser scribing, *J. Carbon Res. C*, 7(2) (2021), 48; 10.3390/c7020048

163. Z. CIplak, N. Yildiz, A. Calimli, Investigation of graphene/Ag nanocomposites synthesis parameters for two different synthesis methods, *Fuller., Nanotub. Carbon Nanostruct.*, 23(4) (2015) 361–370.

164. X. Cai, L. Lai, Z. Shen, J. Lin, Graphene and graphene-based composites as Li-ion battery electrode materials and their application in full cells, *J. Mater. Chem. A* 5 (2017) 15423–15446. DOI: 10.1039/C7TA04354F.

165. Aurbach, in The Role of Surface Films on Electrodes in Li-Ion Batteries. *Advances in Lithium-Ion Batteries*, eds. W. A. van Schalkwijk and B. Scrosati, Springer US, Boston, MA, (2002), pp. 7–77.

166. R.A. Fong, U. von Sacken, J.R. Dahn, Studies of lithium intercalation into carbons using nonaqueous electrochemical cells, *J. Electrochem. Soc.* 137 (1990) 2009–2013.

167. L. Cheng, X.-L. Li, H.-J. Liu, H.-M. Xiong, P.-W. Zhang, Y.-Y. Xia, *J. Electrochem. Soc.* 154 (2007) A692–A697.

168. D. Aurbach, Review of selected electrode-solution interactions which determine the performance of Li and Li ion batteries, *J. Power Sources*, 89 (2000) 206–218.

169. Buiel, A. E. George, J. R. Dahn, On the reduction of lithium insertion capacity in hard-carbon anode materials with increasing heat-treatment temperature, *J. Electrochem. Soc.* 145 (1998) 2252–2257.

170. Pollak, B. S. Geng, K. J. Jeon, I. T. Lucas, T. J. Richardson, F. Wang, R. Kostecki, The interaction of Li^+ with single-layer and few-layer graphene, *Nano Lett.* 10 (2010) 3386–3388.

171. L. J. Zhou, L. F. Hou, L. M. Wu, First-principles study of lithium adsorption and diffusion on graphene with point defects, *J. Phys. Chem. C*, 116 (2012) 21780–21787.

172. P. C. Lian, X. F. Zhu, S. Z. Liang, Z. Li, W. S. Yang, H. H. Wang, Large reversible capacity of high quality graphene sheets as an anode material for lithium-ion batteries, *Electrochim. Acta* 55 (2010), 3909–3914.

173. O. Vargas, A. Caballero, J. Morales, G. A. Elia, B. Scrosati, J. Hassoun, Electrochemical performance of a graphene nanosheets anode in a high voltage lithium-ion cell, *Phys. Chem. Chem. Phys.* 15 (2013) 20444–20446.

174. Y. Shi, L. Wen, F. Li, H.-M. Cheng, Nanosized $Li_4Ti_5O_{12}$/graphene hybrid materials with low polarization for high rate lithium ion batteries, *J. Power Sources* 196 (2011) 8610–8617.

175. Y. Ding, G. R. Li, C. W. Xiao, X. P. Gao, Insight into effects of graphene in $Li_4Ti_5O_{12}$/carbon composite with high rate capability as anode materials for lithium ion batteries, *Electrochim. Acta* 102 (2013) 282–289.

176. H.-C. Tao, L.-Z. Fan, X. Yan, X. Qu, In situ synthesis of TiO_2–graphene nanosheets composites as anode materials for high-power lithium ion batteries, *Electrochim. Acta* 69 (2012) 328–333.

177. Y. Qiu, K. Yan, S. Yang, L. Jin, H. Deng, W. Li, Synthesis of size-tunable anatase TiO_2 nanospindles and their assembly into anatase@titanium oxynitride/titanium nitride–graphene nanocomposites for rechargeable lithium ion batteries with high cycling performance, *ACS Nano* 4 (2010) 6515–6526.

178. T. Hu, X. Sun, H. Sun, M. Yu, F. Lu, C. Liu, J. Lian, Flexible free-standing graphene–TiO_2 hybrid paper for use as lithium ion battery anode materials, *Carbon* 51 (2013) 322–326.

179. Xiang, B. Tian, P. Lian, Z. Li, H. Wang, Sol–gel synthesis and electrochemical performance of $Li_4Ti_5O_{12}$/graphene composite anode for lithium-ion batteries, *J. Alloy. Compd.* 509 (2011) 7205–7209.

180. D. Cai, P. Lian, X. Zhu, S. Liang, W. Yang, H. Wang, High specific capacity of TiO_2-graphene nanocomposite as an anode material for lithium-ion batteries in an enlarged potential window, *Electrochim. Acta* 74 (2012) 65–72.

181. J. Chang, X. Huang, G. Zhou, S. Cui, P. B. Hallac, J. Jiang, P. T. Hurley, J. Chen, Multilayered Si nanoparticle/reduced graphene oxide hybrid as a high-performance lithium-ion battery anode, *Adv. Mater.* 26 (2014) 758–764.

182. Tang, Y. J. Zhang, Q. Q. Xiong, J. D. Cheng, Q. Zhang, X. L. Wang, C. D. Gu, J. P. Tu, Self-assembly silicon/porous reduced graphene oxide composite film as a binder-free and flexible anode for lithium-ion batteries, *Electrochim. Acta* 156 (2015) 86–93.

183. Jiang, X. Zhou, G. Liu, Y. Zhou, H. Ye, Y. Liu, K. Han, Free-standing Si/graphene paper using Si nanoparticles synthesized by acid-etching Al-Si alloy powder for high-stability Li-ion battery anodes, *Electrochim. Acta* 188 (2016) 777–784.

184. F. Lin, D. Nordlund, T.-C. Weng, Y. Zhu, C. Ban, R. M. Richards, H. L. Xin, Phase evolution for conversion reaction electrodes in lithium-ion batteries, *Nat. Commun.* 5 (2014) 3358–3367.

185. Cabana, L. Monconduit, D. Larcher, M. R. Palacan, Beyond intercalation-based Li-ion batteries: the state of the art and challenges of electrode materials reacting through conversion reactions. *Adv. Mater.* 22 (2010) E170–E192.

186. N. Nitta, G. Yushin, High-capacity anode materials for lithium-ion batteries: choice of elements and structures for active particles, *Part. Part. Syst. Charact.* 31 (2014) 317–336.

187. S. Yang, X. Feng, S. Ivanovici, K. Mullen, Fabrication of graphene-encapsulated oxide nanoparticles: towards high-performance anode materials for lithium storage, *Angew. Chem.* 49 (2010) 8408–8411.

188. Y. Su, S. Li, D. Wu, F. Zhang, H. Liang, P. Gao, C. Cheng, X. Feng, Two-dimensional carbon-coated graphene/metal oxide hybrids for enhanced lithium storage, *ACS Nano* 6 (2012) 8349–8356.
189. Luo, J. Liu, Z. Zeng, C. F. Ng, L. Ma, H. Zhang, J. Lin, Z. Shen, H. J. Fan, Three-dimensional graphene foam supported Fe_3O_4 lithium battery anodes with long cycle life and high rate capability, *Nano Lett.* 13 (2013) 6136–6143.
190. Y. Sun, X. Hu, W. Luo, F. Xia, Y. Huang, Reconstruction of conformal nanoscale MnO on graphene as a high-capacity and long-life anode material for lithium ion batteries, *Adv. Funct. Mater.* 23 (2013) 2436–2444.
191. S. Whittingham, Lithium batteries and cathode materials, *Chem. Rev.* 104 (2004) 4271–4302.
192. Y. Wang, G. Cao, Developments in nanostructured cathode materials for high-performance lithium-ion batteries, *Adv. Mater.* 20 (2008) 2251–2269.
193. F.-Y. Su, C. You, Y.-B. He, W. Lv, W. Cui, F. Jin, B. Li, Q.-H. Yang, F. Kang, Flexible and planar graphene conductive additives for lithium-ion batteries, *J. Mater. Chem.* 20 (2010) 9644–9650.
194. J. Yang, J. Wang, Y. Tang, D. Wang, X. Li, Y. Hu, R. Li, G. Liang, T.-K. Sham, X. Sun, LiFePO4–graphene as a superior cathode material for rechargeable lithium batteries: impact of stacked graphene and unfolded graphene, *Energy Environ. Sci.* 6 (2013) 1521–1528.
195. Wang, A. Liu, W. A. Abdulla, D. Wang, X. S. Zhao, Desired crystal oriented LiFePO4 nanoplatelets in situ anchored on a graphene cross-linked conductive network for fast lithium storage, *Nanoscale* 7 (2015) 8819–8828.
196. C.-C. Yang, S.-P. Luo, $Li_3V_2(PO_4)_3$/C composite with nitrogen-doped-carbon coating and nanoporous graphene oxide, *Mater. Chem. Phys.* 173 (2016) 412–420.
197. Cheng, X.-D. Zhang, X.-H. Ma, J.-W. Wen, Y. Yu, C.-H. Chen, Nano-$Li_3V_2(PO_4)_3$ enwrapped into reduced graphene oxide sheets for lithium-ion batteries, *J. Power Sources* 265 (2014) 104–109.
198. Y. Li, W. Q. Bai, Y. D. Zhang, X. Q. Niu, D. H. Wang, X. L. Wang, C. D. Gu, J. P. Tu, Synthesis and electrochemical performance of lithium vanadium phosphate and lithium vanadium oxide composite cathode material for lithium ion batteries, *J. Power Sources* 282 (2015) 100–108.
199. B. Pei, Z. Jiang, W. Zhang, Z. Yang, A. Manthiram, Nanostructured $Li_3V_2(PO_4)_3$ cathode supported on reduced graphene oxide for lithium-ion batteries, *J. Power Sources* 239 (2013) 475–482.
200. R. Rajagopalan, L. Zhang, S. X. Dou, H. Liu, Lyophilized 3D lithium vanadium phosphate/reduced graphene oxide electrodes for super stable lithium ion batteries, *Adv. Energy Mater.* 6 (2016) 1501760–1501768.
201. R. Dominko, Li_2MSiO_4 (M = Fe and/or Mn) cathode materials, *J. Power Sources* 184 (2008) 462–468.
202. J. Yang, L. Hu, J. Zheng, D. He, L. Tian, S. Mu, F. Pan, Li_2FeSiO_4 nanorods bonded with graphene for high performance batteries, *J. Mater. Chem. A* 3 (2015) 9601–9608.
203. J. Yang, X. Kang, D. He, A. Zheng, M. Pan, S. Mu, Graphene activated 3D-hierarchical flower-like Li_2FeSiO_4 for high-performance lithium-ion batteries, *J. Mater. Chem. A* 3 (2015) 16567–16573.
204. H. Xu, B. Cheng, Y. Wang, L. Zheng, X. Duan, L. Wang, J. Yang, Y. Qian, Improved electrochemical performance of LiMn2O4/graphene composite as cathode material for lithium ion battery, *Int. J. Electrochem. Sci.* 7 (2012) 10627–10632.
205. X. Zhao, C. M. Hayner, H. H. Kung, Self-assembled lithium manganese oxide nanoparticles on carbon nanotube or graphene as high-performance cathode material for lithium-ion batteries, *J. Mater. Chem.* 21 (2011) 17297–17303.

206. S.-M. Bak, K.-W. Nam, C.-W. Lee, K.-H. Kim, H.-C. Jung, X.-Q. Yang, K.-B. Kim, Spinel LiMn2O4/reduced graphene oxide hybrid for high rate lithium ion batteries, *J. Mater. Chem.* 21 (2011) 17309–17315.

207. K.-C. Jiang, S. Xin, J.-S. Lee, J. Kim, X.-L. Xiao, Y.-G. Guo, Improved kinetics of LiNi1/3Mn1/3Co1/3O2 cathode material through reduced graphene oxide networks, *Phys. Chem. Chem. Phys.* 14 (2012) 2934–2939.

208. Venkateswara Rao, A. Leela Mohana Reddy, Y. Ishikawa, P. M. Ajayan, LiNi1/3Co1/3Mn1/3O2 graphene composite as a promising cathode for lithium-ion batteries, *ACS Appl. Mater. Interfaces* 3 (2011) 2966–2972.

209. J.-r. He, Y.-f. Chen, P.-j. Li, Z.-g. Wang, F. Qi, J.-b. Liu, Synthesis and electrochemical properties of graphene-modified LiCo1/3Ni1/3Mn1/3O2 cathodes for lithium ion batteries, *RSC Adv.* 4 (2014) 2568–2572.

210. H. R. Byon, B. M. Gallant, S. W. Lee, Y. Shao-Horn, Role of oxygen functional groups in carbon nanotube/graphene freestanding electrodes for high performance lithium batteries, *Adv. Funct. Mater.* 23 (2013) 1037–1045.

211. H. Kim, Y.U. Park, K.-Y. Park, H.-D. Lim, J. Hong, K. Kang, Novel transition-metal-free cathode for high energy and power sodium rechargeable batteries, *Nano Energy* 4 (2014) 97–104.

212. D.-W. Wang, C. Sun, G. Zhou, F. Li, L. Wen, B. C. Donose, G. Q. M. Lu, H.-M. Cheng, I. R. Gentle, The examination of graphene oxide for rechargeable lithium storage as a novel cathode material, *J. Mater. Chem. A* 1 (2013) 3607–3612.

213. Xiong, X. Li, Z. Bai, H. Shan, L. Fan, C. Wu, D. Li, S. Lu, Superior cathode performance of nitrogen-doped graphene frameworks for lithium ion batteries, *ACS Appl. Mater. Interfaces* 9 (2017) 10643–10651.

214. B. Lung-Hao Hu, F.-Y. Wu, C.-T. Lin, A. N. Khlobystov, L.-J. Li, Graphene-modified LiFePO$_4$ cathode for lithium ion battery beyond theoretical capacity, *Nat. Commun.* 4 (2013) 1687.

12 Advanced Nanocomposites Materials for Supercapacitor Applications

Characteristics and Properties

L. Syam Sundar and Feroz Shaik
Prince Mohammad Bin Fahd University, Kingdom of
Saudi Arabia

12.1 INTRODUCTION

Electrochemical capacitors (ECs), also known as supercapacitors or ultra-capacitors, are energy storage devices that store energy as charge on the electrode surface or sub-surface layer rather than in the bulk material as in batteries. As a result, they can produce high power because energy can be released more readily from the surface or sub-surface layer than from the bulk. Supercapacitors have good cycling ability because charging-discharging occurs on the surface, which does not cause electro-active materials to undergo significant structural changes. Supercapacitors are seen to be one of the most promising energy storage technologies because of these distinct characteristics. Electrochemical double layer capacitors (EDLCs) and pseudocapacitors are the two categories of supercapacitors.

Unlike pseudocapacitors that store charge through quick redox processes on the electrode surface, EDLCs store energy electrostatically at the electrode-electrolyte interface in the double layer. Metal oxides/hydroxides, carbon-based materials, and conducting polymers are the three main categories of electrode materials used in supercapacitors. Whereas conducting polymers like polyaniline, polypyrrole, and polythiophene or metal oxides like MnO_2, V_2O_5, and RuO_2 are utilized for pseudo-capacitors, electrode active materials in EDLCs are carbon-based materials, including activated carbon, mesoporous carbon, carbon nanotubes (CNTs), graphene, and

DOI: 10.1201/9781003364825-12

carbon fibers. Because EDLCs only rely on the carbon-based materials' surface area for charge storage, they frequently have very high power output and greater cycle capabilities.

Pseudocapacitors use redox active materials to store charge on the surface as well as in the sub-surface layer, but EDLCs have lower energy density values than pseudocapacitors. Although metal oxides/hydroxides, conducting polymers, and carbon-based materials are the most popular electroactive materials for supercapacitors, each type of material has specific benefits and drawbacks. For instance, while carbon-based materials can offer high power density and long life cycles, their low specific capacitance (primarily double layer capacitance) restricts their use in high energy density devices. In addition to double layer capacitance, metal oxides/hydroxides feature pseudocapacitance. They also have a large charge/discharge potential range, but have a small surface area and a short life cycle.

High capacitance, good conductivity, low cost, and ease of manufacture are all benefits of conducting polymers, but their mechanical stability and life cycle aren't very good. In order to control, develop, and optimize the structures and properties of electrode materials to improve their performance for supercapacitors, it is important to combine the distinct advantages of these dissimilar, nanoscale capacitive materials to create nanocomposite electroactive materials. The qualities of nanocomposite electrodes depend not only on the individual components utilized but also on the shape and the interfacial features.

All different types of nanocomposite capacitive materials, including mixed metal oxides, conducting polymers mixed with metal oxides, CNTs mixed with conducting polymers or metal oxides, and graphene mixed with metal oxides or conducting polymers, have recently received a lot of attention. Many factors need to be taken into account during the design and fabrication of nanocomposite electroactive materials for supercapacitor applications, including material choice, synthesis techniques, fabrication process parameters, interfacial properties, electrical conductivity, nanocrystallite size, surface area, etc. The development of nanocomposite electroactive materials for supercapacitor applications has advanced significantly; however, there are still many obstacles to be solved.

This chapter will provide a brief overview of the most recent advancements in this emerging field of study, including the synthesis techniques currently employed for producing nanocomposite electroactive materials, the different types of nanocomposite electroactive materials investigated, the structural and electrochemical characterization of nanocomposites, the special capacitive properties of nanocomposite materials, and the mechanism for performance enhancement of nanocomposite electroactive materials.

12.2   SUPERCAPACITOR WORKING PRINCIPLE

A high-capacity capacitor with a capacitance value significantly higher than conventional capacitors but lower voltage restrictions is known as a supercapacitor (SC), also known as an ultra-capacitor. It fills the void left by rechargeable batteries and electrolytic capacitors. In comparison to electrolytic capacitors, it typically stores 10–100 times more energy per unit volume or mass, accepts and delivers charge considerably more quickly, and can withstand many more charge and dis-charge cycles than rechargeable batteries.

Instead of long-term compact energy storage in vehicles such as cars, buses, trains, cranes, and elevators, supercapacitors are used in applications requiring numerous quick charge/discharge cycles. These applications include regenerative braking, short-term energy storage, or burst-mode power delivery. Smaller devices are utilized as a static random-access memory (SRAM) power backup. Supercapacitors, in con-trast to regular capacitors, use electrostatic double-layer capacitance and electro-chemical pseudocapacitance, both of which add to the overall capacitance of the capacitor. Figure 12.1 indicates the line diagram of the supercapacitor (Drummond et al. [1]). Figure 12.2 is the classification of supercapacitors.

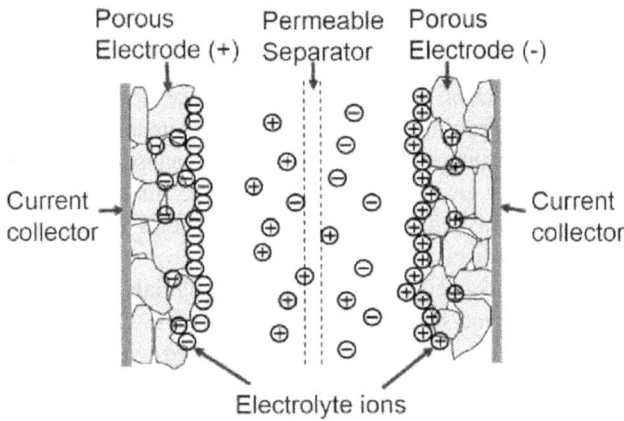

FIGURE 12.1   Line diagram of the supercapacitor (Drummond et al. [1]).

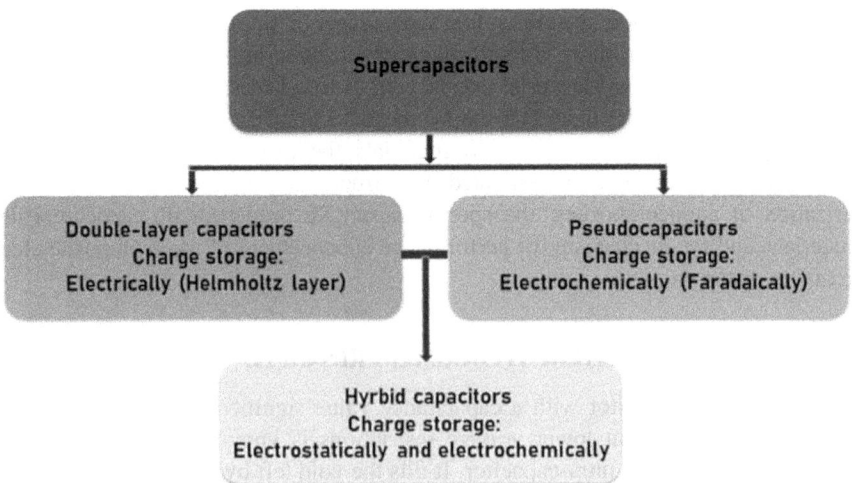

FIGURE 12.2   Classification of supercapacitors.

12.3   FABRICATION AND CHARACTERIZATION OF ACTIVE ELECTRODE NANOCOMPOSITE MATERIALS

12.3.1   FABRICATION METHODS

Several synthesis techniques, such as solid-state reactions (thermal decomposition of mechanically mixed metal salts), mechanical mixing of metal oxides (such as ball milling), chemical co-precipitation, and electrochemical anodic deposition from solutions containing metal salts, have been used to create mixed metal oxide nanocomposites. Reduction of $KMnO_4$ with Pb(II) and Ni(II) salts to form an amorphous mixed oxide precipitant was used to create Mn–Pb and Mn–Ni mixed oxide nanocomposites. Next, anodic deposition was used to directly deposit Mn–V–W oxide and Mn–V–Fe oxide on conductive substrates from an aqueous solution of mixed metal salts.

CNTs and metal oxides were combined mechanically in a mortar to create CNTs-metal oxide nanocomposites, or metal oxides were directly deposited on CNTs using metal-organic chemical vapour deposition (CVD), wet chemical precipitation, or electrochemical deposition. For instance, to create an IrO_2–CNTS nanocomposite, IrO_2 nanotubes were formed on multiwall CNTs using metalorganic CVD using an iridium source of (C6H7)(C8H12)Ir. Thermal CVD was initially used to grow the CNTs on stainless steel plates. The MnO_2–CNTs nanocomposites were created by electrophoretically depositing CNTs on a Ni substrate, followed by the direct current anodic deposition of MnO_2 from the $MnSO_4$ solution. By ruthenium nitrosyl nitrate solution impregnating CNTs, RuO_2–CNTs were created, which were subsequently heated to create composite electrodes.

In-situ polymerization in solutions comprising monomers of the conducting polymer and suspensions of CNTs, metal oxide nanoparticles, or graphene (GN) nanosheets was the primary method used to create nanocomposite materials made of a conducting polymer and metal oxide, CNTs, or graphene (GN). For instance, a CNTs-polyaniline (PANI) nanocomposite was created from a mixture of two materials. CNTs–PANI nanocomposite was created by polymerizing aniline on the surface of CNTs after the addition of an oxidant solution comprising $(NH_4)2S_2O_8$. In order to create MoO_3-poly 3,4-ethylenedioxythiophene (PEDOT) nanocomposites, a suspension of lithium molybdenum nanoparticles was first added with the 3,4 ethylenedioxythiophene monomer; next, iron (III) chloride ($FeCl_3$) was added to the suspension as the oxidizing agent under microwave hydrothermal conditions for polymerization to take place. By employing ammonium peroxydisulfate [(NH_4) $2S_2O_8$)] and $FeCl_3$ as oxidizing agents in a solution that also contained sodium polystyrene sulfonate Na salt, HCl, EDOT monomer, and GN, GN–PEDOT nanocomposite was chemically created. Chemically, G-PANI nanocomposite was produced by oxidatively polymerizing aniline monomers with ammonium peroxydisulfate [$(NH_4)2S_2O_8$] in GN-containing solution.

12.3.2   CHARACTERIZATION (STRUCTURE, ELECTRICAL, CHEMICAL COMPOSITION, AND SURFACE AREA)

The most popular analytical methods used to characterize the morphologies, structures, chemical composition, and surface area of nanocomposite electroactive materials included X-ray diffraction (XRD), scanning electron microscopy/energy-dispersive analysis (SEM/EDX), high-resolution transmission electron microscopy (HRTEM), infrared spectra (IR), and the Brunauer–Emmett–Teller (BET) specific surface areas. The crystallinity and crystal phases of the metal oxide materials were investigated using XRD analysis on the nanocomposite samples. For nanocomposites made of conducting polymers, IR spectra were employed to identify the distinctive bands of a polymer. SEM was utilized to analyze the morphology of nanocomposites, while EDX was employed to ascertain their chemical make-up. Using a four-probe method, the electrical conductivity of nanocomposites was determined. The nanocomposite samples were made into fine powders and subsequently formed into pellets for the measurement. Thermogravimetric analysis (TGA) and differential thermal analysis were used to examine the weight loss of nanocomposite material as well as the heat flow related to the thermal decomposition during synthesis or heat treatment (DTA).

12.3.3 ELECTROCHEMICAL CHARACTERIZATIONS

Typically, an electrochemistry workstation was used to perform cyclic voltammetry (CV) in either aqueous electrolytes or organic electrolytes to characterize the nanocomposite electrode in a three-electrode cell. The working electrode was made of a metal plate or mesh (such as nickel, aluminum, or stainless steel) that had been covered with a binder such as PTFE or polyvinylidenedifluoride and a mixture of nanocomposite and conductive carbon such as acetylene black (PVDF). Saturated calomel electrode (SCE), Ag/AgCl, or other reference electrodes were used. Platinum foil was frequently used as the counter electrode. The CV current value, scan rate, and weight of the nanocomposites were used to calculate the specific current and specific capacitance of the nanocomposite. Using two-electrode systems that had identical electrodes made of the same nanocomposite electroactive material, galvanostatic charge-discharge cycling was carried out.

For charging/discharging the cell in the voltage range of 0–1 V for aqueous electrolytes or 0–2.7 V for organic electrolytes, constant current densities of 0.5 to 10 mA/cm^2 were typically used. The slope (dV/dt) of the linear part of the discharge curve is used to estimate the discharge capacitance (C).

$$C = IdV/dt \qquad (12.1)$$

The weight of the active material of the two electrodes is same in a symmetric supercapacitor. The specific capacitance (Cs) of the single electrode can thus be expressed as:

$$C_s = 2C/m \qquad (12.2)$$

where m is the active material mass of the single electrode. The energy density (Ed) of the capacitor can be expressed as:

$$E_d = \frac{1}{2}(C_s V_{max}^2)$$
(12.3)

The Coulombic efficiency η was evaluated using the following relation:

$$\eta = \frac{t_d}{t_c} \times 100\%$$
(12.4)

where t_d and t_c are the time of charge and discharge, respectively.

A two-electrode system with identical electrodes manufactured of the same nanocomposite active electrode materials was used in electrochemical impedance spectroscopy (EIS) experiments spanning the frequency range of 10 kHz to 10 MHz with a potential amplitude typically of 5 mV.

12.4 DIFFERENT TYPES OF NANOCOMPOSITE ACTIVE ELECTRODE MATERIALS

12.4.1 MIXED PSEUDOCAPACITIVE METAL OXIDE NANOCOMPOSITES

For use in pseudocapacitors, metal oxides such as RuO_2, MnO_2, Co_3O_4, NiO, SnO_2, Fe_3O_4, and V_2O_5 have been used as electroactive materials. The charge storage in pseudocapacitors is facilitated by these metal oxides, which often have many redox states or topologies. RuO_2, which exhibits the greatest specific capacitance values of 720 F g^{-1}, performs remarkably well in supercapacitors, which has sparked a lot of interest in studying metal oxide systems for use in supercapacitors. RuO_2's poisonous toxicity and high cost, however, restrict its practical application. Some drawbacks for other basic metal oxides include poor electrical conductivity, inadequate electrochemical cycle stability, a small voltage working window, and low specific capacitance.

For supercapacitors made of metal oxides to be used commercially, those restrictions must be overcome. Mixed binary or ternary metal oxide systems, including Ni–Mn oxide, Mn–Co oxide, Mn–Fe oxide, Ni–Ti oxide, Sn–Al oxide, Mn–Ni–Co oxide, Co–Ni–Cu oxide, and Mn–Ni–Cu oxide, have demonstrated improved properties as electroactive materials for pseudocapacitors and have provided new insights into this field of study. The progress made recently in the search for electroactive mixed metal oxide nanocomposites for pseudocapacitors is summarized in the section that follows.

12.4.2 MIXED MANGANESE OXIDES

Mn oxides are the most promising novel electroactive materials for pseudocapacitor applications because of their natural abundance, low cost, good energy-storage performance in mild electrolytes, and environmental compatibility. However, Mn oxides have drawbacks such as tiny specific capacitance values, poor electrical conductivity, and low surface areas. Many efforts have been made to include different transition metals into Mn oxides in order to create mixed metal oxide nanocomposites with regulated micro-/nanostructures to improve their electrochemical properties for pseudocapacitors. It is still very difficult to comprehend how they work together synergistically and to develop an integrated material architecture where the attributes of each component can be adjusted and rapid ion and electron transmission is ensured.

MnO_2 nanoflakes-Ni(OH)$_2$ nanowire composites that can be employed in both neutral and alkaline electrolytes and have a very high cycling stability were conceived and created by Jiang et al. [2]. In 1 M Na_2SO_4 neutral aqueous solution, nanocomposites with 70.4 wt% MnO_2 content displayed specific capacitance of 355 F g^{-1} with excellent cycling stability (97.1% retention after 3000 cycles). The MnO_2-Ni (OH)$_2$ nanocomposite with 35.5 wt% MnO_2 content has a remarkable cycle stability and a specific capacitance of 487.4 F g^{-1} in 1 M KOH aqueous alkaline solution. The authors ascribe these exceptional capacitive behaviors to the distinctive MnO_2-Ni (OH)$_2$ core–shell nanostructures shown in Figure 12.3(b). On the surface of Ni(OH)$_2$ nanowires, the interconnected MnO_2 nanoflakes were evenly distributed, resulting in a highly porous surface shape.

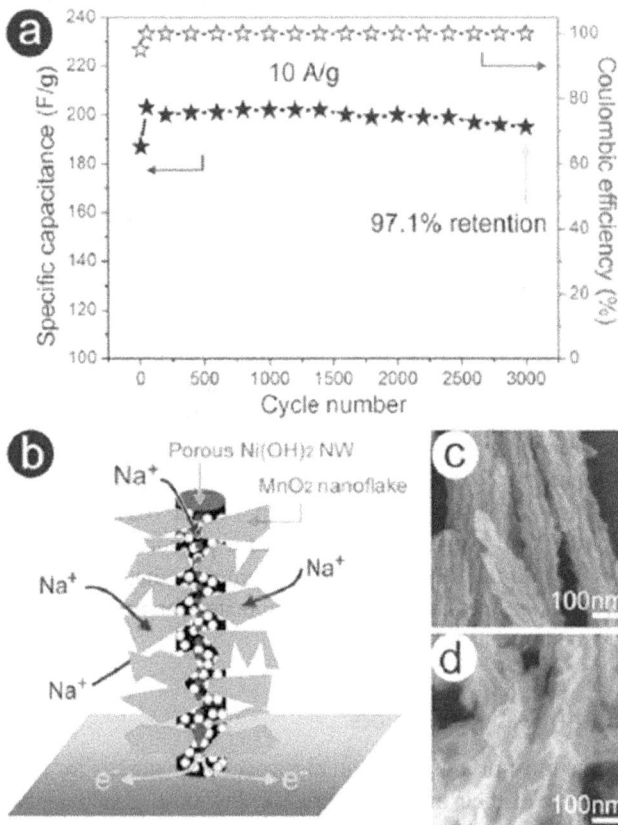

FIGURE 12.3 (a) Specific capacitance as a function of cycle number at 10 A g^{-1}, (b) schematic of the charge storage advantage of the Ni(OH)$_2$–MnO$_2$ core–shell nanowires, (c) and (d) SEM images of the Ni(OH)$_2$–MnO$_2$ core–shell nanowires before and after 3000 cycles (Jiang et al. [2]).

High surface area and more active sites for the redox processes can be provided by this integrated structure. Figure 12.3(a) also displays the MnO_2–$Ni(OH)_2$ nanocomposite's specific capacitance and Coulombic efficiency as a function of cycle number for up to 3000 cycles at a current density of 10 A g^{-1}. The $Ni(OH)_2$–MnO_2 nanocomposites are mostly retained with minimal structural deformation after extensive cycling, as demonstrated in Figure 12.3(c) and (d). MnO_2 was combined with the oxides of Pb, Fe, Mo, and Co to create mixed metal oxide nanocomposites.

By reducing $KMnO_4$ with either lead(II) acetate-manganese acetate or nickel(II) acetate-manganese acetate solutions, Kim et al. [3] created mixed oxides of Mn with Pb or Ni. Using cyclic voltammetry, galvanostatic charge-discharge, XRD, BET analysis, and TGA, the nanocomposite electrodes were characterized. The outcomes demonstrated that adding Ni and Pb to MnO_2 increased the surface area of the mixed oxide because micropores formed. For Mn–Ni and Mn–Pb mixed oxides, the specific capacitance increased from 166 F g^{-1} (for MnO_2) to 210 and 185 F g^{-1}, respectively.

Kim et al. [3] discovered that annealing of the nanocomposites might impact their capacitance: the specific capacitance is decreased when the transition from an amorphous to a crystalline structure occurs at a high temperature (400°C). In a mixed plating solution of $Mn(CH_3COO)_2$ and FeCl3, binary Mn–Fe oxide was electroplated on graphite substrates by Lee et al. [4] at a constant applied potential of 0.8 V vs. SCE. Using cyclic voltammetry in a 2 M KC1 solution, the electrochemical properties of the mixed oxide nanocomposites were studied in both their as-deposited and annealed states.

According to Lee et al. [4], the as-deposited Mn–Fe binary oxide exhibits an amorphous and porous structure. The partially hydrous mixed oxide has optimal ionic and electronic conductivity and produces the highest pseudocapacitive performance after being annealed at 100°C to remove the adsorbed water. The mixed oxide loses its porosity and begins to slowly crystallize at higher annealing temperatures, which causes a reduction in specific capacitance.

Ye et al. [5] studied a number of Mn and Mo mixed oxides (i.e. Mn–Mo–X (X= W, Fe, Co)) and discovered that the specific capacitance of Mo-doped Mn oxides is greater than that of pure Mn oxide. The Mn–Mo–Fe oxide exhibits a rectangular-shaped voltammogram and a high specific capacitance value of 278 F g^{-1} in an aqueous 0.1 M Na_2SO_4 solution. The authors attributed the creation of nanostructure and the presence of poor crystallinity as the causes of the improvement in capacitance of Mn oxides doped with molybdenum. The amorphous structure of mixed metal oxides has a higher specific capacitance than the crystalline structure, according to the results shown above.

In order to develop mixed metal oxide nanocomposite for supercapacitor electroactive materials research, advanced thin-film physical vapour deposition techniques were also applied. In Yang [6] lab, pulsed laser deposition (PLD) was used to create thin coatings of manganese oxide doped with varying amounts of cobalt oxide on silicon wafers and stainless steel substrates.

Yang [7] developed various PLD processing parameters (i.e. temperature, oxygen pressure) to produce various chemical compositions and phases of manganese oxides, such as pure crystalline phases of Mn_2O_3 and Mn_3O_4 as well as amorphous phase of MnOx, before investigating Co-doped manganese oxide film.

The crystalline Mn_2O_3 phase has the highest specific current and capacitance, whereas the values for crystalline Mn_3O_4 films are the lowest. The author next assessed the pseudocapacitance behaviors of these various phases of manganese oxides. The amorphous MnOx films have specific current and capacitance values that fall between Mn_2O_3 and Mn_3O_4. Mn_2O_3 films that are 120 nm thick have a 210 F g^{-1} specific capacitance at 1 mV s^{-1} scan rate with outstanding cycle endurance.

Then, they doped crystalline Mn_2O_3 and amorphous MnOx phases with Co_3O_4, and they used X-ray diffraction and CVs to analyze the mixed Co–Mn oxide films. Undoped and Co-doped amorphous MnOx films' CVs collected at a 20 mV s^{-1} scan rate are presented in Figure 12.4(a), and their specific capacitance calculated from

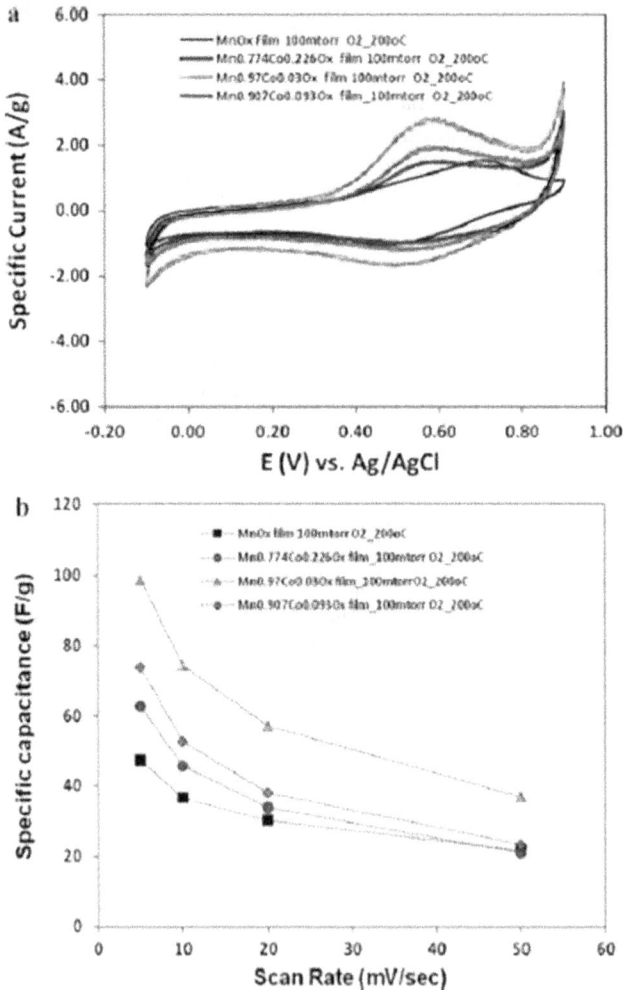

FIGURE 12.4   Cyclic voltammetry (a) and specific capacitance (b) of amorphous MnOx films and various Co-doped amorphous MnOx films deposited by PLD at 200°C in 100 mTorr of O2 (Yang [7]).

the CV curves at scan rates of 5, 10, 20, and 50 mV s^{-1} is shown in Figure 12.4(b). The CVs in Figure 12.2 demonstrate that compared to the undoped amorphous MnOx film, the Co-doped amorphous MnOx film has greater specific currents and capacitances. The highest increase in capacitance was observed at low cobalt doping (3.0 atm%), followed by cobalt doping at 9.3 atm.

12.4.3 OTHER MIXED METAL OXIDES

Many additional mixed metal oxides, in addition to mixed manganese oxides, have also been researched as electroactive materials for supercapacitor applications. By electrochemically depositing Co_3O_4-$Ni(OH)_2$ nanocomposites on a Ti substrate in a solution of $Ni(NO_3)_2$, $Co(NO_3)_2$, and NH_4Cl, followed by heat treatment at 200 °C [8], these materials were created. The high specific capacitance value of 1144 F g^{-1} at 5 mV s^{-1} and long-term cyclability were displayed by Co_3O_4–$Ni(OH)_2$ electrodes. The scientists credited the porous network topologies that facilitate the transit of electrons and ions as well as Faradic redox reactions of both couples of Co^{2+}/Co^{3+} and Ni^{2+}/Ni^{3+} for the good capacitive behaviors of Co_3O_4–$Ni(OH)_2$ nanocomposite.

By anodizing Ti-V alloys with various V compositions using ethylene glycol with 0.2 M HF as the electrolyte and a comparatively high anodization voltage, Y. Yang et al. [9] created mixed V_2O_5-TiO_2 nanotube arrays. For alloys with vanadium contents up to 18 at%, well-defined nanotube structures were formed. When compared to pure TiO_2 nanotube arrays, the combined V_2O_5-TiO_2 nanotube arrays were shown to have significantly improved capacitive characteristics. In repeated cycles, it was discovered that the combined V_2O_5–TiO_2 nanotubes' specific capacitance could reach 220 F g^{-1} and have an energy density of 19.56 Wh kg^{-1}.

SnO_2–Al_2O_3 mixed oxide [10], which exhibits significantly higher electrochemical capacitance than pure SnO_2 and was electrochemically and chemically stable even after cycling 1000 times, is another intriguing mixed oxide.

Compared to NiO and Co_3O_4, spinel nickel cobaltite, whether doped or not (such as $NiCo_2O_4$ and $NiMnxCo_2xO_4y$ (x1.0)), has a substantially higher electronic conductivity. They are used for supercapacitor applications due to their inexpensive cost and variety of oxidation states. By using a co-precipitation technique, Wang et al. [11] created nanostructured $NiCo_2O_4$ spinel platelet-like particles with a restricted size distribution of 5–10 nm. The $NiCo_2O_4$ exhibits outstanding conductivity and demonstrated a high-specific capacitance of 671 F g^{-1} at a current density of 1 A g^{-1} under a mass loading of 0.6 mg cm^{-1}.

A precipitation method was also used by Chang et al. [12] to manufacture $NiCo_2O_4$ and $NiMnxCo_2xO_4y$ (x1.0). They discovered that $NiCo_2O_4$ still has its spinel structure, despite having Mn ions replace 25% of the Co ions. Mn considerably reduces the formation of crystallites during heat processing and greatly increases the specific capacitance of the spinel. The specific capacitance is observed to rise from 30 F g^{-1} for Mn content x = 0–110 F g^{-1} for x = 0.5 at the scan rate of 4 mV s^{-1}. The authors discovered that the surface area of the NiMn0.5Co1.5O$_4$ powder was significantly less than that of the $NiCo_2O_4$ powder. As a result, the $NiMn_{0.5}Co_{1.5}O_4$ electrode's astounding capacitance improvement is not the result of differences in the oxide powders' microstructure.

12.4.4 Carbon Nanotube-Based Nanocomposites

CNTs are excellent electroactive material choices for supercapacitors due to their remarkable material qualities, including high chemical stability, aspect ratio, mechanical strength, and activated surface area. However, there are restrictions on further expanding the effective surface area of the CNTs, together with a relatively high materials cost, which limit the commercial applicability of CNT-based supercapacitors. The electrodes produced from CNTs also exhibit a distinctive pore structure for charge storing. CNTs are combined with conductive polymers and metal oxides to enhance their performance. The most recent advancements in CNT-based nanocomposites for supercapacitor applications will be outlined in this section.

12.4.4.1 Carbon Nanotubes and Polymer Nanocomposites

Arc discharge, chemical vapour deposition, and laser ablation are methods that can be utilized to make CNTs. By employing Ni, Co, and FeS as catalysts in a dc arc discharge of a graphite rod under helium gas, Kay et al. [13] were able to synthesis single-walled CNTs. Next, employing in-situ chemical polymerization of the pyrrole monomer in solution with the single-walled CNT suspension, they created the single-walled CNTs-polypyrrole (PPY) nanocomposite.

As-grown single-walled CNTs, pure PPY, and single-walled CNT–PPY nanocomposite powders created by in-situ chemical polymerization are all depicted in Figure 12.5 using FE-SEM images. As illustrated in Figure 12.5(a), the single-walled CNTs are cross-linked and randomly entangled. Some carbon nanoparticles are also present. The picture of pure PPY produced in Figure 12.5(b) without single-walled CNTs in solution displays a typical granular shape with granules of around 0.2–0.3 mm. Figure 12.5(c) shows that the individual CNT bundles are uniformly coated with PPY, proving that all of the CNTs may be efficiently coated by in-situ chemical polymerization of pyrrole. A maximum specific capacitance of 265 F g^{-1} was obtained from the single-walled CNT-PPY nanocomposite electrode containing 15 wt% of the conducting agent. This electrode was made employing single-walled CNT–PPY nanocomposites as active materials.

As-grown single-walled CNTs, pure PPY, and single-walled CNTs–PPY nanocomposite electrode specific capacitances are depicted in Figure 12.6 as functions of discharge current density. The single-walled CNTs–PPY nanocomposite electrode exhibits extremely high specific capacitance when compared to the pure PPY and as-grown single-walled CNT electrodes. The increase in active surface area of pseudocapacitive PPY by CNTs, according to the scientists, is what caused the improvement in the specific capacitance of the CNTs–PPY nanocomposite, Deng et al. [14] composited CNTs with polyaniline (PANI).

12.4.4.2 Carbon Nanotubes and Other Metal Oxide Nanocomposites

Many additional metal oxides, besides RuO$_2$, have been combined with CNTs to create electroactive materials for supercapacitors. Multi-walled CNTs were grown using thermal CVD by Chen et al. [15] on a stainless steel plate, and IrO$_2$ nanotubes were subsequently coated on top of the CNTs with metal-organic CVD using the

FIGURE 12.5 The FE-SEM images of (a) as-grown single-walled CNTs, (b) pure PPY, and (c) single-walled CNTs-PPY powder [13].

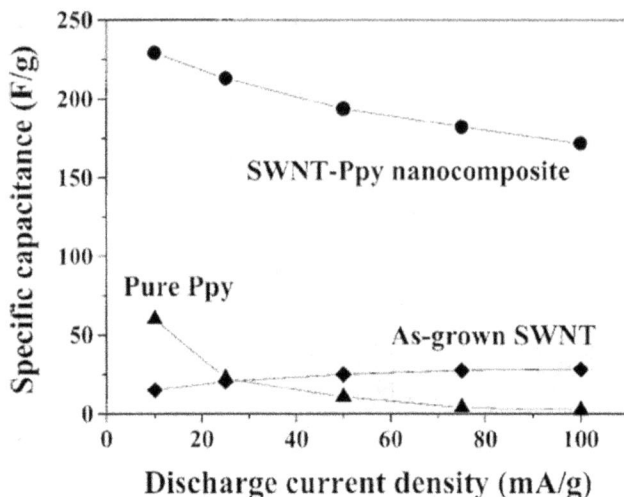

FIGURE 12.6 Specific capacitances of the as-grown single-walled CNTs, pure PPY, single-walled CNTs-PPY, and single-walled CNTs-PPY nanocomposite electrodes as a function of ischarge current density at a charging voltage of 0.9 V for 10 min (Deng et al. [14]).

iridium source of $(C_6H_7)(C_8H_{12})Ir$ at 350°C. On top of the CNTs thin film, crystals of IrO_2 square nanotubes were formed. The morphologies of multi-walled CNTs, IrO_2 nanotubes, and an IrO_2 nanotube-multi-walled CNT nanocomposite are depicted in Figure 12.7. There are upper and lower parts in the CNT thin film, as can be seen from the cross-sectional image of multi-walled CNTs in Figure 12.7(b). In the upper part, which is around 2 m thick, entangled CNTs with no obvious orientation are present. IrO_2 nanotubes grown on a stainless steel substrate are shown in top and cross-sectional views, respectively, in Figures 12.7(c) and (d). IrO_2 nanotubes have been formed over CNTs in Figure 12.7(e) and (f). Figure 12.7(e) shows that the multi-walled CNT wires in the upper part exhibited a high density of formed IrO_2 nanotubes. The resistance is decreased from 90 to 60 and the capacitance is increased by a factor of six compared to multi-walled CNTs by using nanostructured IrO_2–CNTs. A high surface area for storing electrical charge is provided by such a hierarchical structure, along with double-layer capacitance and pseudocapacitance.

The amorphous MnOx–CNTs nanocomposite electrode was created by Lee et al. [16] by electrochemical anodic deposition of MnOx•nH₂O films from MnSO₄•5H₂O solution on CNTs coated Ni substrates. Prior to the deposition of MnOx•nH₂O, the CNTs were electrophoretically deposited on the Ni substrate by applying a 20 V dc voltage. The MnOx–CNTs nanocomposite electrodes had significantly more energy storage capacity than MnOx placed directly on the Ni substrate: the specific capacitances were 415 as measured by CV at a scan rate of 5 mV/s, and it maintained 79% of its initial capacitance value after 1000 cycles. The authors ascribed the improvement to the nanocomposite electrodes' low resistance and substantial surface area. NiO–CNTs [17], V_2O_5–CNTs [18], and SnO_2–V_2O_5–CNTs [19] are a

FIGURE 12.7 SEM image of CNTs top view (a), CNTs cross-sectional view (b), IrO_2 nanotubes top view (c), IrO_2 cross-sectional view (d), IrO_2–CNTs top view (e), IrO_2–CNTs cross-sectional view (f). The inset of (b) and (d) are magnified images. (Chen et al. [15]).

few more metal oxide CNT nanocomposites that are being researched. V_2O_5 was electrochemically deposited on vertically aligned, multi-walled CNTs to create the V_2O_5–CNT composite, which has a specific capacitance of 713.3 F g^{-1} at 10 mV s^{-1}.

12.4.5 GRAPHENE-BASED NANOCOMPOSITES

A two-dimensional monolayer of carbon atoms with sp2 bonds is known as graphene (GN). Because to its exceptionally high electrical and thermal conductivities, outstanding mechanical strength, big specific surface area, and perhaps inexpensive manufacturing cost, it has gained more and more attention in recent years. High specific surface area (2675 m^2 g^{-1}) and high electrical conductivity are two of its best qualities, making it an ideal material for supercapacitor applications. A maximum specific capacitance of 117 F g^{-1} in an aqueous H2SO4 electrolyte has been recorded when thermally exfoliated GN nanosheets are used as supercapacitor electrode materials. It has been reported that supercapacitors formed of chemically altered GN have a specific capacitance of 135 F g^{-1} in an aqueous KOH electrolyte.

The irreversible agglomeration and restacking of GN caused by van der Waals interactions to create graphite, however, while drying GN during the electrode preparation process, represents a significant issue for GN-based supercapacitors. The agglomeration reduces the ability of electrolyte to penetrate into the layers, which has a negative impact on the performance of supercapacitors. The addition of spacers to the GN layers will help to solve this issue. The spacers can be CNTs, metal oxides, or conducting polymers. High electrochemical use of GN layers can be ensured by spacers, and electroactive spacers also add to the total capacitance. Recent advancements in GN-based nanocomposite materials for supercapacitor applications will be discussed in this section.

12.4.5.1 Graphene and Metal Oxide Nanocomposites

In order to create an advanced nanocomposite for supercapacitor applications, metal oxides like CeO_2, RuO_4, V_2O_5, and SnO_2 were combined with GN. The enhanced conductivity of metal oxide and greater GN usage are expected to have a synergistic impact that contributes to increasing the pseudocapacitance. In a Teflon-lined autoclave, Jaidev et al. [20] generated $RuO_2 \cdot xH_2O$–GN nanocomposite by hydrothermally treating GN nanosheets that were created through the exfoliation of graphite oxide in a hydrogen atmosphere. Using electrodes made by combining as-prepared RuO_2 xH_2O–GN, activated carbon, and Nafion (binder) on conducting carbon fabric, a symmetrical supercapacitor was created.

With a specific discharge current of 1 A g^{-1} (20 wt% Ru loading), the hybrid nanocomposite exhibits a maximum specific capacitance of 154 F g^{-1} and an energy density of around 11 Wh kg^{-1}. Moreover, the composite exhibits Coulombic efficiency of 97% for a particular discharge current of 10 A g^{-1} and a maximum power density of 5 kW kg^{-1}. By chemically precipitating 3D GN materials containing Ce $(NO_3)_3$ solution with KOH, CeO_2 was also added to the 3D GN [21].

Despite having a lower specific surface area than pure GN, CeO_2–GN nanocomposite produced high specific capacitance (208 F g^{-1} or 652 mF cm^{-2}) and had a long life cycle. By combining the V_2O_5 sol with the GN/ethanol dispersion and stirring for several days, Bonso et al. were able to create the GN–V_2O_5 nanocomposite [21]. The GN-V2O5 composite electrode that was created in this manner had a specific capacitance of 226 F g^{-1} in 1 M LiTFSI in acetonitrile. Whereas the specific capacitance of V_2O_5 alone was 70 F g^{-1} and that of GN alone was 42.5 F g^{-1}, respectively, the two materials working together had a synergistic effect.

The GNSnO2/CNTs nanocomposite was created by Rakhi et al. [22] by ultrasonically combining chemically functionalized GN and SnO_2–CNTs. SnO_2 was initially chemically precipitated from a $SnCl_2$ solution that contained functionalized multi-walled CNTs to create the SnO_2/CNTs. Figure 12.8 depicts the steps involved in creating a composite made of GN–SnO_2 and CNTs. Figure 12.9(a) and (b) shows

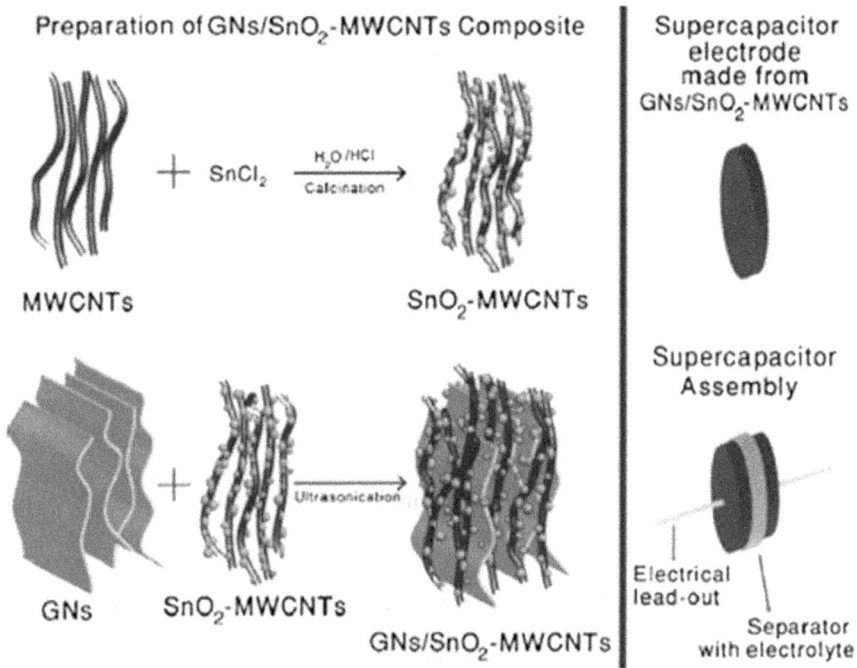

FIGURE 12.8   Schematic of preparation of supercapacitor electrode materials [23].

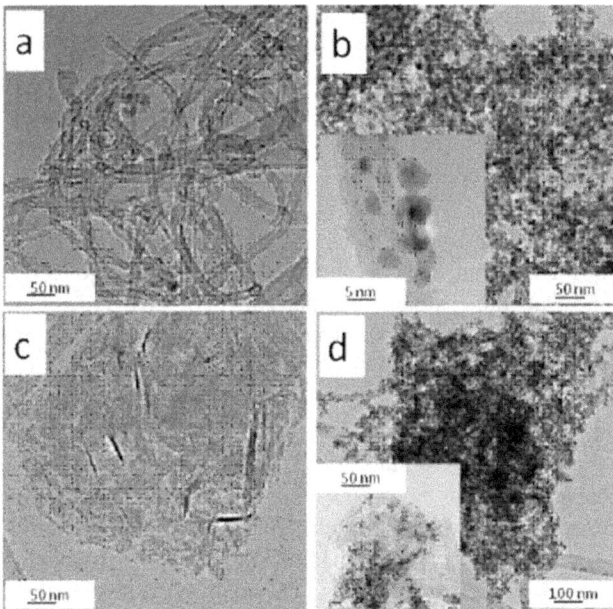

FIGURE 12.9   TEM images of (a) MWCNTs, (b) SnO2–MWCNTs (inset shows HRTEM image), (c) GNs, and (d) GNs/SnO2–MWCNTs composite [23].

the TEM images of multiwall CNTs and SnO_2/CNTs, respectively. The average inner and outer diameters of multi-walled CNTs are 10 nm and 30 nm, respectively, and their average lengths fall between 10 and 30 μm.

SnO_2 nanoparticles appear to be distributed uniformly across the surface of multi-walled CNTs, according to Figure 12.9(b). The inset of Figure 12.9(b) shows a high resolution TEM picture of SnO_2/CNTs, which shows that the SnO_2 nanoparticles are extremely crystalline in nature and have an average particle size of 4–6 nm. Figure 12.9(c) and (d) depicts the TEM images of large-area GN and GN–SnO_2/CNTs composite, respectively. On the surface of GN, SnO_2/CNTs may be seen to be present. The authors created symmetric supercapacitor devices with GN and GN–SnO_2/CNT composite electrodes. With a maximum specific capacitance of 224 F g^{-1}, a power density of 17.6 kW kg^{-1}, and an energy density of 31 Wh kg^{-1}, the latter produced impressive results. The outcomes showed that the capacitance characteristics of GN were enhanced by the dispersion of metal oxide-loaded multi-walled CNTs. With 81% of the initial specific capacitance still present after 6000 cycles, the manufactured supercapacitor device had remarkable life cycle.

12.5   CONCLUSION AND FUTURE SCOPE

The current state of nanocomposite electroactive materials has shown great promise for supercapacitor applications. Different processes, such as solid state reactions, mechanical mixing, chemical coprecipitation, electrochemical anodic deposition, sol-gel, in-situ polymerization, and other wetchemical synthesis, can be used to create different kinds of nanocomposite electroactive materials such as mixed metal oxides, polymers mixed with metal oxides, CNTs mixed with metal oxides or polymers, and graphene mixed with metal oxides or polymers. It has been demonstrated that employing nanocomposite electroactive materials can significantly improve the specific surface area, electrical and ionic conductivities, specific capacitance, cyclic stability, and energy and power density of supercapacitors.

This is explained by the complementary and synergistic behaviors of the constituent parts of the material, special interface properties, a large increase in surface areas, and nanoscale dimensional effects. The two most typical supercapacitor architectures are EDLCs, which use electroactive materials based on carbon, and pseudocapacitors, which use metal oxide or conducting polymers as electroactive materials. Pseudocapacitors provide a number of benefits over EDLCs, including high energy density and inexpensive material costs, but they also have poorer power density and poor cycle stability. The current research focuses on asymmetric supercapacitors that have one electrode made of carbon-based double capacitive materials and one electrode made of pseudocapacitive materials. Applications for asymmetric supercapacitors hold considerable promise for nanocomposite pseudocapactive materials. The main challenge is to maximize the good intrinsic features of nanocomposites, particularly their high surface area and high conductivity, and to enhance the synergistic interaction of various electroactive components.

REFERENCES

1. Drummond, R., Huang, C., Grant, P.S., & Duncan, S.R., Overcoming diffusion limitations in supercapacitors using layered electrodes, *Journal of Power Sources* 433 (2019) 126579.
2. Jiang, H., Li, C., Sun, T., & Ma, J. (2012). High-performance supercapacitor material based on $Ni(OH)_2$ nanowire-MnO_2 nanoflakes core–shell nanostructures. *Chem. Commun.*, 48, 2606–2608.
3. Kim, H., & Popov, N. (2003). Synthesis and characterization of MnO_2-based mixed oxides as supercapacitors. *Journal of the Electrochemical Society*, 150(3), D56–D62.
4. Lee, M. T., Chang, J. K., Hsieh, Y. T., & Tsai, W. T. (2008). Annealed Mn-Fe binary oxides for supercapacitor applications. *Journal of Power Sources*, 185, 1550.
5. Ye, Z. G., Zhou, X. L., Meng, H. M., Hua, X. Z., Dong, Y. H., & Zou, A. H. (2011). The electrochemical characterization of electrochemically synthesized MnO_2 based mixed oxides for supercapacitor applications. *Advanced Materials Research*, 287 (290), 1290–1298.
6. Yang. (2012). Pulsed laser deposition of cobalt-doped manganese oxide thin films for supercapacitor applications. *Journal of Power Sources*, 198, 416.
7. Yang, D. (2011). Pulsed laser deposition of manganese oxide thin films for super-capacitor Applications. *Journal of Power Sources*, 196, 8843.
8. Zhong, J. H., Wang, A. L., Li, G. R., Wang, J. W., Ou, Y. N., & Tong, Y. X. (2012). $Co_3O_4/Ni(OH)_2$ composite mesoporous nanosheet networks as a promising electrode for supercapacitor applications. *J. Mater. Chem.*, 22, 5656.
9. Yang, Y., Kim, D., Yang, M., & Schmuki, P. (2011). Vertically aligned mixed $2O5$-$TiO2$ nanotube arrays for supercapacitor applications. *Chem. Commun.*, 47, 7746–7748.
10. Jayalakshmi, M., Venugopal, N., Phani, R. K., & Mohan, R. (2006). Nano $SnO2$-$Al2O3$ mixed oxide and $SnO2$-$Al2O3$-carbon composite oxides as new and novel electrodes for supercapacitor applications. *Journal of Power Sources*, 158, 1538–1543.
11. Wang, C., Zhang, X., Zhang, D., & Yao, C. (2012). Facile and low-cost fabrication of nanostructured $NiCo2O4$ spinel with high specific capacitance and excellent cycle stability. *Electrochimica Acta*, 63, 220.
12. Chang, S. K., Lee, K. T., Zainal, Z., Tan, K. B., Yusof, N. A., Yusoff, W., Lee, J. F., & Wu, N. L. (2012). Structural and electrochemical properties of manganese substituted nickel cobaltite for supercapacitor application. *Electrochimica Acta*, 67–72.
13. An, K. H., Jeon, K. K., Heo, J. K., Lim, S. C., Bae, D. J., & Lee, Y. (2002). High-capacitance supercapacitor using a nanocomposite electrode of single-walled carbon nanotube and polypyrrole. *Journal of the Electrochemical Society*, 149(8), A1058–A1062.
14. Deng, M., Yang, B., & Hu, Y. (2005). Polyaniline deposition to enhance the specific capacitance of carbon nanotubes for supercapacitors. *Journal of Materials Science Letters*, 40, 5021–5023.
15. Chen, Y. M., Cai, J. H., Huang, Y. S., Lee, K. Y., & Tsai, D. (2011). Preparation and characterization of iridium dioxide-carbon nanotube nanocomposites for super-capacitors. *Nanotechnology*, 22(115706), 7.
16. Lee, C. Y., Tsai, H. M., Chuang, H. J., Li, S., Lin Yi, P., & Tseng, T. (2005). Characteristics and electrochemical performance of supercapacitors with manganese oxide-carbon nanotube nanocomposite electrodes. *Journal of the Electrochemical Society*, 152(4), 716A–A720.
17. Lee, J. Y., Liang, K., An, K. H., & Lee, Y. (2005). Nickel oxide/carbon nanotubes nanocomposite for electrochemical capacitance. *Synthetic Metals*, 150, 153–157.
18. Liu, M. Y., Hsieh, T. F., Hsieh, C. J., Chuang, C. C., & Shu, C. M. (2011). Performance of vanadium oxide on multi-walled carbon nanotubes/titanium electrode for super-capacitor application. *Advanced Materials Research*, 311-313, 414–418.

19. Jayalakshmi, M., Mohan, Rao. M., Venugopal, N., & Kim, K. B. (2007). Hydrothermal synthesis of SnO2-V2O5 mixed oxide and electrochemical screening of carbon nanotubes (CNT), V2O5, V2O5-CNT, and SnO2-V2O5-CNT electrodes for supercapacitor applications. *Journal of Power Sources*, 166, 578–583.

20. Jaidev, J. R. I. (2011). Ramaprabhu S. Hydrothermal synthesis of RuO2.xH2O/ graphene hybrid nanocomposite for supercapacitor application. International Conference on Nanoscience, Technology and Societal Implications, 8–10.

21. Wang, Y., Guo, C. X., Liu, J., Chen, T., Yang, H., & Li, C. (2011). CeO2 nanoparticles/ graphene nanocomposite-based high performance supercapacitor. *Dalton Transactions*, 40, 6388–6391.

22. Bonso, J. S., Rahy, A., Perera, S. D., Nour, N., Seitz, O., Chabal, Y. J., Balkus Jr, K. J., Ferraris, J. P., & Yang, D. (2012). Exfoliated graphite nanoplatelets-V2O5 nanotube composite electrodes for supercapacitors. *Journal of Power Sources*, 203, 227–232.

23. Rakhi, R. B., & Alshareef, H. N. (2011). Enhancement of the energy storage properties of supercapacitors using graphene nanosheets dispersed with metal oxide-loaded carbon nanotubes. *Journal of Power Sources*, 196, 8858.

13 Metals and Metal Oxides-Based Nanomaterials as Solid Nanostructures and Their Applications

M. Amin Mir, M. Waqar Ashraf, and Kim Andrews
Prince Mohammad Bin Fahd University, Kingdom of
Saudi Arabia

Anuj Kumar, Shailendra Prakash, and Anita Bisht
Govt. (PG) College, India

13.1 INTRODUCTION

Because of their unique chemical, physical, and biological features, nanoscale particles have generated a great deal of interest over the past few decades. Many different types of nanoparticles (NPs) have already been created and studied in a wide field of sciences, including material sciences, electronics, physics, medicine, and chemistry [1–5]. Moreover, they are made of different materials, including organic dendrimers, liposomes, gold, carbon, semiconductors, silicon, and iron oxide. In this regard, several solvent-derived mixture methods have been established for metal and metal oxide nanostructure synthesized materials that can be used for solid-state electronic devices and high-temperature devices. However, the appropriate solid-state methods for producing nanostructured materials are still needed [6–11]. For instance, recent research indicated that 3D Au superstructures with properties different from those of Au nanostructures form from evaporated solvents to create solid-state Au nanoparticles (solid-state device incorporation) [12]. As a result, the solvent-free synthesis of nanostructures is very important since it is affordable, environmentally friendly, and industrially feasible. Then, a never-ending problem is the development of novel solid-state techniques to create metallic nanostructured materials. Using the heating method for metallic and organometallic derivatives of poly and oligophosphazene under air at 8000°C, a new solid-state synthesis approach is implemented to generate metallic nanostructure materials [13–17]. Depending on the nature of the metal,

DOI: 10.1201/9781003364825-13

245

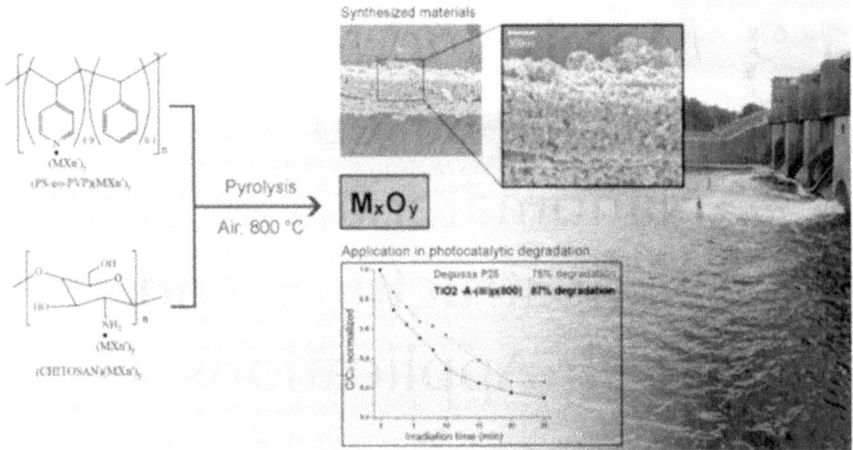

FIGURE 13.1 Solid-state and environmental remediation methods.

nanometal structures (M), metallic oxides (MxOy), and metal salts (MxPyOz, where P = succinates, malonates, etc.) are formed. When it is impossible to prepare the appropriate metallic or organometallic derivative, another approach is to employ mixes like MLn/[NP(O2C12H8)]3 [18–20]. Here, metallic nanoparticles of the pure phase are produced. Nonetheless, a phosphate phase often follows the M or MxOy phase in some of these systems. Several articles [13–23] contain in-depth discussions of these techniques. We will talk about a brand new technique for making nanostructured metallic and metal oxide particles in this chapter, starting with the macromolecule chitosan. The MXn precursor and PS-co-4-PVP under solid-state pyrolysis at 8000°C in air include environmental cleanup applications as shown in Figure 13.1. A brief overview of current solid-state techniques for producing metallic oxide and metallic nanoparticles is provided.

Several research techniques that have been gathered from previously cited literature for generating solid-state nanomaterials are offered in this chapter. An overview of the outcomes from the nanomaterial preparation employing our created solid-state approach is also provided. The outcomes for the various nanostructured metal oxides will next be examined in relation to their use in photocatalysis.

13.2   METAL OXIDE NANOPARTICLES PREPARATION REVIEW

13.2.1   THE KORGEL METHOD

Metal-thiolate complexes are thermolyzed in this process. The end-products are monodispersed, shape-distributed metal-sulfide nanoparticles. In order to increase the solubility of the metal cationic in a solution of chloroform and water, sodium octanoate is used as a biphasic transfer catalyst reaction between dodecanethiol and a metallic salt to create the precursor (Figure 13.2). The organic solvent is evaporated while the aqueous phase is discarded. To get rid of the remainder of the dodecanethiol stabilizer, the material is then redispersed in chloroform. This technique was used to create various metal-sulfide nanoparticles, including Cu_2S

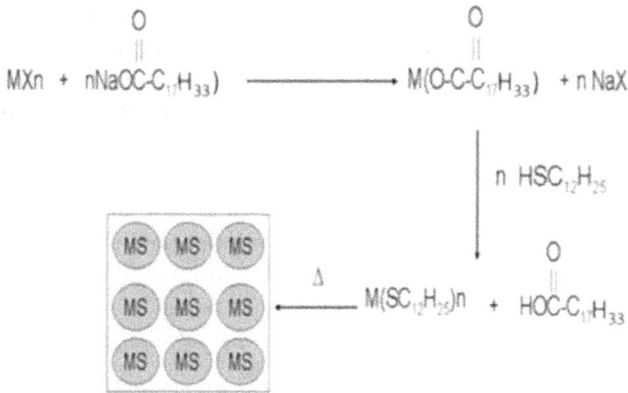

FIGURE 13.2 Simplified illustration of Korgel's approach. The preparation of the solid precursor is accomplished in the first two phases, which are in solution. The $M(SC_{12}H_{25})n$ precursor is solidly thermolyzed in the third stage.

[24,25], NiS [26], and Bi_2S_3 [27]. In a different method, the relevant metal oxides are produced by thermolysis of the related stearate M(OOCC17H33)n [28].

13.2.2 MACROMOLECULAR AND MOLECULAR COMPLEX DECOMPOSITION METHODOLOGY

A molecular complex is simply thermally decomposed solid-state under air at temperatures of 8000°C in this approach. There are several instances of [Ni(en)3](NO3)2 breakdown [29], [bis(2-hydroxy-1-naphtaldehyde) manganese (II)], and nickel dimethylglyoximate, among others. NiO and Mn_3O_4 are produced using MnOOH and Mn_3O_4 correspondingly [30–32]. In addition, ZnO and Co_3O_4 were produced through bis(acetylacetonate)Zn(II) and [Co(NH3)6](NO3)2 complex thermal decompositions [33,34]. Using the solid-state breakdown of the metal transition succinates $M(CH_2)$ $2C_2O_4.xH_2O$ and transition malonates $MCH_2C_2O_4$ xH_2O, a technique to manufacture oxides from the following metals, Fe, Mn, Ni, Co, Cu, and Zn reported [35]. Nanostructured metal oxides have also been synthesized using metal polymeric compounds as precursors. For instance, the [Ru(CO)4]n polymer can be pyrolyzed to produce nanofibers made of ruthenium [36].

When poly(ferrocenylsilanes) are pyrolyzed, nanoparticles made of Fe are produced [37]. Moreover, at temperatures between 500 and 10,000°C, the pyrolysis of poly(ferrocenylsilanes) produces intriguing ferromagnetic ceramic composites that contain a SiC/C matrix with Fe particles within the matrix [38]. It has also been suggested that metal ligand coordination polymers are advantageous precursors of nanostructured metal oxides. [Ga(–OH)(–O2CCH3)2] is a Ga-acetate polymer that pyrolyzes at 8000°C. Ga_2O_3 with a nanostructure is produced via HOAc.H2On [39].

Such clusters can be pyrolyzed to produce Os nanoparticles with a size range of 1–10 nm. Ru nanoparticles were produced using a similar method utilizing ruthenium clusters. This approach was also suggested for producing other perovskites. For instance, $LaMnO_3$ was created by heating the precursors of LaMnOx/C [40]. Several intriguing perovskite oxides, such as $LaMnO_3$, $LaFeO_3$, $LaNiO_3$, $LaCoO_3$, and La0.5Sr0.5Co0.5Fe0.5O3, were added to the solid-state gelation synthesis process.

By calcining CsBr/PbBr2/AIP = 1:1:30 mixtures at 8000°C for 10 min with nitrogen, several composites, including perovskites like $CsPbBr3–Al_2O_3$, were created. The as-obtained perovskites have remarkable thermal stability, a finely defined emission line width of 25 nm, and a high quantum yield of up to 70% [41].

Moreover, on a large scale, perovskites of the nanostructured $BaTiO_3$ and $SrTiO_3$ type have been created by reacting TiO_2 with strontium or barium oxalate at 820°C in the anatase phase [42]. Furthermore, suitable precursors for nanostructured metal oxides have been found in supramolecular metal structures. Take Na6 [Fe_2(–O) (–CO_3)–(chnida)2] as an example.

Composites of Co_3O_4/ZnO created by heating the Co_3O_4/Zn(OH)$_2$ precursor are another example of heterostructures. The so made heterostructure displays greater photocatalytic activity than ZnO toward the dye Rhodamine B. Fe_3O_4@M nanoparticles were generated on a large scale in the laboratory using a solid-state high-temperature approach, where M is equal to Au, Ag, and Au-Ag alloy [43]. The procedure entails heating a solid-state combination of the relevant metallic salts and Fe_3O_4. The Fe3O4@M nanocomposites as they were created demonstrated catalytic activity in the production of H_2 from NH_3BH_3 and $NaBH_4$. Another common method for creating solid-state nanostructured metal oxides is through the thermal breakdown of organometallic precursors. For instance, heating ferrocene carboxaldehyde results in the formation of Fe_2O_3 nanoparticles (hematite phase) with the size of 5 nm [44].

The solid-state thermolysis of metal organic frameworks (MOFs) leads to another method of synthesis [45]: Mn$_2$(hfbba)2(3-mepy), Zn(ADA)(4,40-bipy)0.5. (H2O)] [Cd(ADA)(4,40-bipy)0], [Mg$_3$(O$_2$CH)$_6$I[NH(CH$_3$)$_2$]0.5.5] [Cd(tdc)(bpy)(H$_2$O)], [(DMF)] n\s[Cu$_3$(TMA)$_2$(H$_2$O)$_3$] Cu/CuO, Co/Co3O4, ZnO, Mn2O3, MgO, and CdS/CdO can all be nanostructured using [Co6(BTC)2(HCOO)$_6$(DMF)$_6$]. Depending on the setting for MOF thermolysis, these nanoparticles distributed in a carbon matrix demonstrated promising H_2 and CO_2 adsorption characteristics.

The materials as prepared demonstrate favorable selective electrochemical evolution of oxygen. Finally, polynuclear clusters have been used in direct thermal breakdown of [Cr$_3$O(CH$_3$CO$_2$)6(H$_2$O)$_3$]NO$_3$ CH$_3$COOH ([Cr$_3$O]) in argon [46] to create Cr_2O_3 nanoparticles utilizing a solid-state approach. The carbon-embedded nanoplates exhibit improved electrochemical performance.

An additional comparable example is the solid-state production of $CoFe_2O_4$/C from a thermally treated heterometallic trinuclear [CoFe$_2$O(CH$_3$COO)$_6$(H$_2$O) 3.2H$_2$O] [47] complex that exhibited a 50 nm average particle size with a carbon surface coating.

13.2.3 A NEW APPROXIMATION FOR SOLID STATE

The following two steps make up our most recent innovative solid-state approximation [48]:

1. Making macromolecular precursors of the molecules with the generic formula chitosan MXn and
2. PS-co-4-PVP MXn were pyrolyzed, respectively.

The process of making PS-co-4-PVP and macromolecular chitosan MXn is carried out at ambient temperature with a straight forward coordination process employing dichloromethane as the solvent produces MXn precursors (see Equations (13.1) and (13.2)).

$$Mxn + Chitosan \xrightarrow{CH_2Cl_2} Chitosan - Mxn \qquad (13.1)$$

$$Mxn + PS - co - 4-PVP \xrightarrow{CH_2Cl_2} PS - co - 4-PVP.Mxn \qquad (13.2)$$

The reaction is slow because the polymer has many coordination sites (approximately 350 for chitosan, Mw = 61.000, for example) and because some of the reactants are insoluble. In order to ensure a favorable high level of coordination, around two weeks of reaction time is required.

The insoluble can be simply decanted and dried in a vacuum. Materials are solid, stable, and have the metallic salt's distinctive hue (e.g., green for Cr, white for Ti and Ni, etc.). Moreover, pyrolyzing results in a solid form of chitosan. Metal oxides MXOY are formed by heating MXn and PS-co-4-PVP-MXn precursors at 8000°C without air in a crucible.

13.3 TRANSITION METAL (FIRST SERIES)

13.3.1 TITANIUM

Titanium oxide (TiO$_2$) are only a few of the eight polymorphic forms of titanium dioxide (TiO$_2$), a well-known semiconductor (bronze). Anatase is a solar photocatalyst that is particularly effective at photodegrading different dye pollutants [49]. We chose the following antecedents for this study:

(Chitosan). (I) (Cp2TiCl2) (PS-co-4-PVP).

Chitosan, TiOSO$_4$ (III), (Cp$_2$TiCl$_2$) (II), and (PS-co-4-PVP).(IV) (TiOSO$_4$) and (chitosan).V and (Ti(acac)$_2$) (PS-co-4-PVP). (Ti(acac)$_2$) (VI) (VI). The polymer and the metallic precursors both play a role in determining the resulting polymorph of the pyrolytic products, as shown in Table 13.1.

It is interesting that the chitosan polymer allows for the formation of pure rutile and pure anatase phases employing Cp$_2$TiCl and TiO^{+2} as metallic salts linked to the polymeric chain, respectively. Rutile and anatase phases are mixed via the polymer PS-co-4-PVP.

It's interesting to note that the (TiOSO$_4$)(chitosan) precursor's result always produces a pure anatase phase. In contrast, phase mixes like anatase/rutile or anatase/brookite are nearly always produced via solution synthesis techniques.

TABLE 13.1

Eg Values of NiO and NiO Included in SiO$_2$, TiO$_2$, Al$_2$O$_3$, and Na$_4$.2 Ca$_2$.8 (Si$_6$O$_{18}$) Matrices [105]

Composite	Precursor Formula	Eg(eV)
NiO	Chitosan_NiCl2	5.2
NiO	PSP-4-PVP_NiCl2	5.2
NiO/SiO$_2$	Chitosan_NiCl2	5.0
NiO/SiO$_2$	PSP-4-PVP_NiCl2	5.5
NiO/TiO$_2$	Chitosan_NiCl2	5.2
NiO/TiO$_2$	PSP-4-VP_NiCl2	5.2
NiO/Al$_2$O$_3$	Chitosan_NiCl2	5.4
NiO/Na4.2Ca2.8(Si6O18)	Chitosan_NiCl2	5.6

Photocatalytic interaction with blue methylene was also used to investigate TiO$_2$ made from precursors; hence, this will be covered in more detail later in this chapter.

The anatase is considered the best photocatalyst produced from chitosan. The (TiOSO$_4$) precursor at 8000°C, increased the productivity of the conventional Degussa P25 photocatalyst without additional dopants or phases, thus achieving a discoloration rate of 98% at a pH solution of 9.5 in only 25 minutes.

13.3.2 VANADIUM METAL

Vanadium contains a variety of metal oxides due to its many oxidation states, with VO, V$_2$O$_3$, VO$_2$, and V$_2$O$_5$ being the most prevalent [50]. The nucleation in lamellar form is the most frequent morphology for V$_2$O$_5$. Many morphologies have been observed, including spindle-like bundles [51], V$_2$O$_5$ macroplates, nanoribbons, nanowires, and nanorods [52]. We have chosen the predecessors of chitosan and PS-co-4-PVP for our study. The raw materials at molar ratios of 1:1 and 1:5 produced the same pyrolytic V$_2$O$_5$ product [53]. The chitosan precursor yielded the smallest (1:5 ratio) nanoparticles as seen in Figure 13.3, with particles as tiny as 8 nm.

Only a few solid-state approaches have been published [54], despite the fact that various solution methods to create V$_2$O$_5$ nanoparticles have been documented. For instance, single-crystalline V$_2$O$_5$ nanoparticles are produced by solid-state thermal decomposition of [NH$_4$V$_3$.(OH)$_6$(SO$_4$)$_2$], which is employed as a cathode material for Li-ion batteries [54].

13.3.3 TUNGSTEN, MOLYBDENUM, AND CHROME

The VI group's most prevalent oxides are Cr$_2$O$_3$, MoO$_3$, and WO$_3$. The same pyrolitic process can be used to produce these metal oxides from their respective precursors, including (chitosan) (CrCl3)x, PS-co-4-PVP.(CrCl3)x, PS-co-4-PVP.(MoCl4)n, and chitosan (MoCl4)n, as well as (chitosan) WCl4 and PS-co-4-VP WCl4 [55]. The

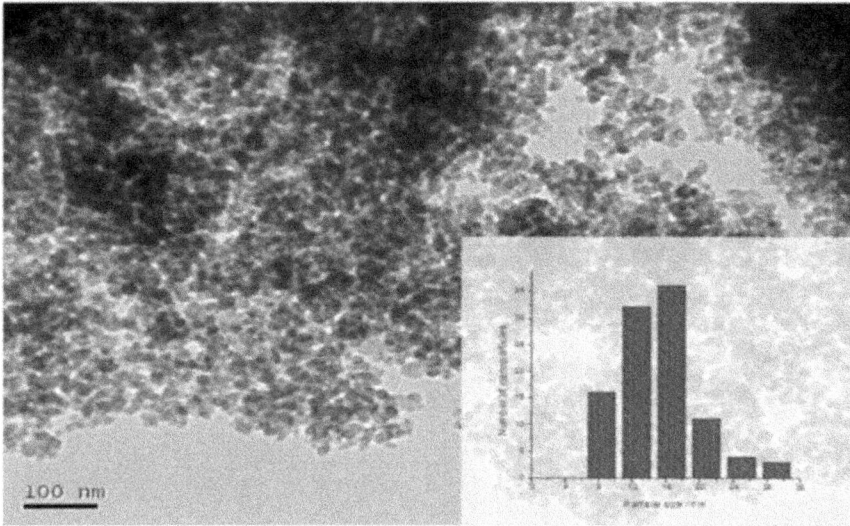

FIGURE 13.3 TEM of pyrolytic product from VCl3.chitosan (1:5) and histogram (inset) [53].

metal/polymer ratio can be adjusted for Cr_2O_3 in the order of 1:1 > 1:5 > 1:10 nanoparticles.

13.3.4 MANGANESE METAL

The precursors PS-co-4-PVP $(MnCl_2)n$ and chitosan $(MnCl_2)n$ were chosen for this work. Mn_2O_3 is evident based on XRD analysis [56]. SEM analysis is shown in Figure 13.4 and demonstrates the existence of thick grains, some of which are united in a three-dimensional pattern. Manganese oxide production was confirmed by an EDS scan.

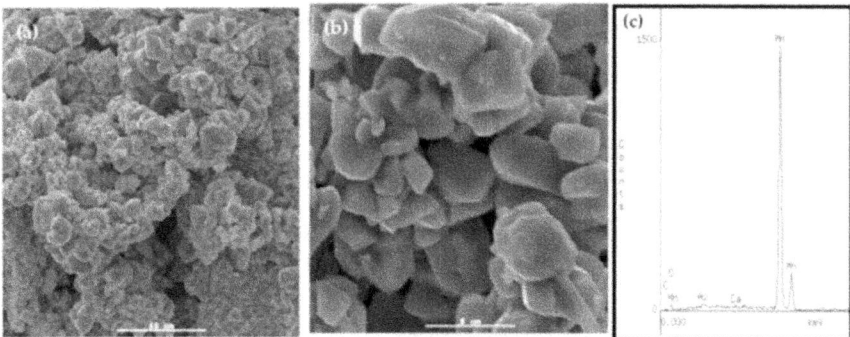

FIGURE 13.4 SEM image: two levels of magnification (a, b) for Mn_2O_3 and EDS analysis (c).

13.3.5 IRON METAL

Hematite (Fe_2O_3) is one of the more prevalent iron oxides [57]. Chitosan ($FeCl_2$)n, chitosan ($FeCl_3$)n, PS-co-4-PVP ($FeCl_2$)n, and PS-co-4-PVP.($FeCl_3$)n with molar ratios of 1:1, 1:5, and 1:10 were produced for the research with Fe. The kind of the polymer, the iron salts' degree of oxidation, and the metal/polymer ratio were all taken into account in a thorough analysis.

Fe_2O_3 (hematite) was obtained in each case. Figure 13.5 displays TEM, SEM, and XRD for the chitosan ($FeCl_3$)n 1:1 pyrolytic product. The hematite structure

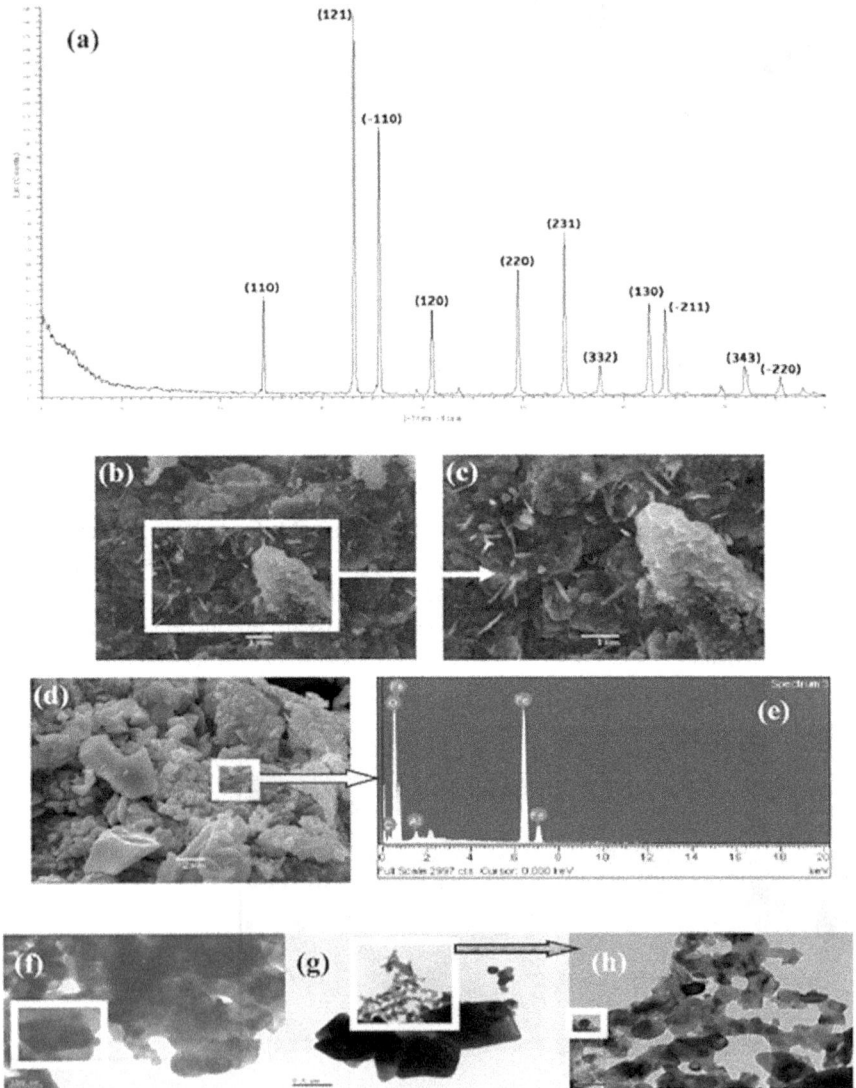

FIGURE 13.5 XRD (a), SEM analysis (b–d), EDS (e), and TEM image (f–h) of pyrolytic product from chitosan. ($FeCl_3$)n 1:1, hematite–Fe_2O_3.

was used to index the XRD data [58]. Figure 13.5b–d shows SEM images with a variety of irregular shapes typical of solid-state techniques, but they also accurately depict the Fe and O atomic content as illustrated by the EDS study shown in Figure 13.5e. The TEM investigation reveals agglomerates with the tiniest nanos-tructures attached in linear arrangements (Figure 13.5f–h). The smallest Fe_2O_3 nanoparticle size is produced by chitosan when combined with the $FeCl_2$ salt, but the smallest nanoparticles were produced by the PS-PVP polymer when combined with $FeCl_3$. The tiniest nanoparticles for both polymers may be seen in the 1:1 molar ratio precursor.

13.3.6 COBALT METAL

The most prevalent cobalt oxides are Co_3O_4 and CoO [59], with Co_3O_4 serving as an example of an electrode material for supercapacitors. In this instance, chitosan $(CoCl_2)n$ and PS-co-4-PVP.$(CoCl_2)n$ and precursors were used. As seen in Figure 13.6a, XRD examination demonstrates the presence of Co_3O_4 pure phase

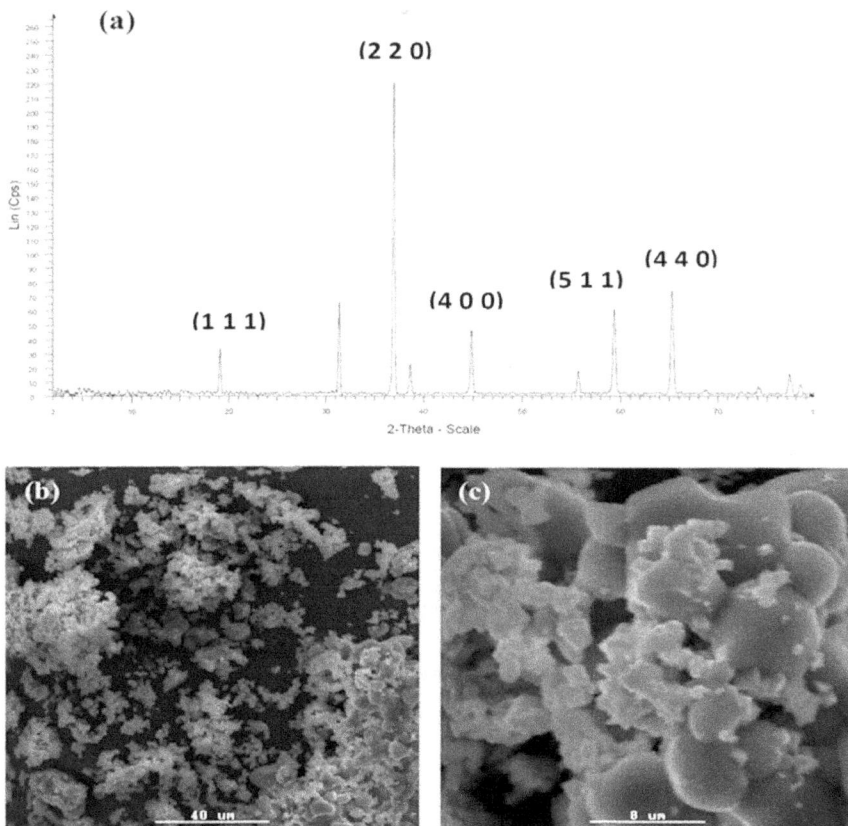

FIGURE 13.6 XRD pattern (a) and SEM images (b, c) of chitosan.(CoCl2)n. pyrolytic product.

[60]. After pyrolysis, SEM images show a thick morphology made up of connected grains forming a 3D pattern. There have been reports of a related thermal synthetic technique [61]. As is typical for solid-state approaches, various forms and sizes are noticed in the morphology. There has also been information given about a solid-state approach beginning with Co (II) salts [59] as shown in Figure 13.6b and c.

13.3.7 NICKEL METAL

The precursors PS-co-4-PVP. $(NiCl_2)n$ and chitosan were used. $(NiCl_2)n$ was used for this work. NiO pure phase is definitely present, according to XRD analysis. Figure 13.8a illustrates the porous morphology of a SEM image. Ni and O are confirmed by an EDS study. Precursor PS-co-4-PVP. (NiCl2)n showed additional porosity in shape as depicted in Figure 13.7. Furthermore, Figure 13.7d shows that their EDS analysis also reveals the presence of Ni and O.

The bandgap of NiO, a p-type semiconductor, is in the range of 3.6–4 eV [62]. Several solution procedures [60] have been described for their preparation. Also presented is a solid-state thermal approach commencing with $[Ni(en)_3][NO_3]_2$ [29]. We were able to make NiO using Ni salts and a cheap, readily accessible polymer thanks to the approach described here. The pure NiO phase was produced using the precursors PS-co-4-PVP. $(NiCl2.6H_2O)$ and chitosan $(NiCl_2 \ H_2O)x$ and by heating

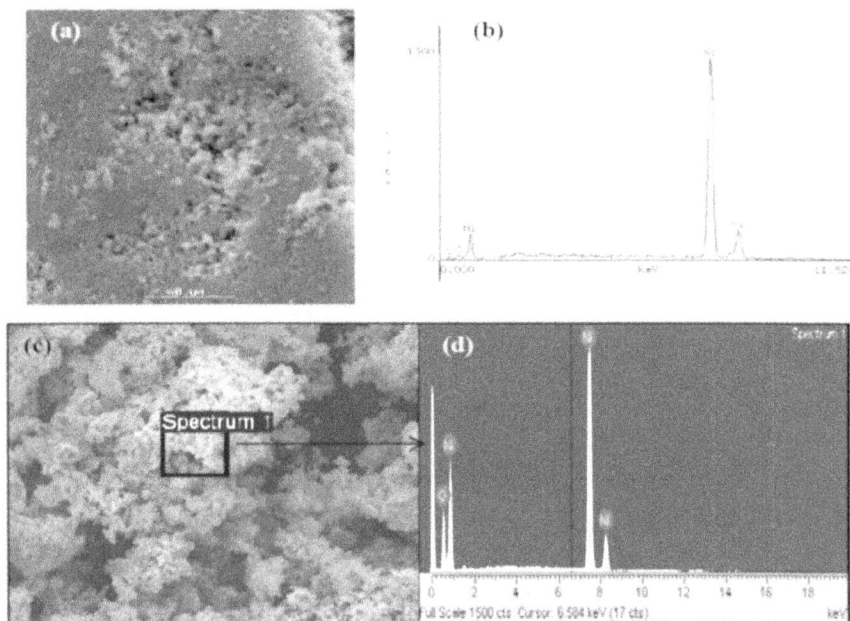

FIGURE 13.7 SEM image (a) and EDS analysis (b) of pyrolytic product 1:1 chitosan. $(NiCl_2)n$ precursor (b) and SEM image (c) and EDS analysis of (d) the 1:1 PS-co-4-PVP. $(NiCl_2)n$ precursor.

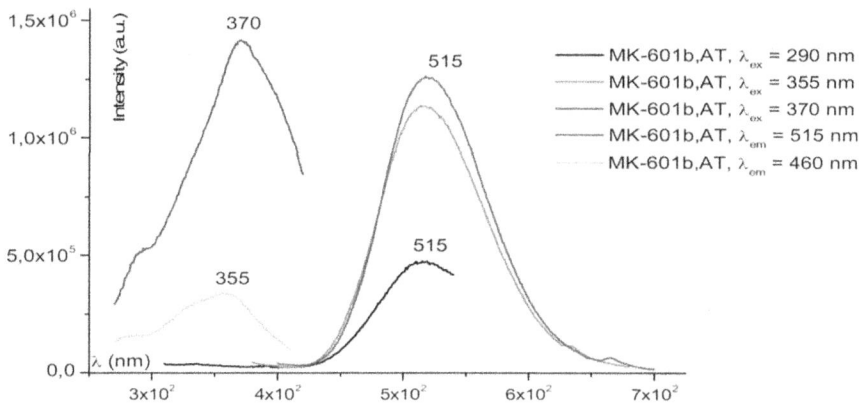

FIGURE 13.8 Luminescence for PS-4-co-PVP.AuC_6F_5 spectrum at various excitation wavelengths.

at 800°C under air. Using both chitosan and PS-co-4-PVP precursors, the NiO was produced with a bandgap of 4.15 eV [62]. The bandgap value of semiconductor metal oxides controls their photocatalytic activity [63]. In actuality, methyl and PS-co-4-PVP precursors were implemented [62].

13.3.8 PRECIOUS AND NOBLE METALS

The same technique can be used to obtain noble metals like Au, Pt, and Ag, as well as metal oxides of valuable metals like Ir, Rh, and Re. By utilizing the proper metallic salts to pyrolyze the corresponding macromolecule chitosan (MLx) and PS-co-4-PVP (MLx) precursors, these metal and metallic nanoparticles can be produced. ML_x = AuCl, $AuCl_3$, AuC_6F_5, $Ag(CF_3SO_3)_2$, $PtCl_2$, $IrCl_3$, $RhCl_3$, and $ReCl_3$ for Au, Ag, Pt, and Re, respectively.

13.3.9 GOLD METAL

The PS-4-co-PVP polymer and the Au metallic salts, such as AuCl, $AuCl_3$, AuC_6F_5, and $Au(PPh_3)Cl$, easily react to form the luminous macromolecule PS-4-co-PVP $AuCl_3$ and PS-4-co-PVP.PS-4-co-PVP, AuCl.PS-4-co-PVP Au(PPh3)Cl and AuC_6F_5 complexes. According to Figure 13.8, the highest luminescence for PS-4-co-PVP.AuC_6F_5 occurs around 550 nm and is most likely the result of Au (I)–Au(I) interactions [64]. It's interesting to note that this luminous activity is not present in the macromolecular complex PS-4-co-PVP.$(AuCl_3)n$.

These macromolecule precursors can be pyrolyzed to produce metallic Au_ foams, with varying pore dimensions according to the Au salts bonded to the polymer [65] as illustrated in Figure 13.9.

Such materials are generated using a variety of solution processes such as dealloying M/M0 alloys and creating metal organic composites and removing the

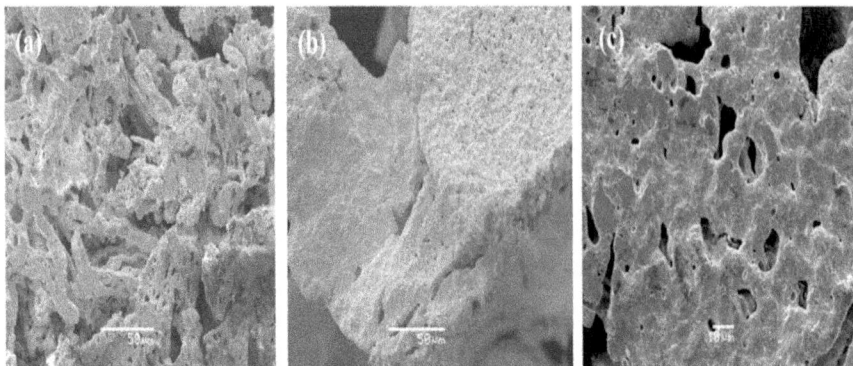

FIGURE 13.9   SEM images for different $(AuXn)n$: (a) $AuCl$, (b) $AuCl_3$, and (c) $Au(C_6F_5)$. Metallic sponges of macro-porous metals demonstrate mechanical strength and stiffness properties [65].

organic components through calcination or dissolving. There haven't been any reports of solid-state approximations, though. The procedure outlined here can be a creative and dependable technique to make metallic foams out of several noble metals.

13.3.10   SILVER METAL

Although there are very few solid-state methods for producing $Ag0$ nanoparticles, several solution approaches have been published [66]. Ag was produced from precursors such as $Ag(CF_3SO_3)$ chitosan that contained Ag, as evidenced by the XRD pattern [65]. Their topography points to a form that resembles foam. Ultimately, EDS analysis verified that Ag was present in a single pure phase.

13.3.11   PLATINUM METAL

Due to significant catalytic stability and activity, nanostructured Pt particles are crucial in the reaction of fuel cells, generating sensors, and usage in the manufacturing and automobile sectors [67]. Pt nanoparticles were produced from the precursors PS-co-4-PVP. $(PtCl_2)n$ and chitosan. (PtCl2)n [68].

The chitosan $(PtCl_2)n$ pyrolytic precursor product generated in a molar ratio concentration of 1:1 yields the lowest particle size (6 nm). Despite the fact that many Pt nanoparticle fabrication techniques have been published [69], few solid-state techniques have emerged. The "sponge-like" topography seen for the pyrolytic products from the precursor's 1:5 ratio PS-co-4-PVP (PtCl2)n is another intriguing feature. There have been few reports of metallic Pt sponge materials using the solid-state approach [65].

13.3.12 IRIDIUM METAL

One of the most catalytically active precious metals on the periodic table is Ir [70]. At the nanoscale, their activity is much increased [71]. This metal has a strong catalytic activity, as do its metal oxides [72]. IrO_2 nanostructured production techniques in isolated solutions are widely reported [73], but no general solid-state techniques have been demonstrated. Ir salt is typically used to create IrO_2. Temperature affects the relative percentage of IrO_2/Ir, with temperatures above 600°C creating pure IrO_2 and temperatures between 250 and 400°C yielding Ir_2O_3 [74]. We have produced a special nanostructured phase of IrO_2 using our approach [63]. The macromolecular precursors PSP-4-PVP.(IrCl3)X and chitosan.(IrCl3)X were heated to form the IrO_2 nanoparticles. The resulting IrO_2 nanoparticles are roughly 15 nm in size, with the polymeric initiator functioning as a solid-state template and influencing the dimensions of the IrO_2-complex but without shape.

13.3.13 RHODIUM METAL

Another valuable metal that is catalytically active is rhodium [75]. At the micro level, their activity is also much increased [76]. Rhodium, one of the noble metals, is crucial for a number of catalytic processes [77]. Rhodium-containing compounds' catalytic mechanism, however, is still a mystery. Rhodium oxide rather than rhodium may constitute the active centers, according to recent research [78].

Rhodium oxides Rh_2O_3 and RhO_2 are the most prevalent types. Although there are numerous uses for these rhodium oxides in catalysis, there are few preparation techniques for nanostructured Rh_2O_3 and RhO_2, and nothing is known about how to control their morphology and size [79]. The macromolecular chitosan $(RhCl_3)X$ precursor was easily heated to produce a nanostructured Rh/RhO_2 combination, whereas the PSP-4-PVP $(RhCl_3)X$ precursor was pyrolyzed to produce pure Rh_2O_3. The shape of the Rh and their metal oxide is not greatly influenced by the type of polymeric precursor used as a solid-state template. The as-obtained products range in size from 20 nm to less.

13.3.14 REPRESENTATIVE METALS

Two of the most used materials for sensors are SnO_2 and ZnO [80]. Wide bandgap n-type semiconductor SnO_2 is crucial for many technological applications, including solar cells, Li-ion batteries, and gas sensing [81]. Its primary use as a sensor is in the detection of H_2 and CO. The nanostructural zinc oxide is among the widely used nanomaterials used in sensors because of its stability in chemical and photochemical reactions, biocompatibility, maximum specific surface area, transparent optical devices, electrochemical feasibility, and more electron mobility. The detection of biological substances has been done using ZnO [81]. By pyrolyzing macromolecule complexes PS-co-4-PVP (ZnCl2)n and PS-co-4-PVP. (SnCl2)n varying the molar ratios at 800°C in air, SnO2 and ZnO were produced.

With a predecessor mixture ratio of 1:1, the corresponding hexagonal and cubic structures for ZnO agglomerates were seen, with average diameters of 30–50 nm.

FIGURE 13.10 The SEM image of pyrolytic precursor from PS-co-4-PVP. (ZnCl2)n in ratio 1:5 used from reference [60].

The pyrolytic predecessor from PS-co-4- PV. (ZnCl2)n in a 1:5 ratio SEM image is shown in Figure 13.10. Ranges with "metallic sponges" and cubic morphological structures are seen here. The PS-co-4-PVP. $(SnCl_2)$n precursor for SnO_2 generated the oxide. The various morphologies and particle sizes of the nanostructured SnO_2 are dependent on the $SnCl_2$:PS-co-4-PVP molar ratios. For each oxide, bigger crystalline crystals are formed when the inorganic salt is utilized as a precursor at a higher weight fraction (1:1).

The morphology of the SEM image appears uneven along with some "foam" structures, as illustrated in Figure 13.11. The Sn and O atoms were verified by EDS analysis. The chitosan macromolecule, PSP-co-4-PVP and the (ZnCl2)n complexes were also used to produce nanostructured ZnO [29].

13.3.15 RARE METALS

Due to their distinctive qualities and potential uses in things like UV-shielding, luminous displays, optical communications, biological probes, and medical diagnostics, rare earth compounds have attracted interest [82]. Amid rare earth compounds, metal oxides demonstrated various properties for applications in a number of

FIGURE 13.11 The SEM image (a) and EDS analysis (b) of SnO_2 obtained from PS-co-4-PVP.(SnCl2)n precursor.

technological industries. Few universal preparation techniques, however, have been published [83].

Rare earth oxide nanostructured compounds can be created using the approach previously described, albeit this includes production of lanthanum oxyhalides in place of the anticipated oxide Ln_2O_3. Electrochemical, magneto-chemical, optical, and fluorescent properties of LnOX compounds (Ln = lanthanide, X = halide) are exceptional and one-of-a-kind [84].

Depending on the type of lanthanide salt used, the PSP-co-4-PV (MLn)n macromolecule and chitosan (MLn)n complexes yield products with the chemical composition M_2O_3 or MOCl. When MLn is $M(NO_3)3$ or $M_2(SO_4)2$, the corresponding M_2O_3 is produced. When MLn is MCl_3, the result is the oxychloride MOCl. For instance, CeO_2 is produced via the transmutation of the macromolecular precursor PSP-co-4-PVP $(Ce(NO_3)_3)n$.

XRD was used to distinguish between the various materials of preparation. Figure 13.12a illustrates how the TEM image reveals the normal configurations of nanoparticles which are somewhat agglomerated. The HRTEM images in Figure 13.12b show that the technique permitted aggregations of CeO_2 single-crystal nanoparticles. Yet, as shown by XRD [60] chitosan $NdCl_3$ and PSP-co-4-PVP $NdCl_3$ pyrolysis resulted in NdOCl in both cases. A combination with Nd_2O_3 oxide was occasionally seen as well. The pyrolytic products from chitosan $NdCl_3$ are shown in a SEM image to have a three-dimensional metallic foam shape. According to the suggested formula, EDS validated the occurrence of Nd, O, and Cl. Figure 13.13c illustrates uneven forms and varying sizes that were visible in the TEM image [62].

The amazing 3D metallic foam structure of the Eu+3 doped $NdOCl/Eu_2O_3$ materials produced by pyrolyzing chitosan $NdCl_3/EuCl_3$ is shown in Figure 13.13 [85]. It's interesting that we found Eu in the corresponding EDS along with the usual Nd, O, and Cl atoms (Figure 13.15b). Hence, the primary emission line for this system's solid-state luminescence is attributed to the 5D0 2F2 transition which is seen around 566 nm [85].

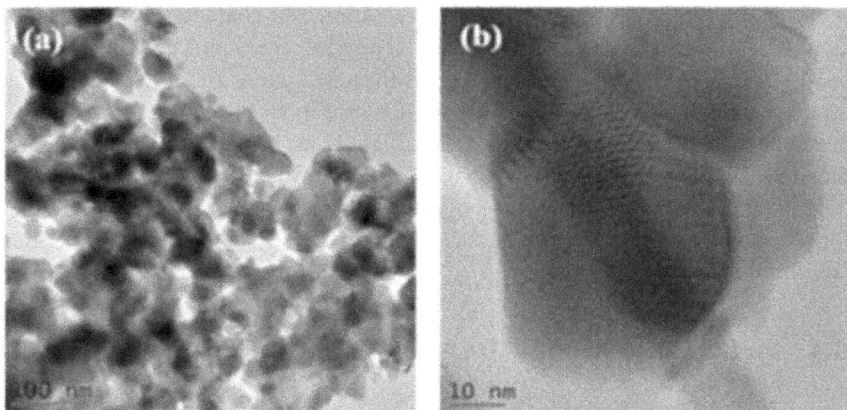

FIGURE 13.12 The TEM image (a) HRTEM (b) image of transmuted material from precursor PSP-co-4-PVP. $(Ce(NO_3)3)n$.

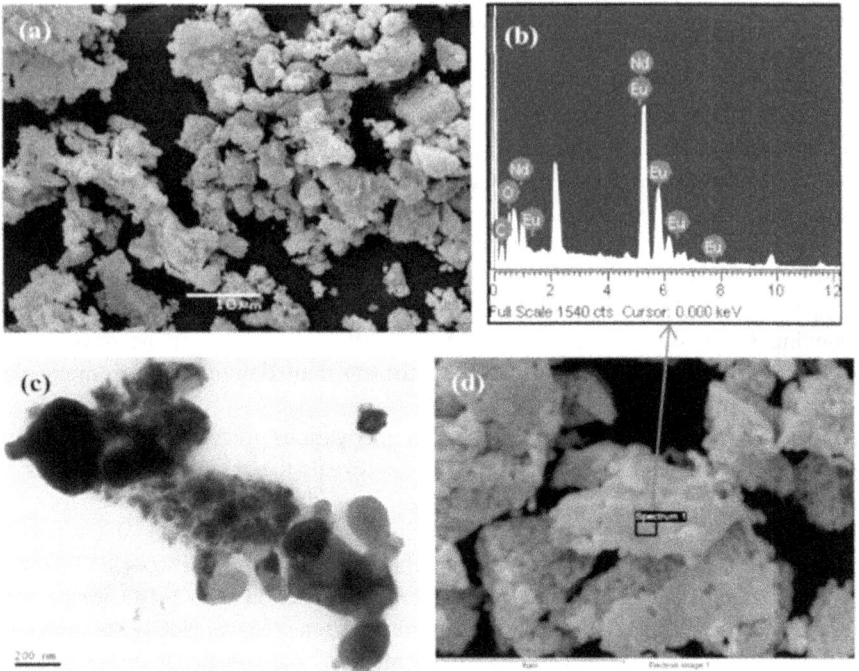

FIGURE 13.13   SEM (a, d), EDS (b), and TEM image (c) of transmuted material from Eu^{3+} doped, chitosan.NdCl$_3$.

13.3.16   ACTINIDES

Actinides are the significant and promising materials among the several actinide oxides that are used as a ceramic sensor catalyst in electrical and optical substances as well as in conventional nuclear factory [86]. In reference to this, there are not many publications describing the creation and characteristics of nanostructured ThO$_2$ that have been published. Using the chitosan Th(NO$_3$)$_4$ and PS-co-4-PVP Th(NO3)4 precursors, we synthesized nanostructured ThO$_2$. The precursor, which is composed of polymeric chitosan and PS-co-4-PVP material, determines the shape and average size of ThO$_2$ as it is formed. The average diameters of the polymer chitosan and PS-co-4-PVP precursors were 50 and 40 nm, respectively. The precursor polymer's characteristics and the as-obtained thoria's intensity emission maxima determine whether they exhibit the anticipated luminescence.

13.4   METALLIC AND METAL OXIDE INTEGRATION IN SOLID MATRIX

SiO$_2$, TiO$_2$, Al$_2$O$_3$, glasses, and other solid-state materials are used to create nanoparticles and/or nanostructures for a variety of practical applications, including catalysis [87]. Solid-state technology was developed to create various composites, including M/M0/xO0 y and MxOy/M0 xO₀y with M0 xO₀y solid matrix, i.e., SiO_2, TiO_2, and Al_2O_3, in air and thermal conditions to treat the chitosan MLn/M0 xO0 y and PS-co-4-PVP.MLn/M0 precursors. We have created many metallic nanoparticles and metal oxides incorporated into solid matrices using this synthetic technique.

As is well known in catalysis, adding M_0 and MxOy to solid matrices causes the catalytic material to be more stable and increases catalyst surface area within the solid matrix [87]. Silica was combined with gold and the bimetallic Au/Ag to create Au/SiO_2 [65] and $Au/Ag/SiO_2$ [88]. Figure 13.14 shows nanoparticles of roughly 5 nm in size. SEM-EDS mapping was used to look at how the Au nanoparticles were distributed inside the silica as illustrated in Figure 13.15.

A constant dispersion of Au nanoparticles within silica is seen in the above images. Moreover, bimetallic Au^0/Ag^0 nanoparticles were included by pyrolyzing the PSP-4-PVP under air at 800°C. Chitosan and precursors to $(AuCl_3/AgSO_3CF_3)$ $n.SiO_2$ $(AgSO_3CF_3)$ $n.SiO_2$).

Pyrolysis of the relevant materials was also used to create the Ag nanostructures inside SiO_2 (PS-co-4-PVP). chitosan, $(AgNO_3)n$, $(SiO_2)n$, precursors of $(AgNO_3)n$ and $(SiO_2)n$ [89]. For the Ag/SiO_2 composites made from chitosan, well-dispersed Ag nanoparticles were seen inside SiO_2. Particle sizes of $(AgNO_3)n.(SiO_2)m$ and $(PS-co-4-PVP).(AgNO_3)n$ $(SiO_2)m$ are 5 nm and 6 nm, respectively.

In addition, silica contained ZnO and SnO_2 [90] $ZnCl_2$, chitosan, SiO_2, and $SnCl_2$ pyrolyzation. Zn_2SiO_4 and SiO_4 mixes and pure SnO_2 can be produced from SiO_2 precursors at 800°C and air, respectively. The silica matrix contains a consistent distribution of SnO_2 nanoparticles.

Using the sol-gel technique, a mixed solution of the chitosan and PVP precursors was used to incorporate IrO_2 into SiO_2 [91]. Pyrolysis of the isolated solid-state chitosan occurred after that. IrO_2/SiO_2 nanocomposites are created via the reactions of $(IrCl_3)x(SiO_2)$ with PSP-4-PVP. As a result of the homogeneous distribution of IrO_2 particles throughout the SiO_2 matrix, porous stable materials suitable for high-thermal catalytic purposes are produced.

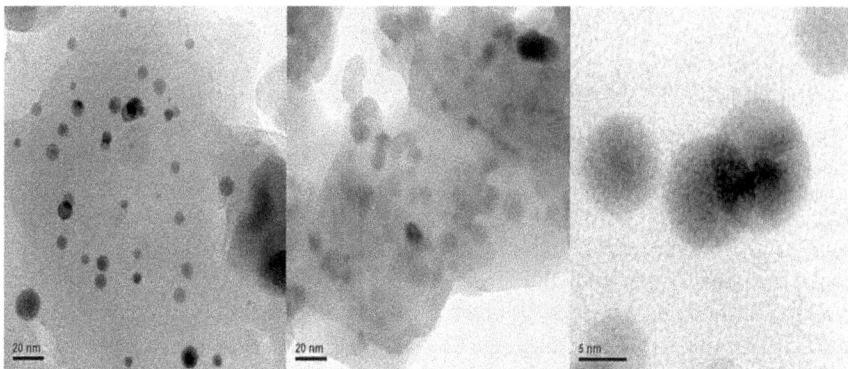

FIGURE 13.14 HRTEM images of Au nanoparticles inside SiO_2 from (PS-co-4-PVP)_ $(AuCl_3)n\_(SiO_2)n$ precursor [65,94].

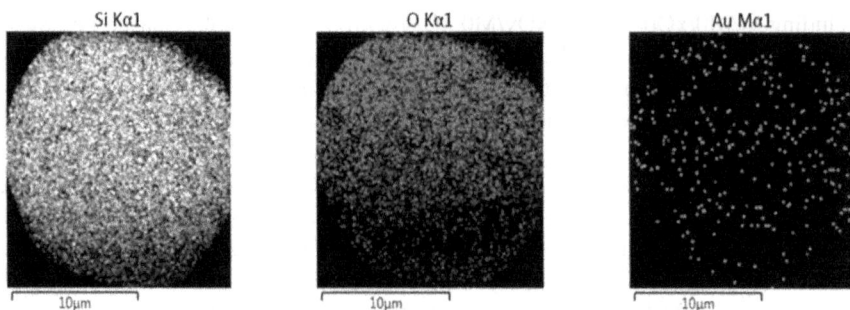

FIGURE 13.15 EDS images of elements of the Au/SiO$_2$ nanocomposite from chitosan. (AuCl$_3$)n, (SiO$_2$)n in 1:1 molar ratio polymer/metal, obtained from reference [65].

Re and Th underwent the same processes. Using the sol-gel technique, a mixture of precursor's chitosan and PSP-4-PVP were used to incorporate ReO$_3$ into SiO$_2$ [92]. Pyrolysis of the solid chitosan occurred after that. The precursors PSP-4-PVP (ReCl$_3$)X (SiO$_2$)y and (ReCl$_3$)X (SiO$_2$)y produce ReO$_3$/SiO$_2$ nanocomposites.

SiO$_2$ contains ReO$_3$ nanoparticles as small as 1 nm in size. As a result of the homogeneous distribution of ReO$_3$ nanoparticles throughout the SiO$_2$ matrix, semiporous materials ideal for high-thermal catalytic purposes are produced. Similar solid-state techniques were used to incorporate ThO$_2$ into SiO$_2$ and TiO$_2$ [93]. Chitosan was pyrolyzed at 800°C under air to create the ThO$_2$/SiO$_2$ composites, precursors of Th(NO$_3$)$_4$/SiO$_2$ and PS-co-4-PVP Th(NO$_3$)$_4$/SiO$_2$ [94]. On the other hand, ThO$_2$/TiO$_2$ composites were made in a similar way by pyrolyzing PS-co-4-PVP Th(NO$_3$)$_4$/TiO$_2$ and chitosan Th(NO$_3$)$_4$/TiO$_2$ precursors at 800°C. With respect to the precursors of PS-co-4-PVP polymer or chitosan, respectively, ThO$_2$ particles demonstrate a reasonable dispersal within the silica. The dispersion consisted of 250 nm and 950 nm, respectively.

General dispersion of thoria into SiO$_2$ and TiO$_2$ matrices is revealed by SEM-EDS mapping analysis. ThO$_2$/SiO$_2$ and ThO$_2$/TiO$_2$ composites' luminous characteristics exhibit a dependence on luminescence intensity, with TiO$_2$ matrix exhibiting the highest luminescence intensity.

13.5 USES OF PHOTOCATALYSIS

Growing wastewater production at industrial facilities frequently results in serious environmental issues [95]. The majority of industrial wastewaters contain hazardous chemical compounds that are resistant to direct biological treatment [96].

Organic pollutants come in a plethora of diverse forms such as phenols, dyes, biphenyls fertilizers, insecticides, plasticizers, hydrocarbons, soaps, proteins, medicines, glucose, and more [97]. Thus, it is imperative to create a method that is both effective and affordable for lowering the content of organic contaminants before discharging wastewater into the aquatic environment. Adsorption and coagulation, two commercially available wastewater treatment methods, just cluster or segregate contaminants from water; they are not entirely "eliminated" or "destroyed" into less hazardous and biodegradable components [98]. Alternative treatment techniques for water, like chemical and membrane technologies, frequently have large running costs and occasionally produce additional harmful contaminants [99].

As a result of increased efficiency, simplicity, reproducibility, and handling ease, recently obtained oxidation pathways like photocatalysis are increasingly being used in the breakdown of organic structures among various chemical, physical, and biological technologies used to control pollution [100]. The following significant benefits of heterogeneous photocatalysis make it a viable option for treating wastewater:

i. Operating under ambient thermodynamic conditions;
ii. Complete contaminants mineralization and intermediate molecules without the creation of pollutants from other sources;
iii. Low operational costs [100].

Nanostructured metal oxides are the most popular photocatalyst [10,11]. Moreover, the recent preparation mostly uses in-solution techniques [101]. These techniques pose challenges when it comes to sepration of the solid from the solvent, also including the template and stabilizer [102]. The photocatalytic destruction of organic contaminants is one of the principal uses of nanostructured metal oxides in environmental restoration. The following qualities are essential for a suitable metal oxide photocatalytic system:

1. Good bandgap
2. Good topography
3. Large area and stable
4. Reuse of requirements

When a semiconductor metal oxide's energy gap is close to 3.1 eV, it is Ultraviolet light active, and when it is close to 2.8 eV, it is visible active light [96]. The principal photocatalytic processes used by metal oxides with characteristics, such as Cr, Zn, Ti, Sn and V, include absorption of light, that causes separation of charge process and the ensuing production of positive holes that oxidize the substances of organic nature. Photo-excited electrons are excited from the outer band to conduction band in this process, generating an electron/hole pair (e-/h+). Moreover, Ultra-violet and visible light, or combination of both are used to activate the metal oxide.

A substance that has been adsorbed on the surface of the photocatalyst can be reduced or oxidized by the photogenerated pair (e/h+).

Metal oxides' photocatalytic activity originates from two places:

1. The oxidation of OH- anions produces OH radicals;
2. The reduction of O2 produces O2- radicals.

Nanostructured semiconductor oxides of metals are used expansively in redox photocatalytic reactions (CB) due to their complete valence band (VB) and the hollowness of conduction band electrical arrangement.

It produces an electron-hole pair which were exposed to a photon with energy greater than the bandgap, h > Eg, pumping 1 electron from VB into CB and leaving a hole in VB. Whereas the electrons in the CB have a strong reducing potential, the produced holes in VB have large oxidation energy.

The extreme reactiveness of electrons and the hole formed take part in the photocatalytic destruction of organic materials. As already mentioned, an effective photocatalyst must have the following characteristics: a sufficient bandgap, acceptable shape, large arial surface, stability, and reusability. The way a particular nanostructured metal oxide is prepared will determine whether these qualities are achieved. One of the most popular and effective metal oxide for the photocatalytic decomposition of many dye organic pollutants is TiO_2 [99].

However, the preparation technique determines the morphology of the bandgap, surface area, stability, and recyclability, which thus determines relative efficacy. The solid-state approach implemented in this situation could produce nanoparticle size metal oxides with the above properties, producing a powerful photocatalyst for the oxidation of organic contaminants.

The metal oxides produced in the above-described solid state have the advantage of being directly applicable to photocatalytic heterogeneous catalysis.

Anatase produced by the (chitosan) ($TiOSO_4$) precursor at 800°C was the most efficient degradant of methylene blue for all of the TiO_2 products. If the pH of the solution was 9.5, this substance increased the efficiency of the typical photocatalyst Degussa P25 without the use of additional phases or dopants, achieving a 98% discoloration rate in just 25 minutes. Figure 13.15a plots the TiO_2 produced from (chitosan) ($TiOSO_4$) as a function of temperature and irradiation time. A 25-minute discoloration rate of 86.5% at 800°C was noted. Figure 13.15b illustrates the 98% discoloration at pH 9.5 after pH optimization. The produced utilizing our solid-state technique is one of the most effective TiO_2 materials known to science for methylene blue degradation [103]. Hematite–Fe_2O_3 was a different photocatalytic-assayed system [104]. Figure 13.15 illustrates the high rate of MB degradation by the nanoparticle material made from chitosan (FeCl2)y 1:1 during simulated sunshine (full visible spectrum) irradiation. The MB degradation was measured at 655 nm and reached 73% in 60 minutes and >94% after 150 minutes. The porous shape of the hematite–Fe_2O_3 may contribute to the high photocatalytic effectiveness [104]. In contrast, NiO and its nanocomposites with bandgap values in the range of 5.0–5.6 eV (Table 13.1) indicate that they can be employed as suitable photocatalysts when exposed to UV light [65]. In actuality, NiO and its nanocomposites with SiO_2, TiO_2, and Al_2O_3 show a good and effective photocatalytic action [105]. In actuality, NiO and its nanocomposites with SiO2, TiO_2, and Al_2O_3 show a good and effective photocatalytic action [105]. The NiO/TiO_2 composite formed from the chitosan ($NiCl_2.6H_2O$)x/TiO_2 precursor had a greater methylene blue efficiency (see Table 13.2). This may be as a result of a p-n junction functioning as NiO as p-NiO and TiO_2 as n-TiO_2. As a result, it appears that the matrix is essential for the NiO/TiO_2 composite, and in this instance, NiO serves as the matrix rather than as an

TABLE 13.2

Photodegradation Kinetic Data of MB Processed with NiO and NiO/SiO$_2$, NiO/TiO$_2$, NiO/Al$_2$O$_3$, and NiO/Na$_4$2Ca$_2$.8(Si6O18) Composites [104]

Photo catalyst	Discoloration Rate (%)
NiO-CHITOSAN	71
NiO-PS-4-PVP	68
NiO/SiO$_2$-CHITOSAN	69
NiO/SiO$_2$-PS-4-PVP	48
NiO/TiO$_2$-CHITOSAN	91
NiO/TiO$_2$-PS-4-PVP	81
NiO/Al$_2$O$_3$-CHITOSAN	45
NiO/Na4.2Ca2.8(Si6O18)	75

active semiconductor. For evidence, consider the photocatalyst produced by precursors (chitosan) (TiOSO$_4$) at various temperatures. Figure 13.16 shows the normalized methylene blue in the presence of catalyst and effect of pH on MB discoloration. Figure 13.17 shows normalized methylene blue in the absence of catalyst.

We have also investigated the photocatalytic activity of ReO$_3$, which was produced by pyrolyzing the fore product PSP-4-PVP and chitosan (ReCl$_3$)X. ReO$_3$ produced from chitosan and PSP-4-PVP precursors exhibit moderate and high activity in their as-prepared states, respectively. The composite ReO$_3$/SiO$_2$ made from the solid-state pyrolysis of the precursors chitosan (ReCl$_3$)X(SiO$_2$)y and PSP-4-PVP (ReCl$_3$)X(SiO$_2$)y exhibits moderate photocatalytic activity toward the degradation of methylene blue and is comparable to that of ReO$_3$. This is the first account of the composite ReO$_3$/SiO$_2$/ReO$_3$ photocatalytic activity.

The photocatalytic activity development of IrO$_2$ and their composite with SiO$_2$ made from chitosan that was solidified through pyrolysis was also evaluated.

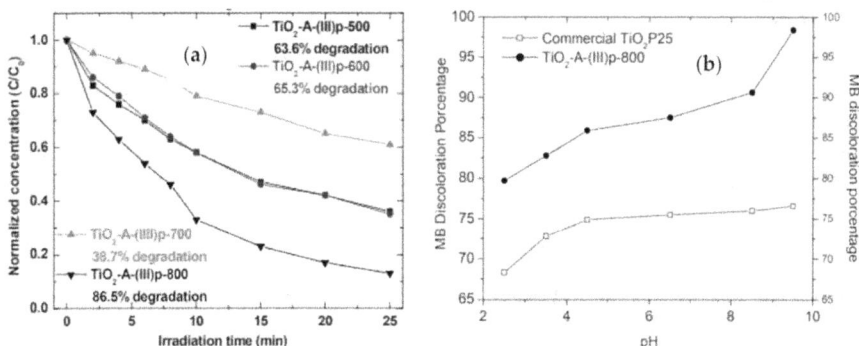

FIGURE 13.16 (a) Normalized MB amount as a function of TiO$_2$ time. (b) Effect of pH on (1 x10^{-5} M) MB discoloration using TiO$_2$.

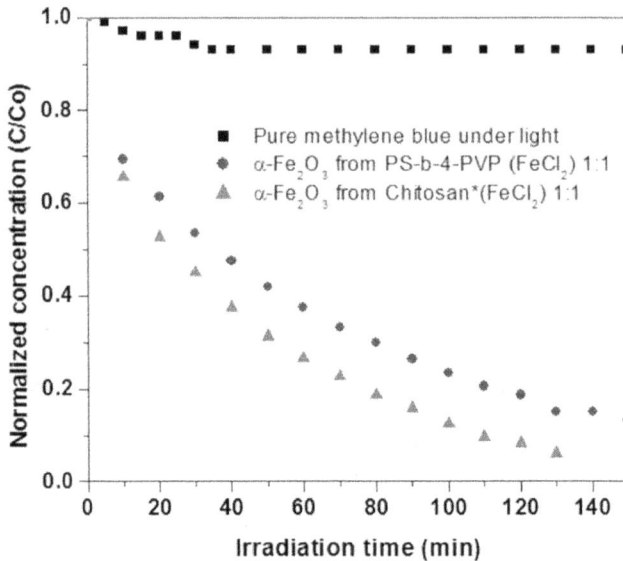

FIGURE 13.17   Normalized MB amount without catalyst, in presence of a-Fe$_2$O$_3$.PSco-4-PVP and a-Fe$_2$O$_3$.chitosan [104].

PSP-4-PVP, (IrCl$_3$)X. (IrCl$_3$)X, PSP-4-PVP, and chitosan. (IrCl$_3$)x(SiO$_2$)y. progenitors of (IrCl$_3$)x(SiO$_2$).

Due to respective bandgap values of 3 and 3.7 eV, the oxide Rh$_2$O$_3$ and combination Rh/RhO$_2$ are capable of photocatalysis when exposed to UV light [106]. In fact, methylene blue degrades by 70% and 78% in equal amounts over the course of 300 minutes in Rh$_2$O$_3$ and the Rh/RhO$_2$ mixture. To the best of our knowledge, no earlier reports of the photodegradation of contaminants employing this form of Rh oxide have been made.

Thoria catalytic activity and ThO$_2$/SiO$_2$ and ThO$_2$/TiO$_2$ composites have also been tested for the first time. Th(NO$_3$)4 precursor showed activity of 66% in 300 minutes, whereas precursor PS-co-4-PVP Th(NO$_3$)4 showed 67% activity during the same period of time that occurred for MB degradation. Moreover, the photocatalytic effectiveness of the ThO$_2$/SiO$_2$ and ThO$_2$/TiO$_2$ composites dramatically declines. ThO$_2$ > ThO$_2$/TiO$_2$ > ThO$_2$/SiO$_2$ is the order in which the photocatalytic activity for the degradation of methylene blue occurs, which may be caused by a reduction in the surface's active sites as a result of the encapsulating of ThO$_2$ in TiO$_2$. The porous shape of ThO$_2$, which is contained within the SiO$_2$ and TiO$_2$ matrices, may be another factor [99].

13.6   METALLIC AND METAL OXIDE NANOSTRUCTURES: POSSIBLE MECHANISM OF PRODUCTION

Although the process by which nanoparticles develop in solutions is well understood [107], solid-state preparation techniques are scarce, and it is uncertain which factors affect the size and outer structure of newly created nanoparticles. With respect to this, it is demonstrated that the pyrolysis of the $NP(OC_8H_{12})_2(OC_6H_4PPh_2-Mn(CO)_2(5-C_5H_4Me)_2$ precursor happens via the preliminary creation of a layered graphite host [16]. Moreover, the synthesis of the nanostructured Mn_2O_3 and Co_2O_3 molecules from their core fore products, (chitosan)(MLn)x, MLn $=MnCl_2$, and $CoCl_2$, was verified to get through a transition state, a layered graphitic carbon matrix, which was seen by HRTEM and Raman studies [108].

In recent times with the development of a graphite intermediate, the pyrolysis of TiO_2 at various temperatures was proven [59,103]. A general process is suggested after taking into account all research. In order to create a thermally stable matrix, a 3D network is formed at the initial stage of heating [15]. Because it counteracted the sublimation, this step is essential. In our system, the chitosan and PS-co-4-PVP polymer cross-linking occurring during the initial heating stage could result in a 3D matrix with O-M-O and H2N-M-NH2 links (for the chitosan polymer) and (pyridine)N-M-N(pyridine) connections (for the PS-co-4-PVP polymer) respectively. The organic carbonization process is started in the subsequent phases, resulting in holes where the nanoparticles can form. This happen at about 400°C for the chitosan matrix and 360°C for the PS-co-4-PVP matrix, according to TG/DSC studies. A graphitic carbon layered host (found in the earlier study [16]) serves as a nanoparticle template for later intermediate stage development. This template vanishes upon complete burning, but carbon leftovers always linger and appear as a thin carbon shell around the nanoparticles [16].

13.7 CONCLUSION REMARKS

Although there are numerous solid-state ways to make particles of nano size, there are few reports of mechanistic investigations, thus making this a challenge. In this regard, it is unknown how to manage factors like the size and topography of newly generated nanoparticles. Utilizing the proposed solid-state technique from the chitosan, the PS-co-4-PVP and MXn is a dependable, all-encompassing, and simple method for creating metal and metal oxides from any element on the periodic table using the MXn complexes as precursors. The impact of including metal and metal oxide nanoparticles inside the M0 xOy matrix hasn't been well studied, though. Notwithstanding this, a number of metal oxides produced using the disclosed solid-state technique show excellent photocatalytic activity toward contaminating dyes like methylene blue. In any case, the majority of them require confirmation before a general assessment of the photocatalytic efficiency of these oxides can be determined. Additionally, it is anticipated that the nanoparticle metal oxides discussed could make a substantial contribution to the decontamination of the environment.

REFERENCES

1. Hornyak, G.; Tibbals, H.F.; Dutta, J.; Moore, J. *Introduction to Nanoscience and Nanotechnology*; CRC Press Taylor and Francis: Boca Raton, NY, USA, 2009.
2. Altavilla, C.; Ciliberto, E. *Inorganic Nanoparticles*; CRC Press Taylor and Francis: Boca Raton, NY, USA, 2011.

3. Rao, C.N.; Muller, A.; Cheetham, A.K. *The Chemistry of Nanomaterials*; Wiley–VCH:Weinheim, Germany, 2004.
4. Edelstein, A.S.; Cammarata, R.C. *Nanomaterials, Synthesis and Applications*; Institute of Physics Publishing: Bristol, UK, 2002.
5. Cao, G. *Nanostructures and Nanomaterial, Synthesis, Properties and Applications*; Imperial College Press: London, UK, 2011.
6. Díaz, C.; Valenzuela, M.L. Metallic nanostructures using oligo and polyphosphazenes as template or stabilizer in solid state. In *Encyclopedia of Nanoscience and Nanotechnology*; Nalwa, H.S., Ed.; American Scientific Publishers: Valencia, CA, USA, 2010; Volume 16, pp. 239–256.
7. Walter, E.C.; Ng, K.; Zach, M.P.; Penner, R.M.; Favier, F. Electronic devices from electrodeposited metal nanowires. *Microelectron. Eng.* 2002, 61–62, 555–561. [CrossRef]
8. Walkers, G.; Parkin, I.P. The Incorporation of noble nanoparticles into hast matrix thin films: Synthesis, characterization and applications. *J. Mater. Chem.* 2009, 19, 574–590. [CrossRef]
9. Scott, C.A. Epitaxial growth and properties of doped transition metal and complex oxide films. *Adv. Mater.* 2010, 21, 219–248.
10. Teo, B.; Sun, X. Silicon-based low-dimensional nanomaterials and nanodevices. *Chem. Rev.* 2010, 107, 1454–1532. [CrossRef] [PubMed]
11. Khomutov, G.B.; Kislov, V.V.; et al. Interfacial nanofabrication strategies in development of new functional nanomaterials and planar supramolecular nanostructures for nanoelectronic and nanotechnology. *Microelectron. Eng.* 2003, 69, 373–383. [CrossRef]
12. Diaz, C.; Valenzuela, M.L.; Bobadilla, D. Bimetallic Au/Ag metal superstructures from macromolecular metal complexes in solid-state. *J. Chil. Chem. Soc.* 2013, 58, 1194–1997. [CrossRef]
13. Amin Mir, M. Synthesis, Catalysis and antimicrobial activity of 5d-metal chelate complex of Schiff base ligands. *Inorg. Chem. Commun.* 2022, 142, 109594. 10.1016/j.inoche.2022.109594
14. Díaz, C.; Valenzuela, M.L. Organometallic derivatives of polyphosphazenes as precursors for metallic nanostructured materials. *J. Inorg. Organomet. Polym.* 2006, 16, 419–435. [CrossRef]
15. Amin Mir, M.; Ashraf, M. W. Synthesis and X-ray crystallographical analysis of 5, 8-dihydroxy-1, 4-naphthoquinonne, cobalt (II), nickel (II) and copper (II) chelate complexes. *Arab. J. Sci. Eng.* 2022, 47, 535–541. 10.1007/s13369-021-05900-4
16. Díaz, C.; Valenzuela, M. L.; Lavayen, V.; O'Dwyer, C. Layered Graphitic Carbon Host Formation during Liquid-free Solid State Growth of Metal Pyrophosphates. *Inorg. Chem.* 2012, 51, 6228–6236. [CrossRef]
17. Amin Mir, M.; Ashraf, M. W., Andrews, K. Synthesis and the formation analysis of Ni (II), Zn (II) and L-glutamine binary complexes in dimethylformamide-aqueous mixture. *Results in Chemistry* 2021, 3, 100188. 10.1016/j.rechem.2021.100188
18. Díaz, C.; Valenzuela, M.L.; Cáceres, S.; O'Dwyer, C. Solution and surfactant-free growth of supported high index facet SERS active nanoparticles of rhenium by phase demixing. *J. Mater. Chem. A* 2013, 1, 1566–1572.
19. Díaz, C.; Valenzuela, M.L.; Cáceres, S.; O'Dwyer, C.; Diaz, R. Solvent and stabilizer free growth of Ag and Pd nanoparticles using metallic salts/cyclotriphosphazenes mixtures. *Mater. Chem. Phys.* 2013, 143, 124–132. [CrossRef]
20. Amin Mir, M.; Kumar, A.; Madwal, S. P.; Jassal, M.M.S. Synthesis, spectral analysis and anti-microbial properties of Cu, Ag, Au complexes of 2, 5-dihydroxy-1, 4-benzoquinone and 3, 6-dichloro-2, 6-dihydroxy-1, 4-benzoquinone. *Results Chem.* 2021, 3, 100209. 10.1016/j.rechem.2021.100209

21. Amin Mir, M. Formation and stability analysis of La (III), Er (III) Ho (III) lomatiol chelate complexes. *Res. J. Chem. Environ.* 2022, 26 (9), 21–25.
22. Díaz, C.; Valenzuela, M.L. A general solid-state approach to metallic, metal oxides and phosphates nanoparticles. In *Advances in Chemical Research*; Nova Science Publishers: New York, NY, USA, 2011.
23. Amin Mir, M.; Ashraf, M. W. Chelate formation and stability analysis of cobalt, nickel and copper with lomatiol. *Arab. J. Chem.* 2021, 14, 102930.
24. Larsen, T.H.; Sigman, M.; Ghezelbash, A.; Christopher-Doty, R.; Korgel, B.A. Solventless synthesis of copper sulfide nanorods by thermolysis of a single source thiolate-derived precursor. *J. Am. Chem. Soc.* 2003, 125, 5638–5639. [CrossRef] [PubMed]
25. Sigman, M.; Ghezelbash, A.; Hanrath, T.; Saunders, A.E.; Lee, F.; Korgel, B.A. Solventless Synthesis of monodisperse Cu2S nanorods, nanodisks, and nanoplatelets. *J. Am. Chem. Soc.* 2003, 125, 16050–16057. [CrossRef] [PubMed]
26. Ghezelbash, A.; Sigman, M.; Korgel, B.A. Solventless synthesis of nickel sulfide nanorods and triangular nanoprisms. *Nano Lett.* 2004, 4, 537–542. [CrossRef]
27. Sigman, M.; Korgel, B.A. Solventless synthesis of Bi2S3 (Bismuthinite) nanorods, nanowires, and nanofabric. *Chem. Mater.* 2005, 17, 1655–1660. [CrossRef]
28. Han, Y.C.; Cha, H.G.; Kim, C.h.W.; Kim, Y.H.; Kang, Y.S. Synthesis of highly magnetized iron nanoparticles by a solventless thermal decomposition method. *J. Phys. Chem. C* 2007, 111, 6275–6280. [CrossRef]
29. Farhadi, S.; Roostaei-Zaniyani, Z. Preparation and characterization of NiO nanoparticles from thermal decomposition of the Ni(en)3(NO3)2 complex: A facile and low-temperature route. *Polyhedron* 2011, 30, 971–975. [CrossRef]
30. Li, X.; Zhang, X.; Li, Z.; Qian, Y. Synthesis and characteristics of NiO nanoparticles by thermal decomposition of nickel dimethylglyoximate rods. *Solid State Commun.* 2006, 137, 581–584. [CrossRef]
31. Davar, F.; Salavati-Niasari, M.; Mir, N.; Saberyan, K.; Monemzadeh, M. Thermal decomposition route for synthesis of Mn3O4 nanoparticles in presence of a novel precursor. *Polyedron* 2010, 29, 1747–1753. [CrossRef]
32. Yanh, Z.; Zhang, Y.; Zhang, W.; Wang, X.; Qian, Y.; Weng, X.; Yang, S. Nanorods of manganese oxides: Synthesis, characterization and catalytic application. *J. Solid State Chem.* 2006, 179, 679–684.
33. Soofivand, F.; Salavati-Niasari, M.; Mohandes, F. Novel precursor-assisted synthesis and characterization of zinc oxide nanoparticles/nanofibers. *Mater. Lett.* 2013, 98, 55–58. [CrossRef]
34. Farhadi, S.; Pouzare, K.; Sadedhinejad, S. Simple preparation of ferromagnetic CO3O4 nanoparticles by thermal dissociation of the CoII(NH3)6(NO3)2 complex at low termperature. *J. NanoStruct. Chem.* 2013, 3, 16. [CrossRef]
35. Randhawa, B.S.; Gandotra, K. A comparative study on the thermal decomposition of some transition metal carboxylates. *J. Therm. Anal. Calorim.* 2006, 85, 417–424. [CrossRef]
36. Chunxiang Li Weng, K.L. Thermolysis of polymeric Ru(CO)4.n to metallic ruthenium: Molecular shape of the precursor affects the nanoparticle shape. *Langmuir* 2008, 24, 12040–12041.
37. Nelson, J.M.; Nguyen, P.; Petersen, R.; Rengel, H.; Macdonald, P.M.; Lough, I.J.; Manners, I.; Raju, N.P.; Greedan, J.E.; Barlow, S.; et al. Thermal ring-opening polymerization of hydrocarbon-bridged 2.ferrocenophanes: Synthesis and properties of poly(ferrocenylethy1ene)s and their charge-transfer polymer salts with tetracyanoethylene. *Chem. Eur. J.* 1997, 3, 573–584. [CrossRef]

38. Tang, B.Z.; Petersen, R.; Foucher, D.A.; Lough, A.; Coombs, N.; Sodhi, R. Novel ceramic and organometallic depolymerization products from poIy(ferrocenyIsiIanes) via pyroIysis. *J. Chem. Soc. Chem. Commun.* 1993, 6, 523–525. [CrossRef]
39. Chang, B.S.; Brijith, T.; Chen, J.; Tevis, I.D.; Karanja, P.; Çınar, S.; Venkatesh, A.; Rossini, A.J.; Thuo, M.M. Ambient synthesis of nanomaterials by in situ heterogeneous metal/ligand reactions. *Nanoscale* 2019, 11, 14060–14069. [CrossRef] [PubMed]
40. Cai, B.; Akkiraju, K.; Mounfield, W.P.; Wang, Z.; Li, X.; Huang, B.; Yuan, S.; Su, D.; Roman-Leshkov, Y.; Shao-Horn, Y. Solid-state gelation for nanostructured perovskite oxide aerogel. *Chem. Mater.* 2019, 31, 9422–9429. [CrossRef]
41. Wang, B.; Zhang, C.; Zheng, W.; Zhang, Q.; Bao, Z.; Kong, L.; Li, L. Large-scale synthesis of highly luminescent perovskite nanocrystals by template-assisted solid-state reaction at 800. *Chem. Mater.* 2020, 32, 308–314. [CrossRef]
42. Mao, Y.; Banerjee, S.;Wong, S.S. Large-scale synthesis of single-crystalline perovskite nanostructures. *J. Am. Chem. Soc.* 2003, 125, 15718–15719. [CrossRef]
43. Nalluri, S.R.; Nagarjuna, R.; Patra, D.; Ganesan, R.; Balaj, G. Large scale solid-state synthesis of catalytically active Fe3O4@M (M =Au, Ag and Au-Ag alloy) core-shell nanostructures. *Sci. Rep.* 2019, 9, 6603. [CrossRef]
44. Dey, A.; Zubko, M.; Kusz, J.; Reddy, V.R.; Banerjee, A.; Bhattacharjee, A. Thermal synthesis of hematite nanoparticles: Structural, magnetic and morphological characterizations. *Int. J. Nano Dimens.* 2020, 11, 188–198.
45. Das, R.; Pachfule, P.; Banerjee, R.; Poddar, P. Metal and metal oxide nanoparticle synthesis from metal organic frameworks (MOFs): Finding the border of metal and metal oxides. *Nanoscale* 2012, 4, 591–599. [CrossRef] [PubMed]
46. Zhu, J.; Jiang, Y.; Lu, Z.; Zhao Ch Xie, L.; Chen, L.; Duan, J. Single-crystal Cr2O3 nanoplates with differing crystalinities, derived from trinuclear complexes and embedded in a carbon matrix, as an electrode material for supercapacitors. *J. Colloid Interface Sci.* 2017, 498, 351–363. [CrossRef] [PubMed]
47. Yuan, Y.; Chen, L.; Yang, R.; Lu, X.; Peng, H.; Luo, Z. Solid-state synthesis and characterization of core–shell CoFe2O4–carbon composite nanoparticles from a heterometallic trinuclear complex. *Mater. Lett.* 2012, 71, 123–126. [CrossRef]
48. Díaz, C.; Valenzuela, M.L.; Lavayen, V.; Mendoza, K.; Peña, O.; O'Dwyer, C. Nanostructured copper oxides and phosphates from a new solid-state route. *Inorg. Chim. Acta* 2011, 377, 5–11. [CrossRef]
49. Ismail, A.A.; Bahnemannb, D.W. Mesoporous titania photocatalysts: Preparation, characterization and reaction mechanisms. *J. Mater. Chem.* 2011, 21, 11686–11707. [CrossRef]
50. Wang, Y.; Cao, G. Synthesis and enhanced intercalation properties of nanostructured vanadium oxides. *Chem. Mater.* 2006, 18, 2787–2804. [CrossRef]
51. Mao, C.J.; Pan, H.C.; Wu, X.C.; Zhu, J.J.; Chen, H.Y. Sonochemical route for self-assembled V2O5 bundles with spindle-like mosphology and their novel application in serum albumin sensing. *J. Phys. Chem.* 2006, 110, 14709–14713.
52. Avansi,W.; Ribeiro, C.; Leite, E.; Mastelaro, V. Vanadium pentoxide nanostructures: An effective control of morphology and crystal structured in hydrothermal conditions. *Cryst. Growth Des.* 2009, 9, 3626–3631. [CrossRef]
53. Diaz, C.; Barrera, G.; Segovia, M.; Valenzuela, M.L.; Osiak, M.; O'Dwyer, C. Crystallizing vanadium pentoxide nanostructures in the solid state using modified block co-polymer and chitosan complexes. *J. Nanomater.* 2015, 2015, 105157. [CrossRef]
54. Fei, H.L.; Liu, M.; Zhou, H.J.; Sun, P.C.; Ding, D.T.; Chen, T.H. Synthesis of V2O5 micro-architectures via in situ generation of single-crystalline nanoparticles. *Solid State Sci.* 2009, 11, 102–107. [CrossRef]

55. Diaz, C.; Valenzuela, M.L.; Zepeda, L.; Herrera, P.; Valenzuela, C. General group VI transition nanostructured metal oxides and their inclusion into solid matrices by a solution-solid approach. *J. Chil. Chem. Soc.* 2021, 66, 5380–5386.

56. Han, Y.; Chen, F.; Zhong, Z.; Ramesh, K.; Chen, L.; Widjaja, E. Controlled synthesis, characterization, and catalytic properties of Mn2O3 and Mn3O4 nanoparticles supported on mesoporous silica SBA-15. *J. Phys. Chem. B* 2006, 110, 24450–24456. [CrossRef]

57. Laurent, S.; Forge, D.; Port, M.; Roch, A. Magnetic iron oxide nanoparticles: Synthesis, stabilization, vectorization, physicochemical characterizations and biological applications. *Chem. Rev.* 2008, 108, 2064–2110. [CrossRef]

58. Liu, Y.; Yu, C.; Dai,W.; Gao, X.; Qian, H.; Hu, Y.; Hu, X. One-post solvothermal synthesis of multi-shelled _-Fe2O3 hollow spheres with enhaced visible-light photocatalytic activity. *J. Alloy. Compd.* 2013, 551, 440–443. [CrossRef]

59. Jiu, J.; Ge, Y.; Li, X.; Nie, L. Preparation of Co3O4 nanoparticles by a polymer contribution route. *Mater. Lett.* 2002, 54, 260–263. [CrossRef]

60. Li, Y.; Tan, B.;Wu, Y. Mesoporous Co3O4 nanowires arrays for lithium ion batteries with high capacity and rate capability. *Nano Lett.* 2008, 8, 265–270. [CrossRef]

61. Salvati-Niasari, M.; Khansari, A.; Davar, F. Synthesis and characterization of cobalt oxides nanoparticles by thermal process. *Inorg. Chem. Acta* 2009, 362, 4937–4942. [CrossRef]

62. Diaz, C.; Carrillo, D.; De la Campa, R.; Soto, A.-P.; Valenzuela, M.L. Solid-state synthesis of LnOCl/Ln2O3 (Ln = Eu, Nd) by using chitosan and PS-co-P4VP as polymeric supports. *J. Rare Earth* 2018, 36, 1326–1332. [CrossRef]

63. Palacios-Hernandez, P.; Hirata-Flores, G.; Contreras-Lopez, O. M. Synthesis of Cu and Co metal oxide nanoparticles from thermal decomposition of tartrate complexes. *Inorg. Chem. Acta* 2012, 392, 277–282. [CrossRef]

64. Rowashden-Omary, M.; Lopez-Luzuriaga, J.M. Golden metallopolymers with an active T state via coordination of poly(4-vinyl)pyridine to pentahalophenyl-gold (I) precursors. *J. Am. Chem. Soc.* 2009, 131, 3824–3825. [CrossRef]

65. Yaqoob, A.A.; Umar, K.; Nasir, M.; Ibrahim, M. Silver nanoparticles, various methods of synthesis, size afecting factors and their potential applications: A review. *Appl. Nanosci.* 2020, 10, 1369–1378. [CrossRef]

66. Tappan, B.; Steiner, S.; Luther, E. Nonporous metal foams. *Angew. Chem. Int. Ed.* 2010, 49, 4544–4565. [CrossRef]

67. Chen, A.; Holt-Hindle, P. Platinum-based nanostructured materials: Synthesis, properties and applications. *Chem. Rev.* 2010, 110, 3767–3804. [CrossRef]

68. Wu, J.; Qi, L.; You, H.; Gross, A. H. Icosahedral platinum alloy nanocrystals with enhanced electrocatalytic activities. *J. Am. Chem. Soc.* 2012, 134, 11880–11883. [CrossRef]

69. Leong, G.J.; Schulze, M.C.; Strand, M.B.; Maloney, D. Richards, R.M. Shape-directed platinum nanoparticle synthesis, nanoscale design of novel catalysts. *Appl. Organomet. Chem.* 2014, 28, 1–17. [CrossRef]

70. Cotton, F.A.; Wilkinson, G. Chapter 22 and 30. In Advanced Inorganic Chemistry; John Wiley and Sons: New York, NY, USA, 1980. 105. Jin, R. The impacts of nanotechnology on catalysis by precious metal nanoparticles. *Nanotechnol. Rev.* 2012, 1, 31–56. [CrossRef]

71. Liu, L.; Corma, A. Metal catalysts for heterogeneous catalysis: From single atoms to nanoclusters and nanoparticles. *Chem. Rev.* 2018, 118, 4981–5079. [CrossRef]

72. Chen, R.S.; Korotcov, A.; Huiang, A.S.; Tsai, D. One-dimensional conductive IrO2 nanocrystals. *Nanotechnology* 2006, 17, 67–87. [CrossRef]

73. Ortel, E.; Reier, T.; Strasser, P.; Kraehnert, R. Mesoporous film template by PEO-PB-PEO block-copolymers: Self-assembly, crystallization behavior and electrocatalytic performance. *Chem. Mater.* 2011, 23, 3201–3209. [CrossRef]

74. Quinson, J. Surfactant-free precious metal colloidal nanoparticles for catalysis. *Front. Nanotechnol.* 2021, 3, 770281. [CrossRef]

75. Fernandez-Garcia, M.; Martinez-Arias, A.; Hanson, J.; Rodriguez, C. Nanostructured oxides in chemistry, characterization and properties. *J. Am. Chem. Rev.* 2004, 104, 4063–4104. [CrossRef]

76. Kim, Y.L.; Ha, Y.; Lee, N.S.; Kim, J.G.; Baik, J.M.; Le, C.; Yoon, K.; Lee, Y.; Kim, M.H. Hybrid architecture of rhodium oxide nanofibers and ruthenium oxide nanowires for electrocatalysts. *J. Alloy. Compd.* 2016, 663, 574–580. [CrossRef]

77. Bai, J.; Han, S.-H.; Peng, R.; Zheng, J.-H.; Jiang, J.-X.; Chen, Y. Ultrathin rhodium oxide nanosheet nanoassemblies: Synthesis, morphological stability, and electrocatalytic application. *ACS. Appl. Mater Interfaces* 2017, 9, 17195–17200. [CrossRef]

78. Shimura, K.; Kawai, H.; Yoshida, T.; Yoshida, H. Simultaneously photodeposited rhodium metal and oxide nanoparticles promoting photocatalytic hydrogen production. *Chem. Commun.* 2011, 47, 8958–8960. [CrossRef] [PubMed]

79. Tricoli, A.; Righettoni, M.; Dupont, L.; Teleki, A. Semiconductor gas sensors, dry synthesis and application. *Angew. Chem. Int. Ed.* 2010, 49, 7632–7659. [CrossRef]

80. Wang, H.; Rogach, A.L. Hierarchical SnO2 nanostructures, Recent advances in design, synthesis and applications. *Chem. Mater.* 2004, 26, 123–133. [CrossRef]

81. Qin, X.; Xu, J.; Wu, Y.; Liu, X. Energy-transfer editing in lanthanide-activated upconversion nanocrystals: A toolbox for emerging applications. *ACS Cent. Sci.* 2019, 5, 29–42. [CrossRef]

82. Binnemans, K. Lanthanide-based luminescent hybrid materials. *Chem. Rev.* 2009, 109, 4283–4374. [CrossRef]

83. Kort, K.R.; Banerjee, S. Shape-controlled synthesis of well-defined matlockite LnOVl (Ln, La, Ce, Gd, Dy) nanocrystals by a novel non-hydrolytic approach. *Inorg. Chem.* 2011, 50, 5539–5544. [CrossRef]

84. Du, Y.P.; Zhang, Y.W.; Sun, L.D.; Yan, C.H. Atomically efficient synthesis of self assembled monodisperse and ultrathin lanthanide oxychloride nanoplates. *J. Am. Chem. Soc.* 2009, 131, 3162–3163. [CrossRef] [PubMed].

85. Allende, P.; Barrientos, L.; Orera, A.; Laguna-Bercero, M.A.; Salazar, N.; Valenzuela, M.L.; Diaz, C. TiO2/SiO2 composite for efficient protection of UVA and UVB rays through of a solvent-less synthesis. *J. Clust. Sci.* 2019, 30, 1511–1517. [CrossRef]

86. Lin, Z.W.; Kuang, Q.; Lian, W.; Jiang, Z.Y. Preparation and optical properties of ThO2 and Eu-doped ThO2 nanotubes by the Sol-Gel method combined with porous anodic aluminum oxide template. *J. Phys. Chem. B* 2006, 110, 23007–23011. [CrossRef]

87. Liu, S.; Han, M.Y. Silica-coated metal nanoparticles. *Chem. Asian J.* 2010, 5, 36–45. [CrossRef] [PubMed].

88. Teja, A.; Koh, P.Y. Synthesis properties and applications of magnetic iron oxide nanoparticles. *Prog. Cryst. Growth Charact. Mater.* 2009, 55, 22–45. [CrossRef].

89. Yaqoob, A.A.; Umar, K.; Nasir, M.; Ibrahim, M. Silver nanoparticles, various methods of synthesis, size affecting factors and their potential applications: A review. *Appl. Nanosci.* 2020, 10, 1369–1378. [CrossRef]

90. Chen, X.; Mao, S.S. Titanium dioxide nanomaterials: Synthesis, properties, modifications, and applications. *Chem. Rev.* 2007, 107, 2891–2959. [CrossRef] [PubMed]

91. Zheng, Y.; Cheng, Y.; Wang, Y.; Bao, F.; Zhou, L.; Wei, X.; Zhang, Y.; Zheng, Q. Quasicubic α-Fe2O3 nanoparticles with excellent catalytic performance. *J. Phys. Chem.* 2006, 110, 3093–3097. [CrossRef]

92. Hua, J.; Gengsheng, J. Hydrothermal synthesis and characterization of monodispere α-Fe2O3 nanoparticles. *Mater. Lett.* 2009, 63, 2725–2727. [CrossRef]

93. Zhang, Y.; Mei, J.; Yan, C.; Liao, T.; Bell, J.; Sun, Z. Bioinspired 2D nanomaterials for sustainable applications. *Adv. Mater.* 2019, 32, 1902806. [CrossRef]

94. Adeyemo, A.A.; Adeoye, I.O.; Bello, O.S. Metal organic frameworks as adsorbents for dye adsorption, overview, prospects and future challenges. *Toxicol. Environ. Chem.* 2012, 94, 1845–1863.

95. Ali, I.; Asim, M.; Khan, T.A. Low cost adsorbents for the removal of organic pollutants from wastewater. *J. Environ. Manage.* 2012, 113, 267–276. [CrossRef] [PubMed]

96. Padmanabhan, O.P.; Sreekumar, K.; Sengupta, P.; Dey, G.; Werrier, K. Nanocrystalline titanium dioxide formed by reactive plasma synthesis. *Vacuum* 2006, 80, 1252–1255. [CrossRef]

97. Gaya, U.I.; Abdullah, A.H. Heterogeneous photocatalytic degradation of organic contaminants over titanium dioxide: A review of fundamentals, progress and problems. *J. Photochem. Photobiol. C* 2008, 9, 1–12. [CrossRef]

98. Chong, M.N.; Jin, B.; Chow, C.W.; Saint, C. Recent developments in photocatalytic water treatment technology: A review. *Water Res.* 2010, 44, 2997–3027. [CrossRef]

99. Ray, C.; Pai, T. Recent advances of metal–metal oxide nanocomposites and their tailored nanostructures in numerous catalytic applications. *J. Mater. Chem.* 2017, 5, 9465–9478. [CrossRef]

100. Preeti, S.; Abdullah, M.M.; Saiga, I. Role of nanomaterials and their applications as photo-catalyst and sensor: A review. *Nano Res. Appl.* 2016, 2, 1–10.

101. Finney, E.; Finke, R. Nanocluster nucleation and growth kinetic and mechanistic studies: A review emphasizing transition-metal nanoclusters. *J. Colloid Interface Sci.* 2008, 317, 351–374. [CrossRef] [PubMed]

102. Mozaffari, S.; Li, W.; Thompson, C.; Ivanov, S.; Seifert, S.; Lee, B.; Kovarik, L.; Karim, A. Colloidal nanoparticle size control, experimental and kinetic modeling investigation of the ligand–metal binding role in controlling the nucleation and growth kinetics. *Nanoscale* 2017, 9, 13772–13785. [CrossRef] [PubMed]

103. Allende, P.; Laguna, M.A.; Barrientos, L.; Valenzuela, M.L.; Diaz, C. Solid state tuning morphology, crystal phase and size through metal macromolecular complexes and its significance in the photocatalytic response. *ACS Appl. Energy Mater.* 2018, 1, 3159–3170. [CrossRef]

104. Guo, H.; Li, H.; Fernandez, D.; Willis, S.; Jarvis, K.; Henkelman, G.R.; Humphrey, S.M. Stabilizer-free CuIr alloy nanoparticle catalysts. *Chem. Mater.* 2019, 31, 10225–10235. [CrossRef]

105. Oska, G. Metal oxide nanoparticles, synthesis, characterization and application. *J. Sol.-Gel. Sci. Technol.* 2006, 37, 161–164. [CrossRef]

106. Allende, P.; Barrientos, L.; Orera, A.; Laguna-Bercero, M.A.; Salazar, N.; Valenzuela, M.L.; Diaz, C. TiO2/SiO2 composite for efficient protection of UVA and UVB rays through of a solvent-less synthesis. *J. Clust. Sci.* 2019, 30, 1511–1517. [CrossRef]

107. Nguyen, T.; Thah, T.K.; Maclean, N.; Mahiddine, S. Mechanisms of nucleation and growth of nanoparticles in solution. *Chem. Rev.* 2014, 114, 7610–7630.

108. Dong, H.; Sun, L.D.; Yan, C.H. Basic understanding of the lanthanide related up conversion emissions. *Nanoscale* 2013, 5, 5703–5714. [CrossRef] [PubMed]

14 Applications of Iron Nanoparticles on Osmotic Membrane Processes to Reduce Energy Consumption and Biofouling

M.M. Armendáriz-Ontiveros
Instituto Tecnológico de Sonora, México

S. Sethi
The University of Sydney, Australia

S.G. Salinas-Rodriguez
IHE Delft Institute for Water Education, Netherlands

G.A. Fimbres-Weihs
The University of Sydney, Australia

14.1 INTRODUCTION

Water scarcity has worsened worldwide in recent years [1]. Desalination via osmotic membrane processes, such as reverse osmosis (RO), is a technology that can help reduce water scarcity problems [2]. Moreover, pressure retarded osmosis (PRO) can assist RO desalination by extracting mechanical energy from a salinity gradient [3]. However, biofouling is a significant problem for both RO and PRO, reducing water production, increasing energy consumption and increasing chemical cleaning, thus increasing desalination process costs [4,5]. This is because biofouling involves the adhesion, growth and reproduction of microorganisms (MOS) on the surface of the RO membrane, which creates extra hydraulic resistance to the passage of water through the membrane [6].

One way to reduce biofouling is to pretreat the feed water before it reaches the membrane module, since it is possible to eliminate a large part of the MOS present

DOI: 10.1201/9781003364825-14

in the feed water, as well as the nutrients required for their reproduction [7]. However, biofouling problems cannot be eliminated 100%; currently, it can only be controlled, delayed or reduced, since even if 99.9% of the MOS present in the feedwater are eliminated, the remainder can still pass through the pretreatment filters in the form of MOS or spores [8], reaching the membrane modules and colonizing the membrane surface, eventually leading to biofouling problems.

Another way to reduce biofouling is the use of antimicrobial materials such as metal nanoparticles (NPs) [9]. Biofouling mitigation helps reduce energy consumption in desalination plants by reducing the hydraulic resistance, thereby allowing higher productivity at the same or lower operating pressure. Therefore, much research has focused on membrane surface modifications with NPs. For example, Yang et al. [10] incorporated AgNPs during the formation of nanochannels in a thin-film (TF) nanocomposite membrane. They found an increased rejection of NaCl, boron, and a set of small-molecular organic compounds. Ngo et al. [11] modified a TF composite polyamide (PA) membrane with poly(ethylene glycol) and CuNPs. They enhanced membrane filtration properties and improved anti-biofouling properties against *Escherichia coli*.

Despite these studies, at the time of writing, there are very little representative biofouling data for RO desalination plants. This is because most biofouling studies in the literature have been performed under operating conditions (i.e., pressure, temperature and salt concentration) that are not typically found in RO desalination plants, or where membrane biofouling tests are performed in direct contact with enriched solutions containing bacteria that are not native from seawater. Thus, these data are not directly applicable for RO desalination plants. For example, Yu et al. [12] modified hydrophilic polyvinylidene fluoride membranes with AgNPs-GO nanosheets to improve biocidal properties, but they tested the modified membranes in an enriched solution of *Pseudomonas aeruginosa*, not at typical RO operating conditions. Khajouei et al. [13] added CuNPs in the PA layer of RO TFC membranes, and anti-biofouling performance was tested in agar plate enriched with *E. coli*, a bacterium not typically found in seawater. Armendariz et al. [14] synthesized TFM membranes adding FeNPs and CuNPs in the active layer, and also tested the anti-biofouling properties of these membranes in an *E. coli* solution.

Recently, iron nanoparticles (FeNPs, also commonly referred to as nano zerovalent iron, nZVI) have been used in osmotic membrane modification due to their easy synthesis process and low costs in comparison to other NPs. For example, Armendariz et al. [15] modified the surface of a commercial membrane with FeNPs, and carried out biofouling experiments using seawater and a high concentration of a native seawater bacteria strain (*Bacillus halotolerans* MCC1) at 6.3 MPa. They found a significant reduction of biofilm formation on the RO membrane without significantly affecting the desalination performance. Given that this recent data set was obtained at typical RO plant operating conditions, it can be used to predict the performance of these modified membranes in an industrial setting. Therefore, the objective of this chapter is to carry out a techno-economic analysis of a SWRO and a SWRO-PRO hybrid plant using modified membranes with FeNPs, to predict the reduction in energy consumption due to the reduction of biofouling, and its effect on water production cost. Hence, this work constitutes the first techno-economic assessment of SWRO and SWRO-PRO plants using FeNP coated RO membranes.

14.2 CASE STUDY

14.2.1 MATERIALS

The chemical reagents used are as follows: nutrient agar (DIFCO TM), D-glucose anhydrous (MEYER), yeast extract (SIGMA) and peptone (Bacto, DIFCO) were used for bacterial growth. Ferric nitrate ($Fe(NO_3)_3 \cdot 9H_2O$; >99%, ACS, Fermont), sodium borohydride ($NaBH_4$; >99%, Fluka), ethanol (C_2H_5OH; > 99%, J.T. Bakerand) and distilled water were used to synthesize FeNPs. A commercial RO membrane (Dow SW30HR) was used for accelerated biofouling and mass transfer tests.

14.2.2 RO MEMBRANE MODIFICATION WITH FENP OR NZVI

FeNPs were synthesized following the methodology of Arancibia-Miranda et al. [16,17]. Afterward, the commercial RO membrane surface was modified with FeNPs via the immersion method according to Armendariz et al. [18]. An aqueous solution of FeNPs (0.3 wt %) was prepared, followed by sonication for 1 h to ensure a homogenous solution and dispersion onto the membranes. The PA carboxyl groups of the membrane active layer were activated by immersing the RO membrane into distilled water for 10 min, and then the RO membrane was immersed into the FeNPs solution for 24 h. The modified membrane is referred to as FeNP, while the unmodified membrane is referred to as UC (uncoated).

14.2.3 MASS TRANSFER TEST ON RO MEMBRANES

Membrane performance was evaluated in terms of flux and hydraulic resistance. The modified and unmodified membranes were evaluated in the CF042 crossflow system (Sterlitech). The pure water permeance test was carried out using distilled water (conductivity < 5 μS cm^{-1}) and a pressure range of 1–6 MPa using a high-pressure pump (Hydra Cell, M03SASGSNSCA). The temperature was kept at 28 ± 1°C using a concentric pipe heat exchanger and a benchtop chiller (PolyScience LS5). The flux behavior is assumed to follow the model of Kedem and Katchalsky [19]:

$$J_V = \frac{m_p}{t \; \rho_p \; A_m} = \frac{TMP - \sigma \Delta \pi_{tm}}{\mu (R_m + R_f)} \qquad (14.1)$$

where J_V is the volumetric permeate flux (m s^{-1}), m_p is the mass of permeate (kg s^{-1}), t is the time elapsed during permeation (s), ρ_p is the permeate density (kg m^{-3}), A_m is the membrane area (m^2), TMP is the transmembrane pressure (Pa), σ is the reflection coefficient, $\Delta \pi_{tm}$ is the osmotic pressure difference across the membrane (Pa), μ is the dynamic viscosity of the feedwater (Pa s), R_m is the clean membrane resistance (m^{-1}) and R_f is the fouling resistance (m^{-1}). The clean membrane resistance (R_m) was determined from the pure water flux data on a clean membrane using equation (14.1).

The observed salt rejection (S_R) was calculated by:

$$S_R = \frac{C_f - C_p}{C_f} \times 100\% \qquad (14.2)$$

where C_f (mg L^{-1}) and C_p (mg L^{-1}) are the salt concentrations of the feed solution and permeate solution, respectively. Finally, the salt permeance (B) can be determined from the relationship between salt mass flux (J_s) and polarization (Γ), which is given by:

$$B = \frac{J_V}{\Gamma}\left(\frac{100\% - S_R}{S_R}\right) \qquad (14.3)$$

14.2.4 ACCELERATED BIOFOULING TEST ON RO MEMBRANES

The accelerated biofouling test was carried out under the same working conditions tested in previous studies [8,15,18,20] using a CF042 crossflow system (Sterlitech Corp.), with effective membrane area of 0.0042 m^2, consisting of two membrane cells connected in parallel as depicted in Figure 14.1. The feed water consisted of samples from the Sea of Cortez in Mexico, pretreated using cartridge filtration (5 μm) and ultrafiltration (0.1 μm). After pretreatment, the seawater was sterilized in autoclave (Felisa FE-399) for 15 minutes at 0.103 MPa and 121°C. A high concentration (10^9 CFU mL^{-1}) of the native gram-negative strain (*Bacillus halotolerans* MCC1) isolated from the Sea of Cortez, Mexico was added to the feed. The strain was fed daily with nutrient broth comprising glucose, peptone and yeast extract. Tables 14.1 and 14.2 show the accelerated biofouling experimental parameters.

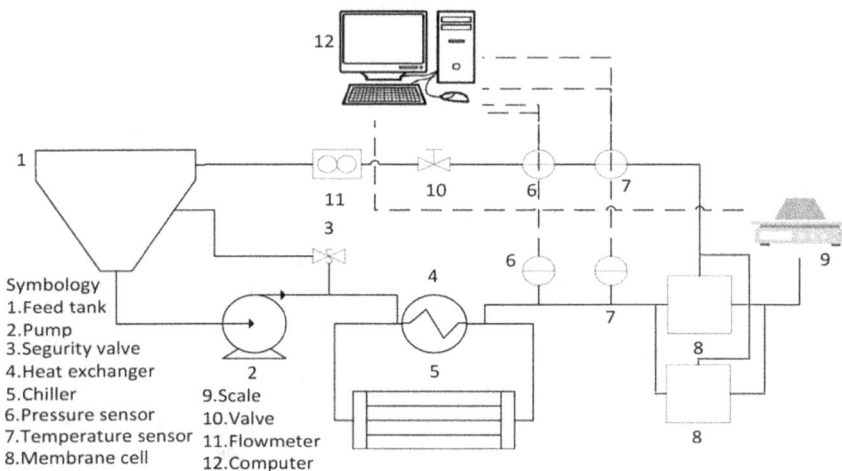

FIGURE 14.1 Schematic diagram of the experimental setup for testing the uncoated and coated membranes.

TABLE 14.1

Biofouling Test Feed Water Parameters [18]

Parameter	Value
Temperature (°C)	27.52 ± 2.48
Electrical conductivity (μS cm^{-1})	48,230 ± 514
Total dissolved Solids (mg L^{-1})	34.51 ± 0.18
Salt fraction (%)	3.534 ± 0.05
Dissolved oxygen (mg L^{-1})	8.21 ± 0.21
pH	7.03 ± 1.01

TABLE 14.2

Biofouling Test Experimental Conditions [18]

Parameter	Value
Operation pressure (MPa)	6.3
Duration of test (h)	90
Temperature (°C)	28–30
pH	6–8
Crossflow velocity (m s^{-1})	0.15
Reynolds number	700
Bacteria concentration (CFU mL^{-1})	10^9
Nutrient broth (mL d^{-1})	40

The fouling resistance R_f (m^{-1}) was calculated as follows:

$$R_f = \left(\frac{TMP - \sigma \Delta \pi_{tm}}{\mu J_V} \right) - R_m \tag{14.4}$$

14.2.5 CHARACTERIZATION OF BIOFOULED MEMBRANES

The biofouled membranes were analyzed in terms of biofilm layer thickness, total cell count, total organic carbon (TOC) and bacterial viability of the biofilm. The biofilm thickness was determined using an inverted microscope (Zeiss Axio). The bacterial viability of the biofilm was evaluated using bacterial viability kit (Molecular Probes, LIVE/DEAD BacLight L7007). The results were analyzed by fluorescence microscopy (Nikon eclipse 50i). The live/dead micrographs were processed using MATLAB (R2017b). The amount of organic matter was determined by the ignition method [21]. The total cell count was determined in a Neubauer chamber using an optical microscope (Unico IP753PL) in 100X.

14.2.6 STATISTICAL ANALYSIS

The theoretical t-student probability distribution for $p \leq 0.001$ was carried out for each of following variables: membrane hydraulic resistance, fouled flux, fouling resistance, salt rejection, salt permeance, biofilm layer thickness, number of total cells, total organic carbon and bacterial viability, using a sample size of $n < 30$. For all analyses, the professional statistical software STATISTIC 10 was used.

14.2.7 TECHNO-ECONOMIC MODELING OF SWRO AND SWRO-PRO PLANT

In order to understand the effect of coating the membranes on a large-scale desalination plant, a techno-economic model of a desalination plant was developed using MATLAB for the following scenarios:

a. Using an uncoated (UC) RO membrane in a RO desalination process.
b. Using uncoated (UC) RO and PRO membranes in a RO-PRO hybrid process.
c. Using a coated (FeNP) RO membrane in a RO desalination process.
d. Using coated (FeNP) RO and PRO membranes in a RO-PRO hybrid process.

14.2.8 MODELING THE REVERSE OSMOSIS PROCESS

In order to simulate the changes occurring within the RO modules, the MATLAB model solves a series of differential equations that reflect various changing parameters within each RO element, including concentration and pressure changes. The pressure, temperature and concentration differential changes along the membrane element are calculated via mass and energy balances across a differential channel cross-section (a node). The differential equations are integrated for all the nodes into which the channel is discretized, and the permeate flow through the membrane at each node is calculated, along with the consequent changes in feed flow rate. The physical properties of the feed channel solution at each point are obtained using the models described by Sharqawy et al. [22] and Nayar et al. [23]. Further details about the differential equation model of the RO and PRO systems can be found elsewhere [24].

The water flux through the membrane (J_V) is calculated by equation (14.1), which describes its variation along the membrane modules. Moreover, the water flux is a function of the total resistance ($R_m + R_f$) which depends on the extent of membrane fouling. To model the evolution of the fouling resistance over time, an exponential fouling model was fitted to experimental fouling data reported by Ruiz-Garcia and Nuez [25]:

$$\left(\frac{R_f}{R_m}\right)_{UC} = 1.449\left[1 - \exp\left(-\frac{t}{1116 \text{ days}}\right)\right] \qquad (14.5)$$

This model is assumed to apply for the uncoated membrane, R_f is the resistance due to the fouling, R_m is the clean membrane resistance and t is the time elapsed since the start of operation (in days). The value for R_m is assumed to be constant and is obtained as described in Section 14.2.3. For the purposes of this model, the values reported by Armendariz-Ontiveros et al. [15] are used. For the uncoated membrane, a $R_{m,UC}$ value of 4.23×10^{14} m^{-1} is assumed, and for the FeNP coated membrane, the $R_{m,FeNP}$ value of 7.11×10^{14} m^{-1} is used, since the coating process increased the membrane resistance. Nonetheless, the coating reduced the fouling rate [8,18,20], so the fouling curve was modified for the FeNP coated membrane, according to the data reported by Armendariz-Ontiveros et al. [15]:

$$\left(\frac{R_f}{R_m} \right)_{FeNP} = 0.185 \left[1 - \exp\left(-\frac{t}{1116 \text{ days}} \right) \right] \tag{14.6}$$

The total resistance is then determined by adding the membrane resistance to the fouling resistance for each particular time of operation. Finally, the operating pressure of the plant is determined such that it maintains a constant permeate flow rate. As the fouling resistance given by equations (14.5) and (14.6) increases over time, the transmembrane operating pressure is increased accordingly, and the pumping energy cost is calculated on a time basis.

14.2.9 MODELING THE PRESSURE RETARDED OSMOSIS PROCESS

The pressure retarded osmosis (PRO) process is simulated in a manner similar to the RO process, with the outputs of the RO process serving as inputs for the PRO process. The key difference between these two membrane processes is the direction and behavior of the water and salt flows across the membrane. The water flux, $J_{w,PRO}$ is given by:

$$J_{w,PRO} = \frac{\Delta \pi_{eff} - \Delta p_{PRO}}{\mu \left(R_m + R_f \right)} \tag{14.7}$$

where the viscosity (m) and the total membrane resistance are calculated as in the RO process, Dp_{PRO} refers to the hydraulic pressure difference between the draw and the feed sides, and Dp_{eff} is the effective osmotic pressure difference between the draw and the feed, which is given by:

$$\Delta \pi_{eff} = \frac{\pi_d \exp\left(-\frac{J_{w,PRO}}{k_d} \right) - \pi_f \exp\left[J_{w,PRO}\left(\frac{1}{k_f} + \frac{S}{D} \right) \right]}{1 + \frac{B}{J_{w,PRO}} \left\{ \exp\left[J_{w,PRO}\left(\frac{1}{k_f} + \frac{S}{D} \right) \right] - \exp\left(-\frac{J_{w,PRO}}{k_d} \right) \right\}} \tag{14.8}$$

In equation (14.8), p_d and p_f refer to the osmotic pressure in the draw and feed channels, respectively, k_d and k_f refer to the mass transfer coefficients for the draw and feed channels, respectively, S represents the structural parameter of the PRO membrane, D is the salt diffusivity, and B is the salt permeance. The flux of salt J_S through the membrane is calculated through an analogous equation:

$$J_S = B\, \Delta c_{eff} \tag{14.9}$$

The effective concentration difference through the membrane Dc_{eff} can be calculated by:

$$\Delta c_{eff} = \frac{c_d \exp\left(-\frac{J_{w,PRO}}{k_d}\right) - c_f \exp\left[J_{w,PRO}\left(\frac{1}{k_f} + \frac{S}{D}\right)\right]}{1 + \frac{B}{J_{w,PRO}}\left\{\exp\left[J_{w,PRO}\left(\frac{1}{k_f} + \frac{S}{D}\right)\right] - \exp\left(-\frac{J_{w,PRO}}{k_d}\right)\right\}} \tag{14.10}$$

where c_d and c_f are the salt concentrations in the draw and feed, respectively. Further details about the differential equation model of the PRO system and its solution procedure can be found elsewhere [26,27].

14.2.10 OVERALL PLANT COSTING

All costs are reported on a real USD, 2022 basis, with the location specified to be in northwest Mexico. The cost of the plant is assumed to be comprising the capital cost (determined using the cost of the membrane, pumps and other plant equipment), the fixed operating costs (estimated through the capital cost) and the variable operating cost, primarily comprising the energy cost. The membrane replacement cost is determined from the model results, whereby the membrane elements are replaced when the required feed pressure reaches the maximum allowable operating pressure for the membrane modules (i.e., 7 MPa). The cost of energy is obtained through determining the pressure, temperature and concentration conditions, and therefore determining pump energy, energy recovered through an energy recovery device (ERD) and energy recovered through PRO where applicable. The inputs to the model are detailed in Table 14.3.

The total water cost is then obtained by dividing the net present value of the project by the total discounted water output over the project life. The specific energy consumption (SEC) is calculated by dividing the energy usage by the water output.

14.3 DISCUSSIONS

14.3.1 MODIFIED MEMBRANE PERMEATION PERFORMANCE

Figure 14.2a shows the flux results under accelerated biofouling conditions. It can be observed that the FeNPs membrane presents statistically significant higher flux with respect to the uncoated membrane (at least 36% higher). This shows that the FeNP membrane has the ability to maintain the flux over a long time regardless of the biofilm conditions. This result also indicates that uncoated membrane is not able

TABLE 14.3

The Inputs to the System Model

Variable	Value	Units	Source
Required water output	200	L/s	–
Number of membrane trains	220	–	–
Number of membrane modules per train	7	–	–
Membrane module area	28	m^2	[28]
Inlet seawater concentration	35	g/kg	–
Module length	1.016	m	[28]
Channel height	1	mm	[29]
Porosity	0.8	–	[30]
Hydraulic diameter	0.9	mm	–
Pump efficiency	0.7	–	[31]
ERD efficiency	0.75	–	[32]
Operational hours	8400	hours	–
Plant life	30	years	[33]
Real discount rate	7	%	

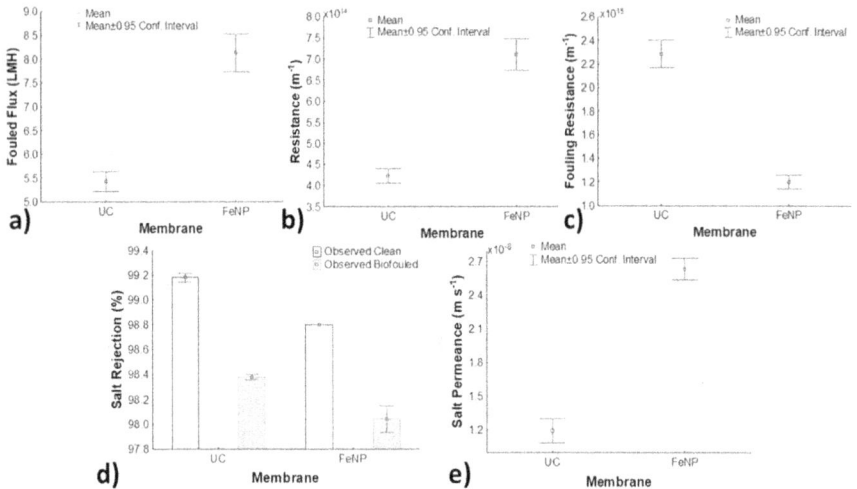

FIGURE 14.2 Desalination performance: (a) fouled flux for coated and uncoated membrane under biofouling conditions; (b) hydraulic resistance determined under clean conditions; (c) average fouling resistance; (d) salt rejection for studied membranes under clean and biofouling conditions and (e) salt permeance under clean conditions.

to maintain the initial flux owing to the formation of the biofilm layer on the membrane. Figure 14.2b and c shows that although the clean membrane resistance is larger for the coated membrane, the lower fouling resistance more than offsets this initial disadvantage. This result can be attributed to the FeNPs preventing the growth of MOS by avoiding the formation of extra resistance to water passage, thereby reducing the fouling resistance. This is beneficial for the desalination process, since the FeNP membrane permeation will be maintained for a longer period under biofouling conditions, in comparison to the uncoated membrane, thereby achieving increased water production or a lower operating pressure over the membrane life. The observed salt rejection results under clean and biofouled conditions are shown in Figure 14.2d. The FeNP coating and the biofouling reduced the salt rejection values. The uncoated membrane presents the highest salt rejection values under clean and biofouled conditions. As regards salt permeance results, Figure 14.2e shows that the FeNP coating increases the salt passage through the membrane (>55% higher mean). This may be due to a phenomenon similar to cake-enhanced and biofilm-enhanced concentration polarization [34,35], related to the increased membrane resistance, as the membrane coating may be acting in a manner similar to a particle fouling layer.

14.3.2   BIOFOULING PERFORMANCE

The results of biofouled membrane characterization after 90 h of accelerated bio-fouling experiments are shown in Figure 14.3. It is possible to observe a highly significant reduction of biofilm cake layer (Figure 14.3a), total cell count (Figure 14.3b), total organic carbon (Figure 14.3c), dead and live cells (Figure 14.3d) due to the FeNP coated effect (52.6, 57.6, 70, 50 and 66.6% less in comparison to uncoated membrane, respectively). This result is attributed to the high biocide effect of FeNPs against the *Bacillus Holotolerans* MMC1. The FeNPs have the ability to

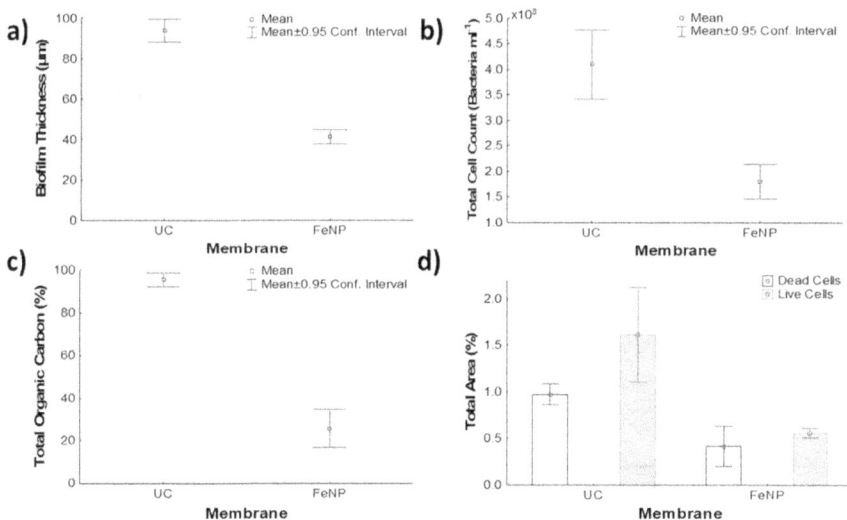

FIGURE 14.3   Characterization of biofouled membranes: (a) biofilm cake layer of studied membranes; (b) total cell for studied membranes; (c) total organic carbon for studied membranes after biofouling experiments and (d) live and dead cell coverage in membranes after biofouling experiments.

produce ROS that provoke cell structural damage, breaking the biomolecules covalent bonds, disrupting the mechanisms of DNA formation, ATP production, protein synthesis, etc., causing MOS death [36–38].

14.3.3 ENERGY CONSUMPTION OF SWRO AND RO-PRO PLANT

The MATLAB model results for the SEC of water production, for all the scenarios mentioned in Section 14.2.7, are presented in Figure 14.4. From these data, it is clear that fouling leads to increased energy consumption for all the scenarios modeled, compared to the cases without fouling (up to 185% for the RO-PRO-UC case). However, the effect of fouling is significantly reduced when the FeNP coated membrane is used. This is because when FeNP coated membranes are used, the desalination process can be operated at a lower transmembrane pressure for a longer time, which reduces the energy operating costs for the high-pressure pumps.

Moreover, it can be observed that the addition of PRO results in a reduction in SEC for all the cases modeled (up to 19% reduction for the UC membrane case without fouling). The effect of PRO is less evident for the UC membrane under fouled conditions (practically no reduction in SEC) because biofouling reduces the PRO water flux, limiting the amount of energy that can be recovered. This is evident, as the cases without fouling and the case with the FeNP membrane under fouling conditions all show significant energy requirement reductions when PRO is implemented (between 4% and 19% reductions).

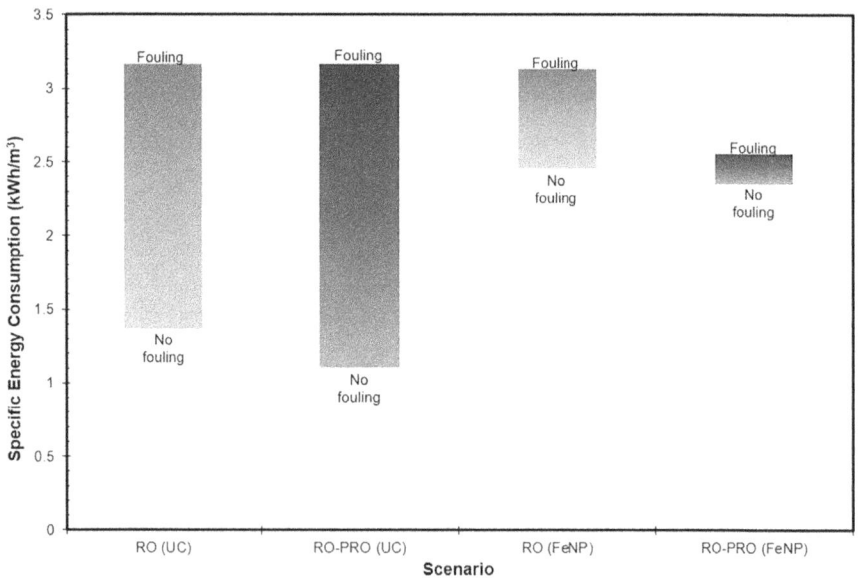

FIGURE 14.4 Specific energy consumption results for water production via RO desalination, showing the effect of fouling (top of bars) compared to the case with no fouling (bottom of bars) and membrane FeNP coating. Conventional RO plant results are shown in brown, and RO-PRO hybrid plant results are shown in blue.

For the cases without fouling, it is clear that utilizing the FeNP coated membrane increases SEC, and this is related to the higher membrane resistance due to coating. Nonetheless, for all the cases with fouling, the SEC is reduced when employing the FeNP coating (1% reduction for standalone RO, and 19% reduction for RO-PRO). This is related to the ability of the coated membrane to maintain a lower operating pressure for a longer time, which also increases the working life of the membrane, thus reducing membrane replacement costs.

14.3.4 WATER PRODUCTION COSTS

Figure 14.5 presents the results of employing FeNP coated membranes for both the standalone RO desalination process and the hybrid RO-PRO process, under scenarios with and without fouling. It can be observed that, just as was the case for SEC, fouling increases the cost of water production due to the increased energy and membrane replacement costs. However, the use of the FeNP coated membranes reduces the impact of fouling, to the extent that the cost of water production for the standalone RO plant is lower when the FeNP coating is used than when the uncoated membrane is used. Nevertheless, for the case without fouling, the cost of water production using an uncoated membrane is slightly lower, as the resistance of the uncoated membrane is lower than its FeNP coated counterpart.

Furthermore, Figure 14.5 also illustrates that despite the energy recovered in the PRO process, the total water production cost is higher when PRO is included than

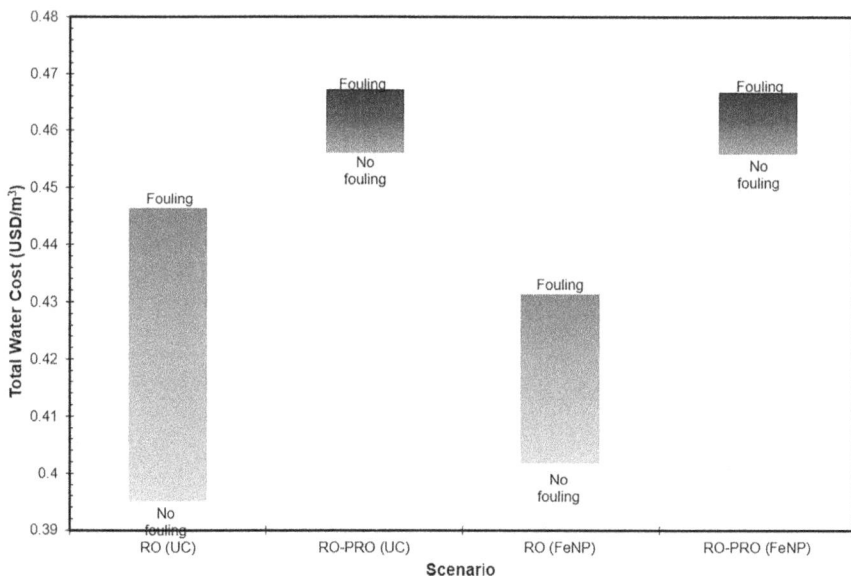

FIGURE 14.5 Total water cost results for water production via RO desalination, showing the effect of fouling (top of bars) compared to the case with no fouling (bottom of bars) and membrane FeNP coating. Conventional RO plant results are shown in brown, and RO-PRO hybrid plant results are shown in blue.

when it is not. This agrees with previous results [26], which shows that currently the capital expenditure associated with the extra ERD necessary for PRO more than offsets any energy cost savings due to the additional energy recovered by PRO. This implies that even though improvements in the permeance of PRO membranes could increase their commercial potential, for PRO to be competitive in the context of seawater desalination, a lower capital cost for ERDs is paramount.

14.4 CONCLUSIONS

Iron nanoparticles (FeNPs) have the potential to reduce the energy requirements for RO desalination plants, due to their anti-biofouling properties. Experimental results from accelerated biofouling tests were used to predict the expected water flux and energy consumption of RO and RO-PRO desalination plants installed in northwest Mexico. These simulated data were in turn used to determine the SEC and total cost of water production. The results show that, for a standalone conventional RO desalination plant, the FeNP coating not only reduces the impact of fouling but also reduces the SEC and cost of water production (by 1% and 3.4%, respectively). Moreover, the benefits of the FeNP anti-biofouling coating are more evident in terms of SEC when PRO is incorporated, as this NP coating allows for an increase in energy recovery from the salinity gradient over the life of the plant.

Although PRO also presents potential to reduce the energy requirements of RO desalination, the results show that incorporating this technology into a RO-PRO hybrid plant increases capital expenditure more than any potential reduction in energy costs, leading to higher water production costs. This is driven by the high cost of the additional ERDs needed. Cost reductions may be possible, however, if the capital cost of ERDs could be reduced, coupled with future improvements in PRO membrane performance.

Overall, this work showcases the potential of FeNPs in reducing the energy intensity of RO desalination, by mitigating one of the most problematic challenges faced by this technology, that is, biofouling. The incorporation of FeNPs can not only decrease energy use but also increase the operating life of membrane elements, leading to an overall lower environmental impact and decarbonization of the desalination industry.

ACKNOWLEDGMENTS

The authors would like to acknowledge funding provided by the Instituto Tecnológico de Sonora through a PROFAPI fund (Project 2022-0012).

REFERENCES

1. Boretti, A. and L. Rosa, Reassessing the projections of the World Water Development Report. *npj Clean Water*, 2019. **2**(1): 15.
2. Ashfaq, M.Y., M.A. Al-Ghouti, H. Qiblawey, N. Zouari, D.F. Rodrigues, and Y. Hu, Use of DPSIR Framework to analyze water resources in Qatar and overview of reverse osmosis as an environment friendly technology. *Environmental Progress & Sustainable Energy*, 2019. **38**(4): 13081.

3. Achilli, A. and K.L. Hickenbottom, *3 - Pressure Retarded Osmosis: Applications*, in *Sustainable Energy from Salinity Gradients*, A. Cipollina and G. Micale, Eds. 2016, Sawston, United Kingdom: Woodhead Publishing. p. 55–75.

4. Goh, P.S., W.J. Lau, M.H.D. Othman, and A.F. Ismail, Membrane fouling in desalination and its mitigation strategies. *Desalination*, 2018. **425**: 130–155.

5. Maddah, H. and A. Chogle, Biofouling in reverse osmosis: Phenomena, monitoring, controlling and remediation. *Applied Water Science*, 2017. **7**(6): 2637–2651.

6. AlSawaftah, N., W. Abuwatfa, N. Darwish, and G.A. Husseini, A review on membrane biofouling: Prediction, characterization, and mitigation. *Membranes*, 2022. **12**(12): 1271.

7. Zhang, H., S. Zhu, J. Yang, and A. Ma, Advancing strategies of biofouling control in water-treated polymeric membranes. *Polymers*, 2022. **14**(6): 1167.

8. Armendáriz-Ontiveros, M.M., G.A. Fimbres Weihs, S. de los Santos Villalobos, and S.G. Salinas-Rodriguez, Biofouling of FeNP-coated SWRO membranes with bacteria isolated after pre-treatment in the Sea of Cortez. *Coatings*, 2019. **9**(7): 462.

9. Agrawal, A., A. Sharma, K.K. Awasthi, and A. Awasthi, Metal oxides nanocomposite membrane for biofouling mitigation in wastewater treatment. *Materials Today Chemistry*, 2021. **21**: 100532.

10. Yang, Z., H. Guo, Z.-k. Yao, Y. Mei, and C.Y. Tang, Hydrophilic silver nanoparticles induce selective nanochannels in thin film nanocomposite polyamide membranes. *Environmental Science & Technology*, 2019. **53**(9): 5301–5308.

11. Ngo, T.H.A., T.D. Nguyen, V.D. Nguyen, and D.T. Tran, Improvement in anti-biofouling property of polyamide thin-film composite membranes by using copper nanoparticles. *Journal of Applied Polymer Science*, 2022. **139**(32): e52739.

12. Yu, Y., Y. Yang, L. Yu, K.Y. Koh, and J.P. Chen, Modification of polyvinylidene fluoride membrane by silver nanoparticles-graphene oxide hybrid nanosheet for effective membrane biofouling mitigation. *Chemosphere*, 2021. **268**: 129187.

13. Khajouei, M., M. Jahanshahi, and M. Peyravi, Biofouling mitigation of TFC membrane by in-situ grafting of PANI/Cu couple nanoparticle. *Journal of the Taiwan Institute of Chemical Engineers*, 2018. **85**: 237–247.

14. Armendariz Ontiveros, M.M., Y. Quintero, A. Llanquilef, M. Morel, L. Argentel Martínez, A. García García, and A. Garcia, Anti-biofouling and desalination properties of thin film composite reverse osmosis membranes modified with copper and iron nanoparticles. *Materials*, 2019. **12** (2081): 1–12.

15. Armendáriz-Ontiveros, M.M., A. García-García, A. Mai-Prochnow, and G.A. Fimbres Weihs, Optimal loading of iron nanoparticles on reverse osmosis membrane surface to reduce biofouling. *Desalination*, 2022. **540**: 115997.

16. Arancibia-Miranda, N., S.E. Baltazar, A. García, A.H. Romero, M.A. Rubio, and D. Altbir, Lead removal by nano-scale zero valent iron: Surface analysis and pH effect. *Materials Research Bulletin*, 2014. **59**: 341–348.

17. Arancibia-Miranda, N., S.E. Baltazar, A. García, D. Muñoz-Lira, P. Sepúlveda, M.A. Rubio, and D. Altbir, Nanoscale zero valent supported by zeolite and montmorillonite: Template effect of the removal of lead ion from an aqueous solution. *Journal of Hazardous Materials*, 2016. **301**: 371–380.

18. Armendáriz-Ontiveros, M.M., A. García García, S. de los Santos Villalobos, and G.A. Fimbres Weihs, Biofouling performance of RO membranes coated with Iron NPs on graphene oxide. *Desalination*, 2019. **451**: 45–58.

19. Kedem, O. and A. Katchalsky, Thermodynamic analysis of the permeability of biological membranes to non-electrolytes. *Biochimica et Biophysica Acta*, 1958. **27**: 229–246.

20. Armendáriz-Ontiveros, M.M., J. Álvarez-Sánchez, G.E. Dévora-Isiordia, A. García, and G.A. Fimbres Weihs, Effect of seawater variability on endemic bacterial biofouling of a reverse osmosis membrane coated with iron nanoparticles (FeNPs). *Chemical Engineering Science*, 2020. **223**: 115753.

21. Drioli, E., A. Criscuoli, and F. Macedonio, *Membrane-Based Desalination: An Integrated Approach (MEDINA)*. 2011, London, UK: Iwa Publishing.
22. Sharqawy, M.H., J.H. Lienhard, and S.M. Zubair, Thermophysical properties of seawater: A review of existing correlations and data. *Desalination and Water Treatment*, 2010. **16**(1–3): 354–380.
23. Nayar, K.G., M.H. Sharqawy, L.D. Banchik, and J.H. Lienhard V, Thermophysical properties of seawater: A review and new correlations that include pressure dependence. *Desalination*, 2016. **390**: 1–24.
24. Toh, K.Y., Y.Y. Liang, W.J. Lau, and G.A. Fimbres Weihs, The techno-economic case for coupling advanced spacers to high-permeance RO membranes for desalination. *Desalination*, 2020. **491**: 114534.
25. Ruiz-García, A. and I. Nuez, Long-term performance decline in a brackish water reverse osmosis desalination plant p redictive model for the water permeability coefficient. *Desalination*, 2016. **397**: 101–107.
26. Yeung, N.O., G.A. Fimbres Weihs, and D. Wiley, Techno-economic assessment of the impact of fouling on SWRO-PRO hybrid desalination systems. in *The International Desalination Association World Congress 2022*. 2022. Sydney, Australia: IDA.
27. Bartholomew, T.V. and M.S. Mauter, Computational framework for modeling membrane processes without process and solution property simplifications. *Journal of Membrane Science*, 2019. **573**: 682–693.
28. Lenntech, *Dow Filmtec SW30-8040*. Lenntech.
29. Xie, P., T.Y. Cath, and D.A. Ladner Mass transport in osmotically driven membrane processes. *Membranes*, 2021. **11**, 29. 10.3390/membranes11010029
30. Liu, F., L. Wang, D. Li, Q. Liu, and B. Deng, A review: The effect of the microporous support during interfacial polymerization on the morphology and performances of a thin film composite membrane for liquid purification. *RSC Advances*, 2019. **9**(61): 35417–35428.
31. Elsey, J., How to define & measure centrifugal pump efficiency: Part 1. *Pumps and Systems Magazine*, 2020.
32. Schunke, A.J., G.A. Hernandez Herrera, L. Padhye, and T.-A. Berry, Energy recovery in SWRO desalination: Current status and new possibilities. *Frontiers in Sustainable Cities*, 2020. **2**: 9.
33. Shahabi, M.P., A. McHugh, M. Anda, and G. Ho, Environmental life cycle assessment of seawater reverse osmosis desalination plant powered by renewable energy. *Renewable Energy*, 2014. **67**: 53–58.
34. Gutman, J. and M. Herzberg, *Cake/biofilm enhanced concentration polarization*, in *Encyclopedia of Membrane Science and Technology*. Wiley: New Jersey, United States. pp. 1–14.
35. Fimbres Weihs, G.A. and D.E. Wiley, CFD analysis of tracer response technique under cake-enhanced osmotic pressure. *Journal of Membrane Science*, 2014. **449**: 38–49.
36. Asl, B.A., L. Mogharizadeh, N. Khomjani, B. Rasti, S.P. Pishva, K. Akhtari, F. Attar, and M. Falahati, Probing the interaction of zero valent iron nanoparticles with blood system by biophysical, docking, cellular, and molecular studies. *International Journal of Biological Macromolecules*, 2018. **109**: 639–650.
37. Jeckelmann, J.-M. and B. Erni, Transporters of glucose and other carbohydrates in bacteria. *Pflügers Archiv: European Journal of Physiology*, 2020. **472**(9): 1129–1153.
38. Morelli, A.M., S. Ravera, and I. Panfoli, The aerobic mitochondrial ATP synthesis from a comprehensive point of view. *Open Biology*, 2020. **10**(10): 200224.

15 Nanotechnology for Wastewater Remediation

Mehreen Shah and Sirajuddin Ahmed
Jamia Millia Islamia University, India

15.1 INTRODUCTION

Wastewater often contains harmful compounds that are extremely harmful to our health. They often contain heavy metals and even carcinogenic compounds [1]. Results of lab analysis showed inhibition of pathogens indicating the bactericidal capacity of the nanoparticles. The degradation period was very short compared to other previous remediation tools, affirming the success of this method. Green nanoparticles leave less polluting residues compared to conventional remediation tools like earlier used in wastewater management. Those conventional remediation techniques often require very high cost for post-processing and often cause pollution once released in water bodies. Hence, scientists and researchers around the globe have now worked on newer remediation tools that work with higher efficiency and also reduce post-processing costs and can later be disposed of without being deleterious to the environment in which they are being released. Green nanoparticles also have an effect on degradation of synthetic dyes and detection of heavy metal ions. Hence, green nanoparticles show great potential for water remediation and are a unique solution for the future.

15.2 NANO-ZEOLITES FOR HEAVY METAL (CD) REMOVAL

Zeolites consist of interlinked silica (SiO_4) and tetrahedra of alumina (AlO_4) and are aluminosilicates. They have high surface area and ion-exchange capacity which enable them to sequester cadmium. Zeolites that are modified have a higher capacity than naturally occurring zeolites [2]. Ghomashi et al. [3] have used modified nano-clinoptilolite zeolite for removing fluoride contamination from industrial wastewater. Goyal et al. have used nano-zeolite socony mobil-5 (NZSM-5) for a 95% removal of bisphenol-5, an industrial pharmaceutical from wastewater. Nano-sized zeolites are more efficient and hence better used and have a high surface to volume ratio and hence possess more absorbent qualities. Zeolites are not toxic and cost efficient, and hence are used for municipal and wastewater treatment in large numbers frequently as found in a review by Renu et al. [4], who also compiled results from different experiments in Table 15.1.

DOI: 10.1201/9781003364825-15

TABLE 15.1

Some Zeolites and Their Absorbance Capacity

Adsorbent	Concentration of Metal (mg/l)	pH	Time (minutes)	Adsorbent (g/l)	Uptake (mg/g)	Absorbance %	Reference
Natural zeolite	9–90	5	1440	–	9	71	[5]
Fly ash zeolite	1124–3372	6.6	1440	10	57–195	96	Izidoro et al., 2013
Synthetic zeolite A	100–2000	–	180	1	316	–	[6]

15.3 GRAPHENE NANOPARTICLES FOR WASTEWATER DECONTAMINATION

A very popular nanomaterial for water metal removal is graphene which is a 2D structure of carbon. It has high efficiency rates. It helps to absorb heavy metals on its surface (Zhang et al., 2017). The adsorption process is highly rapid and spontaneous. Gopalakrishna et al. [7] achieved a 100% removal of chromium using only 70 mg of graphene oxide at an optimum pH of 8. This method involved the addition of functional groups on the surface such as –COOH, –OH, and –C=O; this is a breakthrough advancement in the field of nanotechnology for wastewater decontamination of heavy metals such as chromium. Graphene oxide-based nanofilters are now being increasingly used by scientists for water purification purposes. They have high hydrophilicity and the anti-microbial nature prevents biofouling that is frequently seen in biomembrane filtration. It has high water permeability and ability to reject different molecules based on their molecular charge, hence enabling us to use this to our advantage for water purification technologies at relatively lower costs and ease of access [4]. Lingamdinne et al. [8] used graphene oxide-based spinel nanoparticles as adsorbents for uptake of Cr and Pb heavy metals in wastewater. Tran et al. [9] used engineered graphene-nanoparticle aerogel for capturing phosphate from wastewater. In total, 350 mg/g was removed efficiently. Phosphate is an important nutrient and causes eutrophication in lake bodies. It leads to excess algal growth and lowers the DO (dissolved oxygen) in water. Balasubramani et al. [10] developed graphene oxide nanoparticles for adsorption of the antidiabetic pharmaceutical drug called metformin from water. The uptake of the drug in water was found to be 122.61 mg/g. Reduced graphene oxide-based nanosheets possess excellent adsorption ability and are an edge above carbon nanomaterials that are conventionally used [11]. Mantovani et al. [12] used, odified GO nanosheets for adsorption of heavy metal ions such as Cu^{2+} and Pb^{2+}, and organic dyes such as methylene blue, in wastewater. GO nanosheets have displayed maximum capacity of about 356 mg/g for absorbance of antibiotic ofloxacin (OFLOX), for rhodamine B (RhB) dye = 550 mg/g, 428 mg/g (Khaliha et al., 2021) for methylene blue (MB) dye. White et al. [13–15] used graphene oxide nanoparticles for adsorption of Cu^+ ions in water and in a 1-hour adsorption time achieved a removal efficiency of 99.4%. Mahtab et al. used graphene oxide nanosheets as efficient sorbents for the removal of Pb^{+2}, Zn^{+2}, Cd^{+2}, and Ni^{+2}. The newly synthesized composite displayed an adsorption capacity of 208 mg/g. Mohan et al. [16] synthesized graphene oxide-MgO nanohybrid for the effective removal of Pb^{+2} ions in wastewater, owing to its adsorption nature (190 mg/g) at a dosage of 0.4 g/L at 30 min of contact. Table 15.2 shows graphene and graphene based nanoparticles for adsorbing contaminants in the wastewater.

15.4 FERROUS NANOPARTICLES FOR WATER DECONTAMINATION

Vasiljevic et al. [18–20] used iron titanate (Fe_2TiO_5) nanoparticles for photodegradation of methylene blue (MB) dye in water where photodegradation of the dye occurred after a 4 H exposure with mineralization into end-products H_2O, CO_2. Wasantha et al. [21] used magnetite nanoparticles for adsorption of Pb^{2+} and Cd^{2+} ions for water purification. Iram et al. used Fe_3O_4 hollow nanospheres for the removal of neutral red dye from water via adsorption on the nanosphere and magnetism for a 90% dye removal. Carbon nanotubes (CNTs) containing Fe_2O_3 particles are effective for adsorption of methylene blue and neutral red dyes that were adsorbed in quantities as 42.3 and 77.5 mg/g, respectively, by Song et al. The magnetic nature of the nanoparticles makes it easy to recover the nanoparticles from the water.

15.5 CARBON NANOPARTICLES FOR WATER DECONTAMINATION

Activated carbon had been used for decades for water purification technologies even at municipal level. Treatment with carbon helps in removal of odor and gives clarity to water by removal of suspended soils in it by adsorption [22]. Carbon is highly cost-effective too and offers promising results for water purification goals. Li et al. [23,24] used CNTs for adsorption of heavy metal, cadmium(II). Chen et al. [25] used multiwalled CNTs for uptake of Sr(II) and Eu(III) from water using a formation of complex between the functional groups of the oxidized CNTs and the metals suspended in water. Li et al. [23] were able to achieve simultaneous removal of Pb^{2+}, Cu^{2+}, and Cd^{2+} ions from water by the application of multiwalled CNTs. Sadegh et al. [26] achieved removal of carcinogenic synthetic dye called Malachite Green by the usage of multiwalled CNTs. The CNTs were adsorbents and also had interactions between functional groups on its surface for effective removal of the suspended dye from the wastewater. Ghaedi et al. [27] achieved adsorption of bromothymol blue using multiwalled CNTs that were oxidized. Carbon can be used for water remediation using its various forms such as graphene sheets, multiwalled CNTs, fullerene, and CNTs [28,29].

TABLE 15.2

Graphene and Graphene-Based NPs for Adsorbing Contaminants in Wastewater

Adsorbent	Concentration of Metal (ppm-mg/l)	pH	Time (min)	Uptake by Adsorbent (mg/l)	Adsorbent Dosage (g/l)	Uptake %	Reference
Graphene (GN) with cetyltrimethylammonium bromide	50,100	2	60	21.57	400	**98**	Wu et al., 2013
Graphene sand composite	8–20	1.5	90	2860	10	**93**	[17]
Graphene oxide	52	5	12	44	–	92.6	Yang et al., 2014
Graphene nanosheets with zero valent Fe	15–35	3	90	–	1	70	Li et al., 2016
Spinel Ni-Fe based on graphene oxide	1000	4	120	45	0.12–2.5	>80	[8]

15.6 MECHANISMS EMPLOYED BY NPS FOR WASTEWATER REMEDIATION

15.6.1 PHOTOCATALYSIS

Photocatalysis is an important process that is in motion during nano remediation in wastewater. It refers to the degradation of organic wastes and compounds present in the wastewater by the action of UV or light from a source such as the sun [30]. The photons of light act as photocatalysts, hence helping speed up the rate of the biodegradation reaction. Photocatalysis is an oxidative reaction where oxidative degradation of wastewater contaminants occurs. TiO_2 is a popular choice for this reaction which takes place via oxidative–reduction as particles are irradiated with light of 200–400 nm wavelength causing excitation of the electron and leading to a cascade of oxidative–reduction steps in which wastewater pollutants are biodegradable into harmless compounds such as water and CO_2.

Photocatalysis can be exploited for a higher efficiency rate if we use nanoparticles. Due to large surface area/volume ratio they are more rapidly advanced over their macro counterparts [31]. TiO_2 is the most popular choice for photocatalysis process. The UV irradiation helps in overcoming the energy gap and doping with gas such as nitrogen further reduces the band width. In wastewater, there is also the presence of SO_4^{2-}, HCO_3^{3-}, CO_3^{2-} and Cl^- which too can assist in the generation of radicals. The electron jump to higher level will cause reduction of oxygen [32].

TiO_2-based nanotubes have been effective in the removal of synthetic dyes such as methylene blue, congo red [33] dichlorophenol trichlorobenzene, toulene, and phenols. This removal process can be coupled with the simultaneous removal of heavy metals such as Cd^{+2}. The removal efficiency in one experiment was reported to be 98% for phenol using photocatalysis after a 60 min reaction [34]. Ramandi et al. [35] achieved a total degradation of antibiotic Diclofenac (25 mg/L) within 180 minutes using N-doped TiO_2-NPs synthesized using the sol–gel method with ultrasound irradiation, with sunlight triggering photocatalysis. Diclofenac is extremely harmful to birds if left untreated in wastewater. It causes thinning of egg shells and causes kidney failure, hence declining avian population [36]. Umar et al. [37] synthesized Mn-doped TiO_2 for photodegradation of pharmaceutical pollutant namely chlorothalonil and ketoprofen.

The mechanism of photodegradation is available to nanoparticles that are semiconductors in nature and have a bandgap that allows the excitation of electrons to a higher energy band in the presence of catalyst which is sunlight or irradiation from a light. There is a generation of radicals such as ·OH radicals that then initiate oxidation reactions ending with complete mineralization of CO_2 and H_2O. Vasiljevic et al. [18] used iron titanate (Fe_2TiO_5) nanoparticles for photodegradation of methylene blue (MB) dye in water where both photodegradation of the dye occurred after a 4 H exposure with mineralization into end-products H_2O and CO_2. The mineralization pathway is slower than degradation mechanism.

15.6.2 Photo-Fenton Process

Photo-Fenton process is an advanced oxidation process in which hydroxyl radicals are generated to degrade via oxidation of the hazardous and recalcitrant compounds that are present in wastewater. H_2O_2 or peroxide added is decomposed catalytically by Fe^{+2} and the generated radical of $\cdot OH$ oxidize the contaminant in water. Bello et al. [38] showed a complex formation between SiO_2 and H_2O_2 via a siloxane bridge (Si–O–Si) that enabled dye degradation from wastewater. The coupling with SiO_2 enabled a higher COD removal (20%) than without the semiconductor. In the ferrous cycle, Fe^{3+} is converted to Fe^{2+} by the reduction of H_2O_2 and the Fenton cycle is repeated multiple times [39]. Fe^{+3} ion is a powerful coagulating agent that helps in the sedimentation through coagulation of the pollutants present in the water [40,41]. The radical that is generated in water operates via electron transfer, hydroxyl, and hydrogen addition, hence reacting with compounds in water for their effective oxidation in harmless compounds.

15.6.3 Adsorption

Majority of nanoparticles are able to effect water remediation by their ability to adsorb the contaminants onto their surface. The high surface/volume ratio of nanoparticles due to their extremely small size aids them in this process [42]. An adsorption isotherm is a graphical representation of the relationship between the adsorbate that is remaining in liquid phase and the amount absorbed by the adsorbent in equilibrium. Freundlich and langmuir isotherms are the most studied models to explain the mechanism of metal ion uptake by the adsorbent system. Langmuir isotherms are based on the assumption that there is a monolayer at surface, each site only holds one single molecule, and there are no intermolecular interactions [43]. Freundlich isotherms involve multilayer absorption for heterogeneous molecular systems and as cations and anions both are adsorbed simultaneously. There exists an attractive force between them. There is equilibrium in the solid and liquid phases [44,45].

15.6.4 Catalytic Activity of Nanoparticles

A synthetic compound that is widely indispensable for manufacture of pesticides, insecticides, drugs, fungicides, and for leather tanning is 4-nitrophenol. But if consumed or inhaled, it can cause harmful effects on human health such as cyanosis or blueing of mouth, dizziness, headache, and nausea. It has a toxic and damaging nature on the health of humans and subsequently on ecology. Yet, its need in industrial manufacture is a great issue for environmentalists. 4-Nitrophenol is a cause of concern for scientists and environmentalists worldwide. It is important to cater to the reduction of 4-nitrophenol before it can be properly disposed of without leading to pollution of natural water bodies or seepage into groundwater. When we reduce 4-nitrophenol, the product so formed after the reduction is 4-aminophenol, which is industrially important for the manufacture of sulfur dyes, paracetamol (a therapeutic drug), rubber, natural antioxidants, preparation of black/white films, analgesic and antipyretic usage, and stoppage of corrosion.

Reduction of 4-nitrophenol is done by using $NaBH_4$ and adding Au-NPs as a metal catalyst. Difference of potential between nitrophenolate ions and donor ($H_3BO_3/NaBH_4$) or the acceptor molecule may cause a requirement for a higher activation energy barrier which must be reduced so that the reaction may occur at a lower energy provided at substrate level and rate of reaction be at its utmost optimum speed.

Rate of reaction is significantly increased by increasing the adsorption metallic NPs that have a high surface/volume ratio, thereby lowering the activation energy significantly. 4-Nitrophenol displays a sharp band at 400 nm in its UV-visible spectra, hence affirming the success of usage of metallic nanoparticles for the reduction of 4-nitrophenol as performed in an experiment by Ganapuram et al. Green Ag-NPs are synthesized by plant stem extract. A band at 313 nm on spectra indicates the 4-aminophenol formation, hence proving catalytic reduction potential of Au-NPs.

15.6.5   ANTIOXIDANT ACTIVITY

Gold nanoparticles offer a high speed of catalytic reduction of pollutants due to their small size. Gold nanoparticles display good antioxidant ability, which is referred to the ability to combat the generation and stop the spread of free radicals. Free radicals such as nascent or free oxygen wreaks havoc by subsequent chain of cascade reactions that lead to biologically degraded animal and cellular tissues. How effectively Au NPs demobilize the free radicals is governed by specific surface, size, and concentration as shown in an experiment by Singh et al. Furthermore, it is equally stated that Au NPs that have a high surface to volume ratio exhibit higher antioxidant activity. DPPH assay of Au NPs obtained from *Allium cepa* or common onion showed an effective antioxidant activity of 14.44% at 100 M g/ml as displayed in an experiment by Dr. Sunday Akintelu et al. They can employ various types of absorbent/ascorbate mechanisms, anti-oxidant action, heavy metal detection, antibacterial action, etc.

15.6.6   ANTI-MICROBIAL ACTION OF NPS IN WASTEWATER

Berekaa et al. [46] used N-doped TiO_2 for effectively disarming *Escherichia coli* from wastewater displaying anti-microbial properties of nanoparticles. They exhibit a greater efficiency against gram negative bacteria. The rate of bactericidal activity increases with decrease in the size of nanoparticles. Both size and dosage affect the activity. The smaller the size of the nanoparticles, the more is the rate of efficiency. Microbes are becoming a wider problem particularly the highly contagious ones that are precursors to infections and pandemics. Lot of bacterial strains are gaining resistance through natural selection to antibiotic strains and such strains can lead to occurrences of disease and epidemics. In-vitro studies are performed and metallic nanoparticles effectively cause death to microbes by various pathways.

Nanoparticles cause cell lysis and apoptosis by making holes in the cell wall. They also cause denaturation of cellular proteins, uncoiling, and damage to cellular DNA. They damage the mRNA. They stop cellular and genetic transcription. They cause disruption of enzymes by blocking the receptor sites. There is damage to peptidoglycan and cellular biological cellular membranes [47]. Many bacterial strains have genes that confer penicillin-resistance, sulfonamide-resistance, tryptophan-resistance, etc. When these bacterial strains are treated with anti-biotics, they do not die and only become more resistant over time and pose a great risk to human health and safety. Nanoparticles generate oxygen reactive species (ROS) that further enable the antibacterial action by immediate rupture of the peptidoglycan cell wall, disruption of bacterial replication mechanism, disruption of enzymes, and inhibition of biofilm formation.

Gurunathan et al. [48] disarmed *Pseudomonas aeruginosa* using graphene oxide nanoparticles that released ROS that led to cell death of bacteria, hence promising newer avenues for water filtration. Leung et al. [49,50] performed experiments to create evidences of non-ROS mediated antibacterial action of MgO nanoparticles towards *E. coli*. Nagy et al. [51] created a composite of silver nanoparticles that were deposited in zeolite membrane (AgNP-ZM) which was toxic towards *Staphylococcus aureus* and *E. coli* by generation of Ag^+ ions that caused damage to pathogens.

15.7   CASE STUDY

15.7.1   SYNTHETIC DYE REMOVAL BY NANOPARTICLES IN WASTEWATER

Numerous industries use synthetic dyes in paper, textile, pharmaceutical, food industries, hair coloring, and toy production [52,53]. Methylene blue (MB) and congo red (CR) are both notorious for polluting water. If these dyes are not removed from the discharge of the industries, then they can cause havoc in the delicate water ecology [54]. Scientists have devised newer techniques for the degradation of synthetic dyes in water. Removal of synthetic dyes is highly important to achieve remediation of polluted water as such water if consumed can cause damage to human health and under some conditions may even prove to be carcinogenic, hence leading to fatality [55].

Azo dyes have a double nitrogen bonding which makes them resist degradation in water, hence a challenge to remediate from wastewater [56]. Yao et al. [57] used CNTs for adsorption of synthetic dyes like cationic methyl violet and methylene blue dyes, thus helping remove the dyes from water and achieve decontamination of the wastewater by surface adsorption on the CNT surface. Bouazizi et al. [58] used a nanoscaled-TiO_2 ultrafiltration membrane deposited on a clay bentonite support that displayed a 98% uptake of Direct Red 80 dye in water. Saffaj et al. [59] used $ZnAl_2O_4$–TiO_2 ultrafiltration membrane with a pore size of diameter 5 nm for uptake of synthetic dyes from water. Livani et al. [60,61] used activated carbon adsorbents that were delivered with $NiFe_2O_4$ nanoparticles for the remediation or uptake of synthetic dyes in water, namely Direct Blue 78 and Direct Red 31 dyes, and maximum adsorption capacity was reported to be 299.67 mg/g and 209.13 mg/g for DR31 and DB78, respectively. Ranjithkumar et al. [62] used a deposition of Fe_2O_3 nanocomposite with activated carbon for uptake of Acid Yellow 17 dye from water.

15.7.2 Decolorization by Nickel Oxide Nanoparticles

Due to its nature of absorptivity and magnetism, nickel oxide nanoparticles are effective at water decontamination and can be used for dye absorption. Its catalytic nature allows it to be used for oxidation of numerous compounds that are organic in nature. Song et al. could remove 100 mg/L reactive dye completely after a contact period of 6 hours using nickel oxide nanosheets. Nategi et al. removed 81% of the Orange II dye after 30 minutes of contact time using only 0.6 g/L absorbent. Nickel oxide and nanoparticles derived from it are essential to nanotechnology owing to their high surface to volume ratio, hence having numerous sites that are available for bulk adsorption of the contaminants and effect rapid degradation of the synthetic colorant in the wastewater.

In a study by Jisma et al., gold nanoparticles or Au-NPs were synthesized using biological precursors. Triphenyl methane blue dyes namely Victoria Blue B or VBB and VBR were degraded photocatalytically or using sunlight at 65% and 52% at a time scale of 8 hours. The experiment was eco-friendly and cost-effective, hence worth replicating at a large industrial scale. Toh et al. showed that chlorine and other synthetic derivatives led to carcinogen generation in water, thus posing as a great threat. VB is a chloride salt of triphenylmethane; hence, using bacterial species to degrade the chlorine atom was ineffective as shown by Lade et al. Thus, using conventional bioremediation methods proved to be a failure.

15.7.3 Heavy Metal Ion-Detection in Aqueous Medium

Heavy metals are a serious problem in wastewater management. Heavy metals wreak havoc on both human and animal health. They can cause problems such as methemoglobinemia, oxygen deficiency, cancer among many problems. Heavy metals include Cr, Ni, Zn, Cu, Zn, Hg, and Cd. They can pollute water for several years and such water if consumed can be greatly toxic to health. Such a water must be checked for the presence of heavy metal ions before any consumption can take place. Water that is fit for drinking is called potable water. There are numerous ways to remove heavy metal ions from aqueous solutions, including chemical precipitation, ultrafiltration, reverse osmosis, chemical oxidation/reduction, electrodialysis, and ion-exchangers. But these conventional method-ologies come with a set of challenges and limitations. It's imperative and highly important to develop new methods for cost-effective, high efficiency, and environmentally friendly ways to minimize the chemical by-products and toxin compounds released due to these conventional methods and allow new techniques which lead to protection of the environment and foster ecological balance. Proanthocyanidins-chelated gold nanoparticles display a high efficiency for heavy metal ions and synthetic dyes removal, which indicates the possible advances in devices for bioremediation. Heavy metal can seep into groundwater and local water bodies due to mining activities, plastic, dye industries, vehicular emissions, etc. Even at trace levels or ppm concentrations, heavy metals such as mercury, cadmium, and lead can be highly toxic. Detection is vital for proper remediation to take place.

Optical properties and refractive index of gold nanoparticles are helpful for heavy metal ion detection. Analysis using conventional techniques is time-consuming, expensive, skill-dependent, and highly complicated. Gold nanoparticles offer cheaper and more efficient methods for heavy metal ion detection. Colorimetric methods are usually employed where nanoparticles are helpful for heavy metal detection in water.

15.8 CONCLUSIONS

Bioremediation is mainly incorporated while we practice degradation, attenuation immobilization, and/or detoxification of a plethora of chemical/toxic waste and physical hazardous contaminants from the surrounding site of pollution through the powerful action of the microorganisms. The main principle is to degrade and convert or transform the pollutants to lesser toxic forms. All bioremediation techniques have their own advantages and disadvantages because it has its own specific applications. In purview of these limitations, the latest advent of science has been green nanoremediation that overcomes the shortcomings.

Nanoremediation involves the use of nanoparticles that are extremely small in dimension (>100 nm). Owing to this characteristic, nanoparticles have an extremely high surface/volume ratio which can be exploited to remediate the contaminated site. Nanoparticles can be synthesized by the reduction of the metal ions using plant/microorganism-origin enzymes, extracts, or secondary metabolites. Compared to previously and conventionally used nanoparticle synthesis by chemicals, green nanotechnology offers a much better alternative as it uses natural and nature-derived sources to reduce metal ions into corresponding nanoparticles. Green synthesis helps us to prevent the detrimental loss to the environment that is caused due to the conventional methods that create pollution, thereby defeating the original purpose of using nanoparticles for remediation. These nanoparticles overcome the limitations of conventional strategies of bioremediation that are various in number, hence allowing the researchers to explore new avenues to improve efficiency rate and overcome barriers. Nanoparticles have various mechanisms by which they can overcome these hurdles.

REFERENCES

1. Mohammadi, A. A., Zarei, A, Majidi, S, Ghaderpoury, A. Carcinogenic and non-carcinogenic health risk assessment of heavy metals in drinking water. *Methods* 6, 2019: 1642–1651, ISSN 2215-0161, 10.1016/j.mex.2019.07.017
2. Guth, J.L., Kessler, H. Synthesis of aluminosilicate zeolites and related silica-based materials. In *Catalysis and Zeolites Fundamentals and Applications*; Weitkamp, J., Puppe, L., Eds.; Springer: Berlin/Heidelberg, Germany, 1999; pp. 1–52.
3. Ghomashi, P., Charkhi, A., Kazemeini, M., Yousefi, T. Removal of fluoride from wastewater by natural and modified nano clinoptilolite zeolite. *J. Water Environ. Nanotechnol.* 2020, 5, 270–282.
4. Renu, M. A., Singh, K. Heavy metal removal from wastewater using various adsorbents: A review. *J. Water Reuse Desalin.* 1 December 2017;7 (4):387–419. https://doi.org/10.2166/wrd.2016.104

5. Hamidpour, M., Afyuni, M., Kalbasi, M., Khoshgoft Armanes, A. H., Inglezakis, J. Mobility and plant-availability of Cd (II) and Pb (II) adsorbed on zeolite and bentonite. *Appl. Clay Sci.* 2010;48 (3): 342–348.

6. El-Kamash, A. M. Zaki, A. A. El Geleel, M. A. Modeling batch kinetics and thermodynamics of zinc and cadmium ions removal from waste solutions using synthetic zeolite A. *J. Hazard. Mater.* 2005;127 (1):211–220.

7. Gopalakrishnan, A., Krishnan, R., Sakthivel, T., Venugopal, G., Kim, S.-J. Removal of heavy metal ions from pharma-effluents using graphene-oxide nanosorbents and study of their adsorption kinetics. *J. Indus. Eng. Chem.*. 2015;30:14–19. 10.1016/j.jiec.2015.06.005

8. Lingamdinne, L. P. Koduru, J. R., Choi, Y. L., Chang, Y. Y., Yang, J. K. Studies on removal of Pb (II) and Cr (III) using graphene oxide based inverse spinel nickel ferrite nano-composite as sorbent. *Hydrometallurgy* 2015;165:64–72.

9. Tran, D., Kabiri, S., Wang, L., Losic, D. Engineered graphene-nanoparticle aerogel composites for efficient removal of phosphate from water. *J. Mater. Chem. A.* 2015;3. 10.1039/C4TA06308B

10. Balasubramani, K., Sivarajasekar, N., Naushad, M. Effective adsorption of anti-diabetic pharmaceutical (metformin) from aqueous medium using graphene oxide nanoparticles: Equilibrium and statistical modelling. *J. Mol. Liq*, 2020;301; 112426, ISSN 0167-7322.

11. Baig, N., Ihsanullah, Sajid, M., Saleh, T.A. Graphene-based adsorbents for the removal of toxic organic pollutants: A review. *J. Environ. Manage.* 2019 Aug 15;244:370–382. 10.1016/j.jenvman.2019.05.047. Epub 2019 May 24. PMID: 31132618.

12. Mantovani, S., Khaliha, S., Favaretto, L., Bettini, C., Bianchi, A., Kovtun, A., Zambianchi, M., Gazzano, M., Casentini, B., Palermo, V., Melucci, M.Scalable synthesis and purification of functionalized graphene nanosheets for water remediation, *Chem. Commun.*, 2021, 57, 3765–3768.

13. White, R. L., White, C. M., Turgut, H., Massoud, A., Ryan Tian, Z. Comparative studies on copper adsorption by graphene oxide and functionalized graphene oxide nanoparticles. *J. Taiwan Instit. Chem. Eng.*, 2018;85:18–28, ISSN.

14. Xu, P., Zeng, G. M., Huang, D. L., Feng, C. L., Hu, S., Zhao, M. H., Lai, C., Wei, Z., Huang, C., Xie, G. X., Liu, Z. F. Use of iron oxide nanomaterials in wastewater treatment: A review. *Sci. Total Environ.* 2012 May 1;424:1–10. 10.1016/j.scitotenv.2012.02.023. Epub 2012 Mar 4. PMID: 22391097.

15. Yang, X. et al. Surface functional groups of carbon-based adsorbents and their roles in the removal of heavy metals from aqueous solutions: A critical review. *Chem. Eng. J.* 2019;366:608–621.

16. Mohan, S., Kumar, V., Singh, D. K., Hasan, S. H. Effective removal of lead ions using graphene oxide-MgO nanohybrid from aqueous solution: Isotherm, kinetic and thermodynamic modeling of adsorption. *J. Environ. Chem. Eng.* 2017;5(3):2259–2273, ISSN 2213-3437.

17. Dubey, R., Bajpai, J. & Bajpai, A. K. Green synthesis of graphene sand composite (GSC) as novel adsorbent for efficient removal of Cr (VI) ions from aqueous solution. *J. Water Process Eng.* 2015;5: 83–94.

18. Vasiljevic, Z. Z., Dojcinovic, M. P., Vujancevic, J. D., Jankovic-Castvan, I., Ognjanovic, M., Tadic, N. B., Stojadinovic, S., Brankovic, G. O., Nikolic, M. V. Photocatalytic degradation of methylene blue under natural sunlight using iron titanate nanoparticles prepared by a modified sol–gel method. *R. Soc. Open Sci.* 2020, 7, 200708.

19. Wang, J, Zheng, S, Shao, Y, Liu, J, Xu, Z, Zhu, D. Amino-functionalized Fe_3O_4@SiO_2 core-shell magnetic nanomaterial as a novel adsorbent for aqueous heavy metals removal. *J Colloid Interface Sci.* 2010;349(1):293–299. 10.1016/j.jcis.2010.05.010

20. Lankathilaka, W., de Silva, R. M., Mantilaka, M.M.M.G.P.G., Nalin de Silva, K.M. Magnetite nanoparticles incorporated porous kaolin as a superior heavy metal sorbent for water purification, *Groundw. Sustain. Dev.* 2021;14:100606, ISSN 2352-801X, 10.1016/j.gsd.2021.100606

21. Wasantha Lankathilaka, K. P., de Silva, R. M., Mantilaka, M. M. M. G. P. G., Nalin de Silva, K. M. Magnetite nanoparticles incorporated porous kaolin as a superior heavy metal sorbent for water purification, *Groundw. Sustain. Dev.*, 2021;14:100606, ISSN 2352-801X, 10.1016/j.gsd.2021.100606

22. Chengfeng, R., Rongrong, R. Get rid of chlorine residue active carbon water purification filter core. U.S. Patent CN08791367U, 26 April 2019. *Environ. Prog. Sustain. Energy* 2019;38:13188.

23. Li, Y.-H., Ding, J., Luan, Z., Di, Z., Zhu, Y., Xu, C., Wu, D., Wei, B. Competitive adsorption of Pb2+, Cu2+ and Cd2+ ions from aqueous solutions by multiwalled carbon nanotubes. *Carbon* 2003;41(14):2787–2792.

24. Li, Y.-H., Wang, S., Luan, Z., Ding, J., Xu, C., Wu, D.: Adsorption of cadmium (II) from aqueous solution by surface oxidized carbon nanotubes. *Carbon* 2003;41(5): 1057–1062.

25. Chen, C., Hu, J., Xu, D., Tan, X., Meng, Y., Wang, X. Surface complexation modeling of Sr (II) and Eu(III) adsorption onto oxidized multiwall carbon nanotubes. *J. Colloid Interface Sci.* 2008;323(1):33–41.

26. Sadegh, H., Shahryari-ghoshekandi, R., Agarwal, S., Tyagi, I., Asif, M., Gupta, V. K. Microwave-assisted removal of malachite green by carboxylate functionalized multi-walled carbon nanotubes: Kinetics and equilibrium study. *J. Mol. Liq.* 2015;206, 151–158.

27. Ghaedi, M., Khajehsharifi, H., Yadkuri, A.H., Roosta, M., Asghari, A. Oxidized multiwalled carbon nanotubes as efficient adsorbent for bromothymol blue. *Toxicol. Environ. Chem.* 2012;94(5):873–883.

28. Kotia, A., Yadav, A., Raj, T. R., Keischgens, M. R., Rathore, R., Sarris, I. E. Carbon nanoparticles as sources for a cost-effective water purification method: A comprehensive review. *Fluids* 2020;5(4): 230. 10.3390/fluids5040230

29. Kumari, M., Pittman, C. U., Mohan, D. Heavy metals [chromium (VI) and lead (II)] removal from water using mesoporous magnetite (Fe3O4) nanospheres, *J. Environ. Manag.*, 2022;306. 10.1016/j.jcis.2014.09.012

30. Saravanan, R., Gracia, F., Stephen, A. Basic principles, mechanism, and challenges of photocatalysis. In Nanocomposites for Visible Light-Induced Photocatalysis; Springer, Cham: Berlin/Heidelberg, Germany, 2017; pp. 19–40.

31. Chen, W., Liu, Q., Tian, S., Zhao, X. Exposed facet dependent stability of ZnO micro/nano crystals as a photocatalyst. *App. Surf. Sci* 2019;470:807–816.

32. Gomes, J., Lopes, A., Gmurek, M., Quinta-Ferreira, R.M., Martins, R.C. Study of the influence of the matrix characteristics over the photocatalytic ozonation of parabens using Ag-TiO2. Sci. Total Environ. 2019;646:1468–1477.

33. Raliya, S.R., Avery, C., Chakrabarti, S., Biswas, P. Photocatalytic degradation of methyl orange dye by pristine TiO2, ZnO, and graphene oxide nanostructures and their composites under visible light irradiation. *Appl. Nano Sci.* 2017;7:253–259.

34. Zhang, Y., Xin, Q., Cong, Y., Qi, W., Jiang, B. Application of TiO2 nanotubes with pulsed plasma for phenol degradation. *Chem. Eng. J.* 2013;215–216:261–268. 10.1016/j.cej.2012.11.045

35. Ramandi, S., Entezari, M. H., Ghows, N. Sono-synthesis of solar light responsive S-N-C-tri doped TiO2 photo-catalyst under optimized conditions for degradation and mineralization of Diclofenac. *Ultrason. Sonochem.* 2017;38:234–245.

36. Nethathe, B., Chipangura, J., Hassan, I. Z., Duncan, N., Adawaren, E. O., Havenga, L., Naidoo, V. Diclofenac toxicity in susceptible bird species results from a combination of reduced glomerular filtration and plasma flow with subsequent renal tubular

necrosis. *PeerJ.* 2021 Aug 23;9:e12002. 10.7717/peerj.12002. PMID: 34513332; PMCID: PMC8388555.

37. Umar, K., Ibrahim, M.N.M., Ahmad, A., Rafatullah, M. Synthesis of Mn-doped TiO2 by novel route and photocatalytic mineralization/intermediate studies of organic pollutants. *Res. Chem. Intermediat.* 2019, 45, 2927–2945.

38. Bello, M.M., Abdul Raman, A.A., Asghar, A. (2019), Fenton oxidation treatment of recalcitrant dye in fluidized bed reactor: Role of SiO2 as carrier and its interaction with fenton's reagent. *Environ. Prog. Sustain. Energy*, 38: 13188.

39. Lin, R., Li, Y., Yong, T., Cao, W., Wu, J., Shen, Y. Synergistic effects of oxidation, coagulation and adsorption in the integrated fenton-based process for wastewater treatment: A review. *J. Environ. Manage.* 2022 Mar 15;306:114460. 10.1016/j.jenvman.2022.114460. Epub 2022 Jan 11. PMID: 35026715.

40. Sinha, D. Roy, O. Roy, S. Neogi, S. Removal of organic contaminants from flowback water using Fenton process. *J. Water Process Eng.* 2022;47:Article 102680.

41. Qu, S., Huang, F., Yu, S., Chen, G., Kong, J. Magnetic removal of dyes from aqueous solution using multi-walled carbon nanotubes filled with Fe2O3 particles *J. Hazard. Mater.*, 2008;160(2–3):643–647, ISS0304-3894, 10.1016/j.jhazmat.2008.03.037

42. Apul, O.G., Wang, Q., Zhou, Y., Karanfil, T.: Adsorption of aromatic organic contaminants by graphene nanosheets: Comparison with carbon nanotubes and activated carbon. *Water Res.* 2013;47(4):1648–1654.

43. Langmuir, I. The adsorption of gases on plane surface of glass, mica and platinum. *J. Am. Chem. Soc..* June 1918;40 (9):1361–1402.

44. Freundlich, H. *Kapillarchemie, eine Darstellung der Chemie der Kolloide und verwandter Gebiete.* Akademische Verlagsgesellschaft, 1909.

45. Friedmann, D, Mendiveb, C, Bahnemann, D. TiO2 for water treatment: Parameters affecting the kinetics and mechanisms of photocatalysis. *Appl. Catal. B Environ.* 2010;99:398–406.

46. Berekaa, M.M. Nanotechnology in wastewater treatment: Influence of nanomaterials on microbial systems. *Int. J. Curr. Microbiol. App. Sci.* 2016, 5, 713–726.

47. Hajipour, M. J., Fromm, K.M., Ashkarran, A. A., Jimenez de Aberasturi, D., Ruiz de Larramendi, I., Rojo, T., Serpooshan, V., Parak, W. J., Mahmoudi, M. Antibacterial properties of nanoparticles. *Trends Biotechnol.* 2012;30(10): 499–511.

48. Gurunathan, S., Han, J.W., Dayem, A.A., Eppakayala, V., Kim, J.H. Oxidative stress-mediated antibacterial activity of graphene oxide and reduced graphene oxide in Pseudomonas aeruginosa. *Int. J. Nanomed.* 2012;7:5901.

49. Leung, Y. H., Ng, A. M., Xu, X., et al. Mechanisms of antibacterial activity of MgO: Non-ROS mediated toxicity of MgO nanoparticles towards Escherichia coli. *Small.* 2014;10(6):1171–1183.

50. Li, X., Xia, T., Xu, C., Murowchick, J., Chen, X. Synthesis and photoactivity of nanostructured CdS–TiO2 composite catalysts. *Catal. Today* 2014, 225, 64–73.

51. Nagy, A., Harrison, A., Sabbani, S., Munson Jr, R. S., Dutta, P. K., Waldman, W. J. Silver nanoparticles embedded in zeolite membranes: Release of silver ions and mechanism of antibacterial action. *Int. J. Nanomed.* 2011;6:1833.

52. Hameed, B.H., Ahmad, L., Latiff, K.N. Adsorption of basic dye (methylene blue) onto activated carbon prepared from rattan sawdust. *Dye. Pigm.* 2007;75:143–149.

53. Mahmood, I, Guo, C., Guan, Y., Ishfaq, A., Liu, H. Adsorption and magnetic removal of neutral red dye from aqueous solution using Fe3O4 hollow nanospheres. *J. Hazard. Mater.* 2008;160(2–3):643–647, ISSN 0304-3894, 10.1016/j.jhazmat.2008.03.037

54. Ashrafi, S., Kamani, H., Mahvi, A. The optimization study of Direct Red 81 and methylene blue adsorption on NaOH-modified rice husk. *Desalin. Water Treat.* 2016;57(2):738–746. 10.1080/19443994.2014.979329

55. Affat, S. Classifications, advantages, disadvantages, toxicity effects of natural and synthetic dyes: A review. *University of Thi-Qar Journal of Science (UTsci).* 2021;8:130–135.

56. Chang, M., Shih, Y.H. Synthesis and application of magnetic iron oxide nanoparticles on the removal of reactive black 5: Reaction mechanism, temperature and pH effects. *J Environ Manag.* 2018;224:235–242. 10.1016/j.jenvman.2018.07.021. Chem. Commun., 57 (31) (2021), pp. 3765.

57. Yao, Y., Xu, F., Chen, M., Xu, Z., Zhu, Z. Adsorption of cationic methyl violet and methylene blue dyes onto carbon nanotubes. In: Nano/Micro Engineered and Molecular Systems (NEMS), 2010 5th IEEE International Conference 2010.

58. Bouazizi, A., Breida, M., Achiou, B., Ouammou, M., Calvo, J.I., Aaddane, A., Younssi, S.A. Removal of dyes by a new nano–TiO2 ultrafiltration membrane deposited on low-cost support prepared from natural Moroccan bentonite. *Appl Clay Sci.* 2017;149:127–135. 10.1016/j.clay.2017.08.019

59. Saffaj, N., Persin, M., Alami Younssi, S., Albizane, A., Bouhria, M., Loukili, H., Dach, H., Larbot, A. Removal of salts and dyes by low ZnAl2O4–TiO2 ultrafiltration membrane deposited on support made from raw clay. *Sep. Purif. Technol.* 2005;47(1–2):36–42, ISSN 1383-5866, 10.1016/j.seppur.2005.05.012

60. Livani, M. J., Ghorbani, M. Fabrication of NiFe2O4 magnetic nanoparticles loaded on activated carbon as novel nanoadsorbent for Direct Red 31 and Direct Blue 78 adsorption. *Environ. Technol.* 2018;39(23):2977–993.

61. Ma, L.J., Han, L.N., Chen, S., Hu, J.L., Chang, L.P., Bao, W.R., Wang, J. Rapid synthesis of magnetic zeolite materials from fly ash and iron-containing wastes using supercritical water for elemental mercury removal from flue gas. *Fuel Process. Technol.* 2019;189:39–48.

62. Ranjithkumar, V, Sangeetha, S, Vairam, S. Synthesis of magnetic activated carbon/α-Fe2O3 nanocomposite and its application in the removal of acid yellow 17 dye from water. *J. Hazard. Mater.* 2014;273:127–135. 10.1016/j.jhazmat.2014.03.034

Index

For Product Safety Concerns and Information please contact our EU
representative GPSR@taylorandfrancis.com
Taylor & Francis Verlag GmbH, Kaufingerstraße 24, 80331 München, Germany

www.ingramcontent.com/pod-product-compliance
Lightning Source LLC
Chambersburg PA
CBHW060327220326
41598CB00023B/2632

* 9 7 8 1 0 3 2 4 2 9 0 4 5 *